Since 1973 the Royal College of Obstetricians and Gynaecologists has regularly convened Study Groups to address important growth areas within obstetrics and gynaecology. An international group of eminent scientists and clinicians from various disciplines is invited to present the results of recent research and take part in in-depth discussion. The resulting volume containing the papers presented and also edited transcripts of the discussions is published within a few months of the meeting and provides a summary of the subject that is both authoritative and up to date.

Previous Study Group publications available from Springer-Verlag:

Early Pregnancy Loss
Edited by R. W. Beard and F. Sharp

AIDS in Obstetrics and Gynaecology
Edited by C. N. Hudson and F. Sharp

Fetal Growth
Edited by F. Sharp, R. B. Fraser and R. D. G. Milner

Micturition
Edited by J. O. Drife, P. Hilton and S. L. Stanton

HRT and Osteoporosis
Edited by J. O. Drife and J. W. W. Studd

Antenatal Diagnosis of Fetal Abnormalities
Edited by J. O. Drife and D. Donnai

Prostaglandins and the Uterus
Edited by J. O. Drife and A. A. Calder

The Royal College of Obstetricians and Gynaecologists gratefully acknowledges the sponsorship of this Study Group by William Cook Europe Limited; Hoeschst UK Limited; Organon Laboratories Limited; Serono Laboratories (UK) Limited; HG Wallace Limited.

Infertility

Edited by

A. A. Templeton and J. O. Drife

With 89 Figures

Springer-Verlag
London Berlin Heidelberg New York
Paris Tokyo Hong Kong
Barcelona Budapest

Allan Templeton MD, FRCOG
Professor of Obstetrics and Gynaecology, Aberdeen Maternity Hospital, Foresterhill, Aberdeen AB9 2ZD, UK

James O. Drife, MD, FRCSEd, FRCOG
Professor of Obstetrics and Gynaecology, Clarendon Wing, Leeds General Infirmary, Leeds LS2 9NS, UK

ISBN 3–540–19743–5 Springer-Verlag Berlin Heidelberg New York
ISBN 0–387–19743–5 Springer-Verlag New York Berlin Heidelberg

British Library Cataloguing in Publication Data
Infertility
I. Templeton, A. Allan II. Drife, J. O.
616.6
ISBN 3-540-19743-5

Library of Congress Cataloging-in-Publication Data
Infertility / edited by A.A. Templeton and J.O. Drife.
p. cm.
Includes bibliographical references and index.
ISBN 3-540-19743-5 (alk. paper). --ISBN 0-387-19743-5 (alk. paper)
1. Infertility--Congresses. I. Templeton, A. A. (Alexander Allan) II.Drife, James O., 1947-
[DNLM: 1. Infertility--diagnosis--congresses. 2. Infertility--therapy--congresses. WP 570 143]
RC889.I554 1992
618.1′78--dc20
DNLM/DLC 92-49701
for Library of Congress CIP

Printed in Great Britain

It was not possible to refer all the material back to the authors or discussants, but it is hoped that the proceedings have been reported fairly and accurately.

Typeset by Photo·graphics, Honiton, Devon
Printed by The Alden Press, Osney Mead, Oxford
28/3830-543210 Printed on acid-free paper

Preface

Infertility, as with many aspects of medicine, is at the mercy of rapid technological advance. Many of these developments initially seem attractive to both clinicians and patients, but need to be rigorously assessed if their real value is to be understood and clinical practice is to develop. In this book issues of importance to the management of infertile patients are discussed. The gaps in our knowledge which prevent a better understanding of the condition are identified, and recent developments, both clinical and scientific, are subjected to peer review and discussion. An important feature of the book is an acceptance that training in infertility practice is a real problem. This is perceived not only by the practising clinicians, both doctors and nurses, but particularly by the clinical scientists, including embryologists, who now provide such an essential part of the service. Similarly the provision of the clinical service has been examined in detail from a variety of standpoints, in an attempt to make sensible recommendations which balance real need with limited resource.

The book is based on the papers presented and discussed at the 25th RCOG Study Group held in April 1992. The discussion after each paper was civilised but uncompromising and forms an important part of this publication. The rapid processing of the written and recorded material by the staff at the RCOG, and particularly Miss Sally Barber, has ensured that the book has been produced while the issues are live, the reviews contemporary and the discussion relevant. The publishers, Springer-Verlag, are also to be thanked for their help in this respect.

Finally the recommendations printed at the end of this book were thought through very carefully by the participants, and represent a statement of intent, to which the Study Group felt committed. It is hoped that the recommendations will be of help to those providing the clinical service, as well as those who have responsibility for purchasing services for infertile couples. In addition several of the recommendations are relevant to the present Code of Practice published by the HFEA, and these will no doubt be considered when it is thought timely. As a footnote it is gratifying to know that the last recommendation in the book has been realised and due to pressure from many sources, the confidentiality clause in the Human Fertilisation and Embryology Act has now been amended. Our thanks go to all those who participated in the Study Group and contributed to this book.

May 1992

A. A. Templeton
J. O. Drife

Contents

Back row: Dr R. G. Gosden, Professor J.F. Kerin, Professor S. Franks, Professor R. W. Shaw, Mr R. H. T. Ward, Professor D. H. Barlow, Professor D. T. Baird, Mr H. I. Abdalla, Mr T. A. Sheldon, Dr A. H. Handyside, Professor R. H. Asch, Mr J. Dickson, Dr S. G. Hillier, Professor I. D. Cooke, Dr S. B. Fishel, Mr H. A. Whittall.

Front row: Professor D. J. Cusine, Professor R. J. Aitken, Dr H. J. Leese, Dr A. Dokras, Professor J. O. Drife, Mrs M. M. Inglis, Mrs B. J. Botting, Professor R. M. L. Winston, Professor M. G. R. Hull, Professor P. R. Braude, Dr P. F. Watson, Professor D. G. Whittingham, Professor A. A. Templeton.

Participants

Mr H. I. Abdalla
Director, In Vitro Fertilisation Unit, Fertility & Endocrinology Centre, The Lister Hospital, Chelsea Bridge Road, London SW1W 8RH, UK

Professor R. J. Aitken
Professor/Research Scientist, MRC Reproductive Biology Unit, Centre for Reproductive Biology, The University of Edinburgh, 37 Chalmers Street, Edinburgh EH3 9EW, UK

Professor R. H. Asch
Director, University of California Irvine Medical Center for Reproductive Health and Dean, College of Medicine, 101 The City Drive, Orange, California 92668, USA

Professor D. T. Baird
MRC Clinical Research Professor of Reproductive Endocrinology, Department of Obstetrics & Gynaecology, Centre for Reproductive Biology, The University of Edinburgh, 37 Chalmers Street, Edinburgh EH3 9EW, UK

Professor D. H. Barlow
Nuffield Department of Obstetrics & Gynaecology, John Radcliffe Hospital, Maternity Department, Headington, Oxford OX3 9DU, UK

Dr V. Beral
Director, Imperial Cancer Research Fund, Cancer Epidemiology Unit, University of Oxford, Gibson Building, The Radcliffe Infirmary, Oxford OX2 6HE, UK

Mrs B. J. Botting
Statistician, Office of Population Censuses & Surveys, Medical Statistics Division, St Catherine's House, 10 Kingsway, London WC2B 6JP, UK

Professor P. R. Braude
UMDS Professor of Obstetrics & Gynaecology, 6th Floor, North Wing, St Thomas' Hospital, Lambeth Palace Road, London SE1 7EH, UK

Professor C. M. Campbell
Chairman, Human Fertilisation & Embryology Authority, Paxton House, 30 Artillery Row, London E1 7LS. UK

Professor I. D. Cooke
University Department of Obstetrics & Gynaecology, Jessop Hospital for Women, Leavygreave Road, Sheffield S3 7RE, UK

Professor D. J. Cusine
Head of Department of Conveyancing & Professional Practice of Law, University of Aberdeen, Taylor Building, Old Aberdeen AB9 2UB, UK

Mrs J. Denton
Chairman, Nursing Research Fellow, Multiple Births Foundation, Queen Charlotte's and Chelsea Hospital, Goldhawk Road, London W6 0XG, UK

Mr J. Dickson
Director, ISSUE, Birmingham Settlement, 318 Summer Lane, Birmingham B19 3RL, UK

Dr A. Dokras
Clinical Research Fellow, Nuffield Department of Obstetrics & Gynaecology, John Radcliffe Hospital, Headington, Oxford OX3 9DU, UK

Professor J. O. Drife
Department of Obstetrics & Gynaecology, D Floor, Clarendon Wing, Leeds General Infirmary, Belmont Grove, Leeds LS2 9NS, UK

Dr S. B. Fishel
Senior Lecturer & Scientific Director – "Nurture", Department of Obstetrics & Gynaecology, University of Nottingham, Queen's Medical Centre, Nottingham NG7 2UH, UK

Professor S. Franks
Professor of Reproductive Endocrinology, Department of Obstetrics & Gynaecology, St Mary's Hospital Medical School, London W2 1PG, UK

Dr R. G. Gosden
Senior Lecturer in Physiology, The University of Edinburgh, Medical School Building, Teviot Place, Edinburgh EH8 9AG, UK

Dr A. H. Handyside
Senior Lecturer, Institute of Obstetrics & Gynaecology, Royal Postgraduate Medical School, Hammersmith Hospital, Du Cane Road, London W12 0NN, UK

Dr S. G. Hillier
Senior Lecturer and Director, Reproductive Endocrinology Laboratory, The University of Edinburgh, 37 Chalmers Street, Edinburgh EH3 9EW, UK

Professor M. G. R. Hull
Professor of Reproductive Medicine & Surgery, Department of Obstetrics & Gynaecology, University of Bristol, St Michael's Hospital, Southwell Street, Bristol BS2 8EG, UK

Mrs M. M. Inglis
Counsellor in Obstetrics & Gynaecology, Fifth Floor, The Royal Free Hospital, Pond Street, London NW3 2QG, UK

Professor J. F. Kerin
Professor of Obstetrics & Gynaecology, Director of Reproductive Medicine Programme, Flinders Medical Centre, Bedford Park, South Australia 5042

Dr H. J. Leese
Senior Lecturer, Department of Biology, University of York, Heslington, York YO1 5DD, UK

Dr C. Nezhat
Professor of Clinical Obstetrics & Gynecology, Mercer University School of Medicine, Macon, Georgia and Director, Center for Special Pelvic Surgery™, Fertility & Endoscopy Center, Endometriosis Clinic, Medical Quarters, Suite 276, 5555 Peachtree Dunwoody Road, Atlanta, GA 30342, USA

Professor E. Nieschlag
Direktor, Institut für Reproductionsmedizin, Universität Münster, Steinfurterstrasse 107, D-4400 Münster, Germany

Professor R. W. Shaw
Professor of Obstetrics & Gynaecology, University of Wales College of Medicine, Heath Park, Cardiff CF4 4XN, UK

Mr T. A. Sheldon
Lecturer, Academic Unit of Public Health Medicine, University of Leeds, 30 Hyde Terrace, Leeds LS2 9LN, UK

Dr S. J. Silber
Associate Professor of Obstetrics and Gynecology, Department of Urology & Microsurgery, St Luke's Hospital, 224 S. Woods Mill Road, Suite 730, St Louis, Missouri 63017, USA

Professor A. A. Templeton
Department of Obstetrics & Gynaecology, Aberdeen Maternity Hospital, Foresterhill, Aberdeen AB9 2ZD, UK

Dr P. F. Watson
Senior Lecturer in Physiology, Department of Veterinary Basic Sciences, Royal Veterinary College, Royal College Street, London NW1 0TU, UK

Mr H. A. Whittall
Deputy Chief Executive, Human Fertilisation & Embryology Authority, Paxton House, 30 Artillery Row, London E1 7LS, UK

Professor D. G. Whittingham
Director, MRC Experimental Embryology & Teratology Unit, St George's Hospital Medical School, Cranmer Terrace, London SW17 0RE, UK

Professor R. M. L. Winston
Professor of Fertility Studies, Royal Postgraduate Medical School, Hammersmith Hospital, Du Cane Road, London W12 0HS, UK

Additional Contributors

Dr A. A. Gadir
Research Associate, Honorary Senior Lecturer, Department of Obstetrics and Gynaecology, University of Wales College of Medicine, Heath Park, Cardiff CF4 4XN, UK

Dr H. M. Behre
Senior Registrar, Institut für Reproductionsmedizin, Universität Münster, Steinfurterstrasse 107, D-4400 Münster, Germany

Ms N. I. Boland
Research Worker, Department of Physiology, University Medical School, Teviot Place, Edinburgh EH8 9AG, UK

Dr J. G. Carroll
MRC Experimental Embryology and Teratology Unit, St Georges Hospital Medical School, Cranmer Terrace, London SW 17 ORE, UK

Dr J. K. Critser
Director of Andrology, Methodist Hospital of Indiana Inc., 1701 North Senate Boulevard, Wile Hall 611, Indianapolis, IN 46202, USA

Dr C. J. Dickens
Post Doctoral Fellow, Department of Biology, University of York, Heslington, York YO1 5DD, UK

Dr S. L. Ibbotson
Senior Registrar in Public Health Medicine, Academic Department of Public Health Medicine, University of Leeds, 30 Hyde Terrace, Leeds LS2 9LN, UK

Dr R. A. Margara
Senior Lecturer in Fertility Studies, Royal Postgraduate Medical School, Hammersmith Hospital, Du Cane Road, London W12 0NN, UK

Dr P. Mazur
Corporate Fellow, Biology Division, Oak Ridge National Laboratory, Oak Ridge, TN 37831-8077, USA

Dr F. Nezhat
Associate Professor of Clinical Obstetrics and Gynecology, Mercer University School of Medicine, Macon, Georgia and Director, Center for Special Pelvic Surgery ™, Fertility & Endoscopy Center, Endometriosis Clinic, Medical Quarters, Suite 276, 5555 Peachtree Dunwoody Road, Atlanta, GA 30342, USA

Dr I. L. Sargent
University Lecturer, Scientific Director of IVF Unit, Nuffield Department of Obstetrics and Gynaecology, John Radcliffe Hospital, Oxford OX3 9DU, UK

Mrs J. Timson
Embryologist – "Nurture", Department of Obstetrics and Gynaecology, University of Nottingham, Queen's Medical Centre, Nottingham NG7 2UH, UK

Section I

Epidemiology and Clinical Presentation

Chapter 1

Reproductive Trends in the UK

B. J. Botting

Introduction

The civil registration system is the main source of information about fertility trends and how these have varied over time.

Two Acts of Parliament passed in 1836 established the basic registration procedure. These were "An act for marriages in England" and "An act for registering births, deaths and marriages in England". As a result of two further Acts of Parliament, a registration system was set up in Scotland in 1855 and in Ireland nine years later. Civil registration of births was not compulsory in England until 1874. Stillbirths did not have to be registered until 1927 in England, 1939 in Scotland and 1961 in Northern Ireland.

The decline in the numbers of births was a subject of national concern in the 1930s and early 1940s. This gave rise to a desire to increase the amount of information collected on the subject. Under the Population (Statistics) Act of 1938, additional data including the parents' ages and the mother's parity within marriage were collected at birth registration in England and Wales from mid-1938 onwards and in Northern Ireland and Scotland from the beginning of 1939.

This chapter is mainly based on these registration data but also includes data collected about terminations of pregnancy which take place under the 1967 Abortion Act, about infertility treatment through statistical returns from licensed clinics to the Interim Licensing Authority and from surveys; the ongoing General Household Survey, and the study of triplet and higher order births based on births in 1980 and 1982–85. As a result of using these different sources, some data are presented for the UK as a whole, some for Great Britain (for example, abortion data, as there is no abortion Act in Northern Ireland) and some for England and Wales (for example, specific fertility analyses which are not calculated by the separate Register Offices in Scotland and Northern Ireland).

Overall Fertility

The general fertility rate, the number of live births per 1000 women aged 15–44 years (Fig. 1.1), peaked in 1964 at 93 per 1000 women. Birth rates declined overall between the mid-1960s and 1977 when the general fertility rate reached a minimum of 58. Since 1982 there has been a slight upward trend reaching 64 in 1990.

Comparing the component countries of the UK (Fig. 1.2), fertility rates for Northern Ireland are always much higher than the other countries but mirror the same pattern. The other three countries have rates close to one another. In Northern Ireland the proportion of births occurring outside marriage is lower than in the other countries. Northern Ireland also has the highest mean age at childbirth. This is due, at least in part, to women, on average, having more children. In 1989 36% of births within marriage in Northern Ireland were the woman's third or higher order birth, compared with 23% of births in England and Wales. As a result Northern Ireland has the highest total period fertility rate (average number of children per woman based on the age-specific rates of a given calendar year).

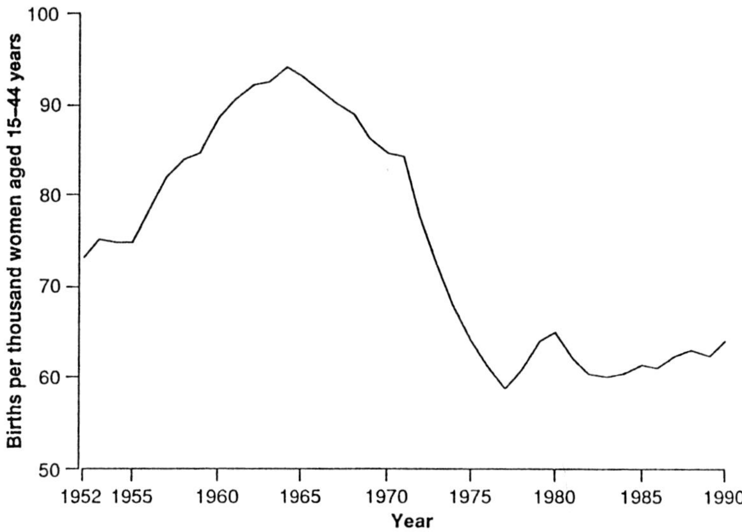

Fig. 1.1. General fertility rate, United Kingdom, 1952–90.

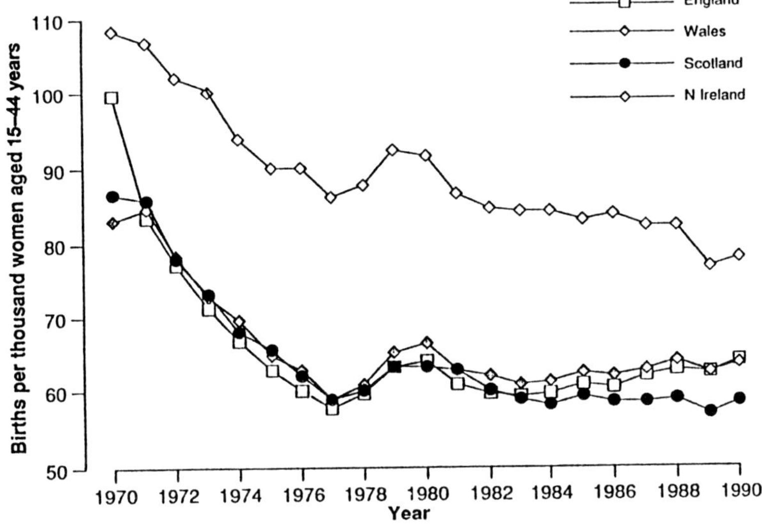

Fig. 1.2. General fertility rates, UK component countries, 1970–90.

Marital Status

One of the most striking trends in recent years has been the dramatic increase in both the number and proportion of all births which occur outside marriage. In 1990 28% of births in the United Kingdom occurred outside marriage. The differences in the level between England, Wales and Scotland were small. The proportion of births outside marriage in Northern Ireland (19%), however, was less than in Great Britain, although the rate of increase over the previous decade was similar [1].

In England and Wales between 1960 and 1980 the number of births outside marriage nearly doubled from 43 000 to 77 000. Since then the number has more than doubled reaching 200 000 in 1990 [2]. In percentage terms this represents an increase from 5% of all births in 1960 to almost 30% in 1990.

The proportion of births which are outside marriage and registered by the mother on her own is not rising, however. The increase is entirely in the proportion of births outside marriage registered by both parents together. Fig. 1.3 shows that, in 1980 when only 12% of all live births were outside marriage, almost half of these were registered solely by the mother. Since then there has been a substantial increase in both the number and the proportion of births registered jointly by both parents. In 1990 73% of all births outside marriage were registered by both parents. In three quarters of these registrations the mother and father stated that, at the time they registered the birth, they were living at the same address.

It also has to be borne in mind that marriage patterns vary from culture to culture. Ethnic origin is not recorded in official registration statistics and

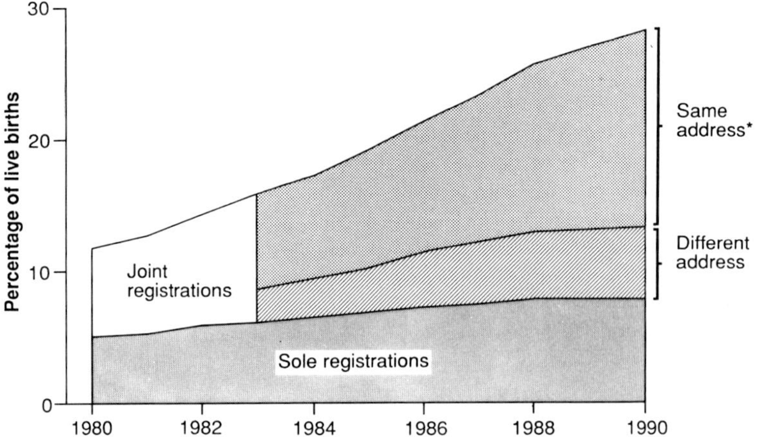

Fig. 1.3. Live births outside marriage as a percentage of total live births, England and Wales, 1980–90. *Estimates of patients address(es) based on a sample of jointly registered births outside marriage for 1983–85, and for all cases for 1986–90. This information is not available for years before 1983.

country of birth is used as a crude approximation to it. While very few women born in Pakistan have babies outside marriage, about half the babies born to women born in the West Indies are outside marriage [2].

Age at Childbirth

Figure 1.4 shows that between 1980 and 1990 fertility levels increased considerably among women in their thirties and forties. In contrast, fertility rates fell among women in their twenties. However, the late twenties remain the peak childbearing years, with fertility levels substantially above those for all other age groups.

The shift to childbearing at older ages is one reason for the steady rise in mean age at childbirth seen since 1975 for births both inside and outside marriage (Fig. 1.5). It is also due to a gradual increase in the proportion of women aged 15–44 years who are in the older part of the age range. An upward shift in the age distribution of women of childbearing ages would, in itself, raise the average age at childbirth if fertility rates were to remain constant. Among married women in 1990 the mean age at childbirth was 28.7 years, the highest since 1952. In contrast, mean age at childbirth for births outside marriage was much lower at 24.6 years, because a far larger proportion of births outside marriage were to women aged under 25 (59% in 1990) compared with births inside marriage (23%).

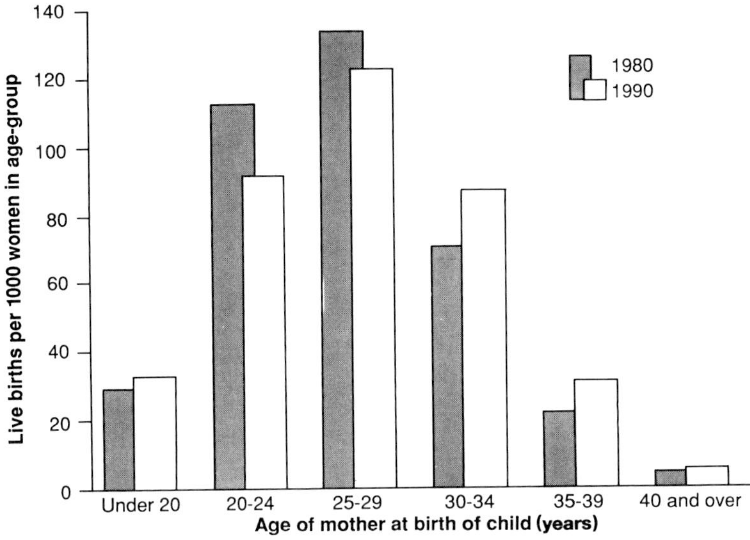

Fig. 1.4. Age-specific fertility rates, England and Wales, 1980 and 1990.

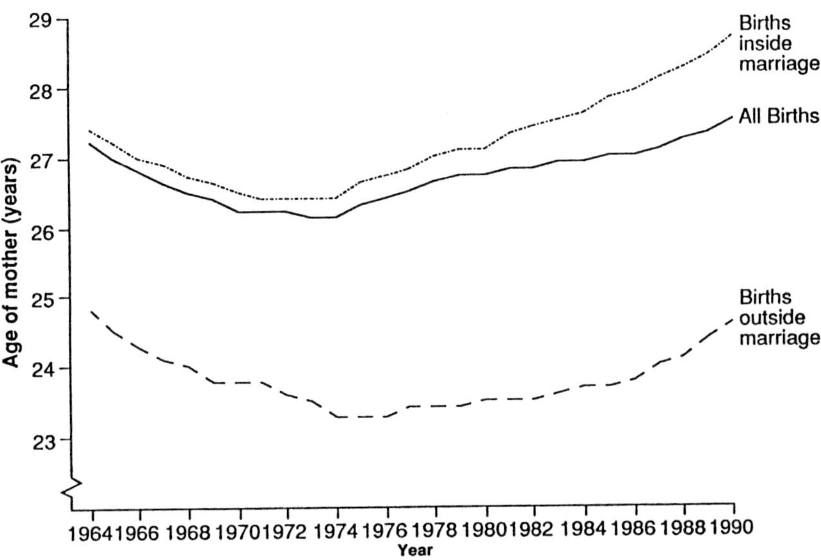

Fig. 1.5. Mean age of mothers at childbirth, England and Wales, 1964–90.

Childlessness

Since the early 1960s there has been an upward trend in the level of childlessness, i.e. those women who had not had a registrable birth by a given age. The proportion of women childless at given ages rose steadily for successive cohorts of women born between 1945 and 1965. There followed a reversal of this pattern for women born in 1970, who are estimated to be following the pattern of the 1960 cohort (Fig. 1.6). Over a third of women born in 1960 reached age 30 without having at least one child. This was twice the proportion for women born in 1945.

It is difficult to determine fully how much is voluntary childlessness and how much is an inability to have children. In Britain it is believed that more than one in ten couples experience difficulty either achieving pregnancy or having a liveborn child [3]. Whether this has varied much over time is not known. Even estimating the current number of these couples is difficult [4].

The current level of childlessness may be explained to some extent by the trend towards later childbearing, later age at marriage and higher levels of marital breakdown. It may also be related indirectly to other factors such as more women being highly qualified. The 1988 General Household Survey (GHS) [5] found that for women born in the late 1950s less-qualified women stated a greater preference for a family with four or more children (12% compared with 6% among more-qualified women) whereas a smaller proportion expected to have no children (5% compared with 10% among more-qualified women). Of course, birth expectations may not necessarily translate into actual childbearing experience, but nevertheless the 1988 GHS showed that by the

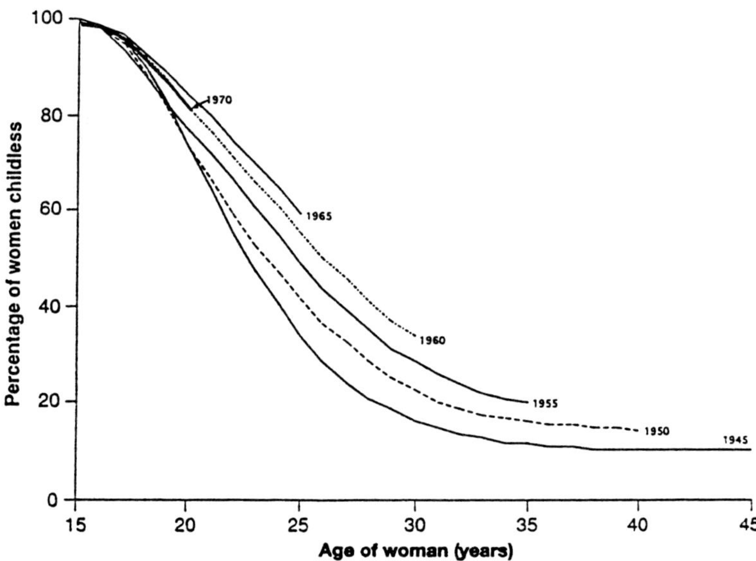

Fig. 1.6. Percentage of women childless at successive ages, 1945–70 cohorts.

time women born in the early 1950s had reached age 30 those with higher academic qualifications had on average 1.2 children whereas those with fewer qualifications had on average 1.7 children.

Infant Mortality

The infant mortality rate (deaths of live-born infants under one year of age per 1000 live births) in England and Wales in 1990 was 7.9, the lowest ever recorded, and 7% lower than the 1989 rate of 8.4 [6]. Between 1989 and 1990, the stillbirth rate (stillbirths per 1000 total births) fell from 4.7 to 4.6, again the lowest rate ever recorded.

Figure 1.7 shows trends in the two components of infant mortality, neonatal deaths (deaths at less than 28 days after live birth) and postneonatal deaths (deaths at ages 28 days or over but under one year) since 1975. These show that the neonatal mortality rate fell continuously from 10.7 per 1000 live births in 1975 to 4.6 in 1990. The postneonatal mortality rate also fell, though less steeply and less regularly, from 5.0 per 1000 live births in 1975 to 3.3 in 1990, after fluctuating around 4.0 between 1983 and 1987. The infant mortality rate has also fallen in each year with the exception of 1986, when there was a small increase due to a rise in the postneonatal mortality rate.

One possible explanation for the slower decrease in postneonatal mortality in recent years compared with neonatal mortality is that improvements in medical knowledge and neonatal intensive care facilities may have prolonged the lives of some babies. The proportion of infant deaths occurring in the postneonatal period has increased from 32% in 1975 to 42% in 1990. Thus

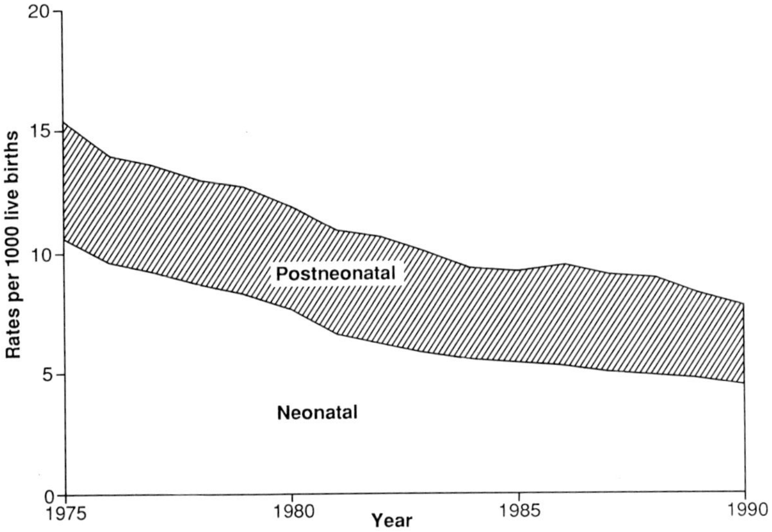

Fig. 1.7. Linked infant mortality rates, England and Wales, 1975–90.

babies who might have died in the neonatal period could now be surviving longer but dying in the postneonatal period.

Pregnancies

Not all pregnancies result in a registrable live or still birth. Fig. 1.8, prepared for the Confidential Enquiry into Maternal Deaths [7], estimated that during 1985–87 76% of pregnancies led to a maternity resulting in one or more registrable live or still births. A further 17% of pregnancies were legally terminated under the 1967 Abortion Act.

The remaining 7% of "other" pregnancies included ectopic pregnancies and spontaneous miscarriages where the mother was admitted to hospital. Hence these figures do not include other pregnancies which miscarry early and where the woman is not admitted to hospital, or indeed where the woman herself may not even know she is pregnant.

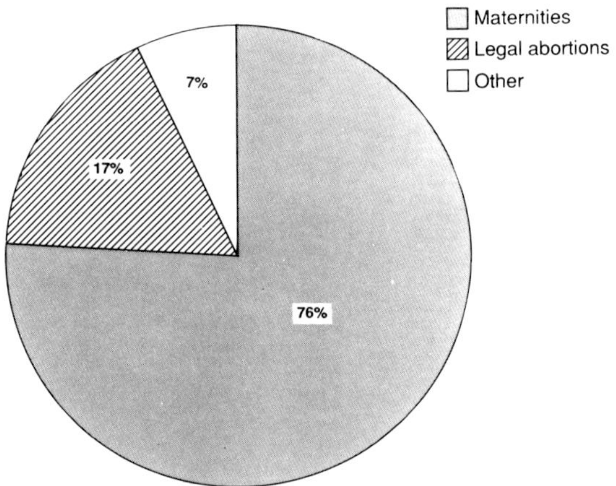

Fig. 1.8. Estimated pregnancies, England and Wales, 1985–87. Source: Confidential Enquiry into Maternal Deaths, 1985–87, Crown copyright.

Conceptions

In England and Wales, data are provided which combine the numbers of maternities resulting in live or still births with the number of legal terminations, adjusted back in time to estimate the year and the women's ages at conception. This adjusted is based on the given period of gestation for terminations and stillbirths and an estimated gestation of nine months for live births. These data are referred to in official statistics as "conceptions" and are based on the estimated year of conception. Miscarriages before 28 completed weeks of gestation are not included in these data.

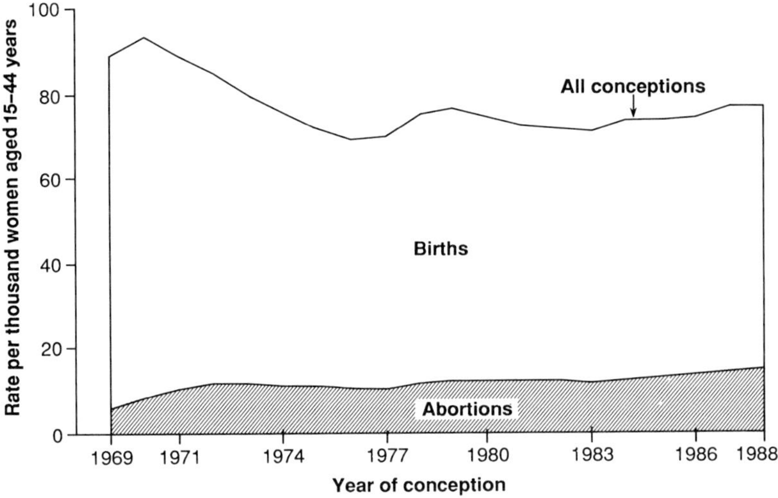

Fig. 1.9. Outcome of conceptions among women aged 15–44 years, England and Wales, 1969–89.

Figure 1.9 shows that conception and birth rates fell between 1970 and 1976, a period when termination rates stabilised between 10 and 12 terminations per 1000 women aged 15–44 years. Birth rates then increased between 1976 and 1979, and these increases were similarly reflected in increases in conception rates. Between 1980 and 1983 both birth and termination rates fell slightly. Fertility and abortion rates then resumed their gradual increase.

In 1989 20% of all conceptions ended in termination. The highest proportions of conceptions ending in termination occur in the youngest and oldest women (Fig. 1.10). Approximately 40% of all conceptions to women aged under 20 or 40 or over are legally terminated. In 1989 the lowest proportion of conceptions

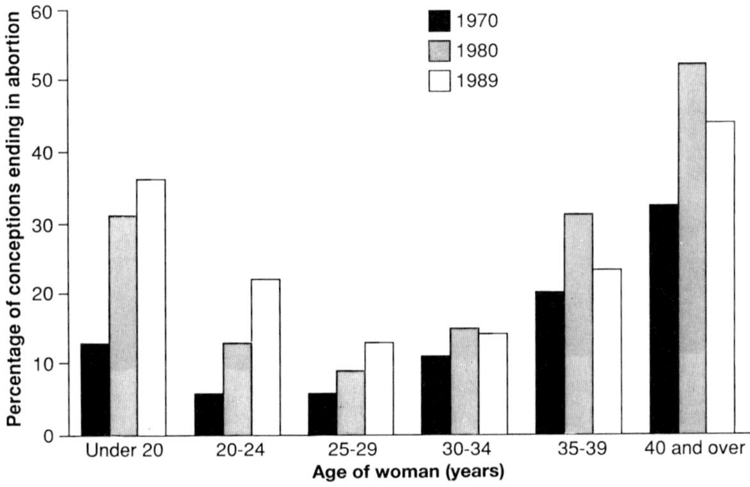

Fig. 1.10. Percentage of conceptions ending in abortion, England and Wales, 1970, 1980 and 1989.

resulting in termination (fewer than 1 in 7) were for women aged between 25 and 34. This is the age group where fertility rates are the highest.

Over time there has been an increase in the percentage of conceptions ending in abortion for age groups under 30. For older age groups there was an increase between 1970 and 1980 followed by a fall between 1980 and 1989, but the level remained above those in 1970.

Abortions

The 1967 Abortion Act came into force on 27 April 1968. Fig. 1.11 shows the abortion rate for residents of England and Wales between 1968 and 1990. It is likely that a major effect of the 1967 Abortion Act was to transfer abortions from the illegal to the legal sector [8]. This would explain the rapid increase in terminations to residents from 22 000 in 1968 to 109 000 in 1972. The number of abortions for residents remained fairly constant until 1978 when there was an increase of over 20 000 terminations over two years. Among other factors, this increase corresponded with an increase in birth rates. The number of terminations for residents rose again between 1979 and 1984, reaching 136 000, and then increased in each successive year reaching 174 000 in 1990.

Residents from elsewhere in the UK may travel to England and Wales for terminations and these are shown in Fig. 1.12. The 1967 Abortion Act does not apply to Northern Ireland where only a few hundred legal terminations are performed each year on medical grounds under the case law which applied in England and Wales before the 1967 Abortion Act [9]. Each year almost

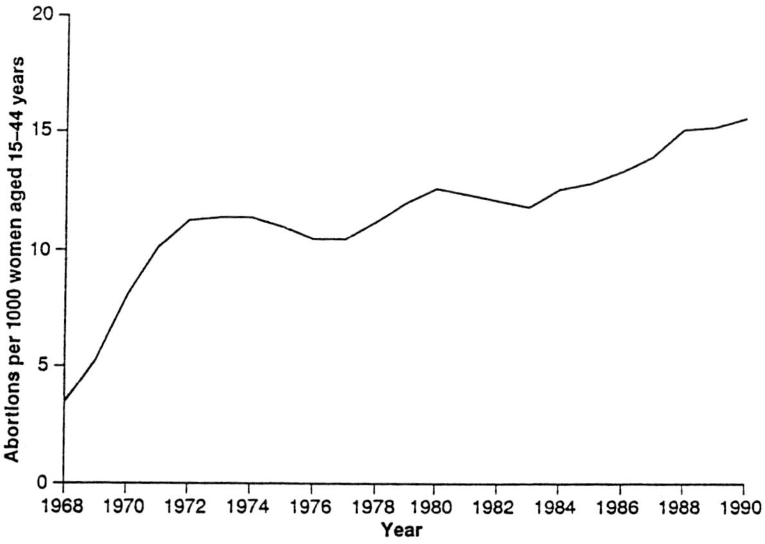

Fig. 1.11. Abortion rate, residents of England and Wales, 1968–90.

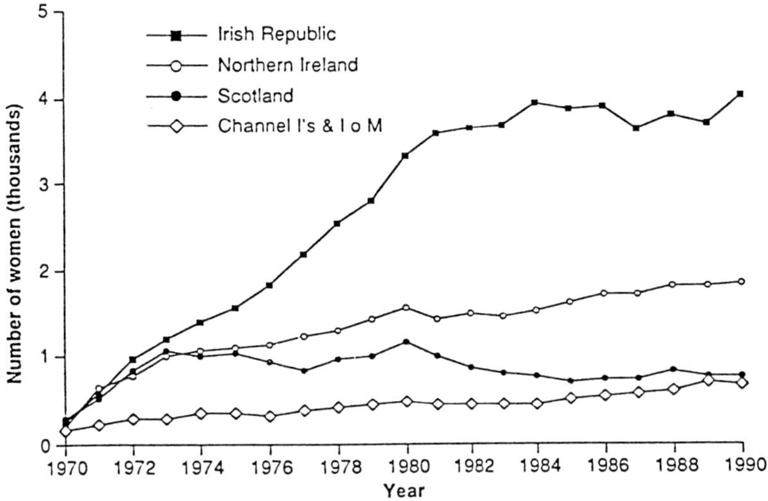

Fig. 1.12. Abortions in England and Wales to residents elsewhere in the UK, 1970–90.

2000 women giving usual addresses in Northern Ireland travel to England and Wales for terminations.

The projected average lifetime abortions presented in Fig. 1.13 is a hypothetical measure which gives the average number of terminations for a cohort of 1000 women who throughout their reproductive years experienced the age-specific abortion rates for the given time period. It is a composite measure as it makes no allowance for women who have more than one termination. Nevertheless, the measure shows that, based on 1970 data,

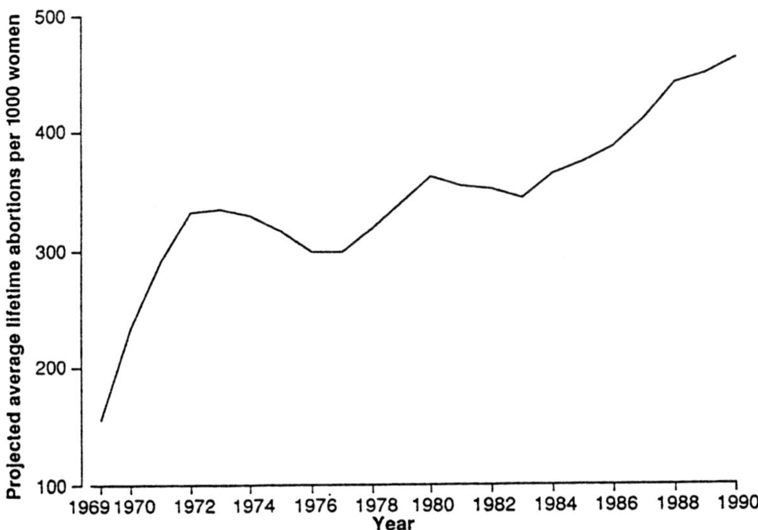

Fig. 1.13. Projected average lifetime abortions, England and Wales, 1969–90.

approximately 20 women in every 100 had a termination, compared with 45 women in every 100 based on 1990 data.

Contraception

Recent patterns of fertility are likely to have been affected by the availability of effective contraception, which means that most people are now able to decide on the number and spacing of their pregnancies.

Although sterilisation should be completely effective in preventing pregnancy, no method of contraception can guarantee that pregnancy will not occur. Since different methods of contraception vary in their effectiveness and their use has changed over time, it is relevant to consider whether changes in contraceptive practices have been associated with changes in the number of conceptions.

Before 1968, contraception services under the NHS were restricted to women who had a medical need to avoid pregnancy. The Family Planning Act of 1967, which came into effect in 1968, enabled local authorities to provide contraceptive advice irrespective of marital status or medical need.

On 1 April 1974 contraceptive supplies prescribed and dispensed through clinics became available free of charge. Since July 1975 contraceptive supplies prescribed by GPs have been free of charge and GPs have been paid "item of service" fees for providing contraceptive services. GPs are not able to prescribe condoms.

Female sterilisation on medical grounds has been available since the NHS began. It also became explicitly available on family planning grounds in 1974. The National Health Service (Family Planning) Amendment Act 1972 gave local authorities the power to provide vasectomy services from November 1972, on the same basis as those for contraception, but some vasectomies had been done under the NHS before this date.

Based on data taken from four surveys carried out by the Office of Population Censuses and Surveys (OPCS) between 1970 and 1983, there was an overall rise in the use of contraception by ever-married women and some marked changes in the popularity of different methods [10–14]. The use of the condom, cap, withdrawal and safe period continued to fall whereas the use of the pill increased. By 1975 the pill, rather than the condom, was the most popular method of contraception among ever-married women. By 1983 the earlier increase in use of the pill had halted and there was a small decrease in usage. This decrease continued to 1986 [13,14].

The main change in the pattern of contraceptive use among ever-married women between 1976 and 1986 was in the proportion of ever-married women aged 16–40 years who were sterilised or who had partners who were sterilised for contraceptive reasons. In 1970, only 4% of these women were sterilised or had partners who were sterilised. By 1986 almost half of all women aged 35–44 years or their partners had been sterilised for contraceptive purposes, compared with approximately 20% in 1976. Fig. 1.14 is based on all women regardless of marital status and shows the use of the contraceptive pill and sterilisation in 1976, 1983, and 1986 by the women's ages. It shows that use of the

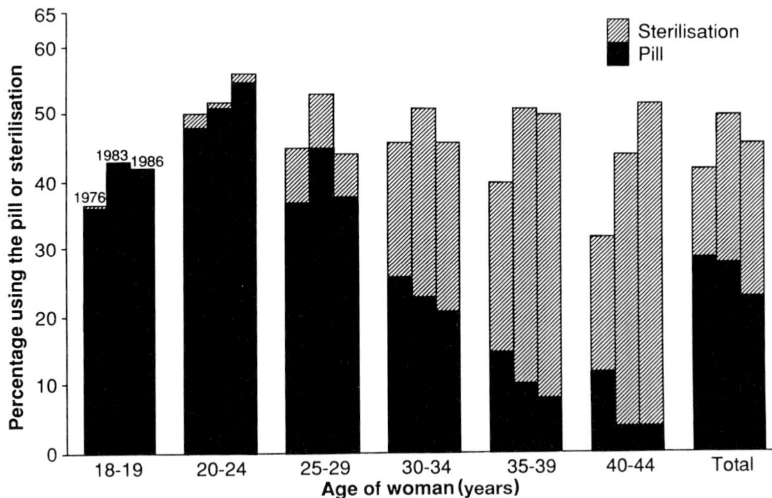

Fig. 1.14. Percentage of women aged 18–44 years using the pill or sterilisation (woman or partner) as a method of contraception by age, Great Britain, 1976, 1983 and 1986.

contraceptive pill was concentrated in the younger age-groups and sterilisation in the older age-groups.

The decrease in conception rates for residents of England and Wales between 1974 and 1977 coincided with the growth in freely available contraceptive services (although many sterilisation operations are paid for privately). The subsequent decrease in pill usage between 1976 and 1983 can be contrasted with increases in conception rates between 1977 and 1980 followed by a fall until 1983.

It is not possible directly to correlate termination rates with contraceptive practices since many of those seeking legal termination may not have been using any form of contraception. A study reviewing the contraceptive practices of patients seeking legal abortion in the Scottish Highlands in 1986 showed that two-thirds of patients seeking a legal termination had not been using contraception [15]. In order to correlate termination rates and contraceptive practice more directly, better information about the use of methods would be required.

Multiple Births

Patterns in the incidence of twin and higher order births have not followed that of overall fertility. The collection of data about multiple births in England and Wales began in July 1938 through the Population (Statistics) Act. Its aim was to collect additional data so that fertility rates could be monitored more fully. The reason for analysing multiple births was given as follows: "Although fertility is usually measured by the number of children born rather than by the number of maternities experienced, it is necessary to remember that the former is a composite total made up of the number of maternities, which is susceptible to voluntary control and the number of extra children born in multiple maternities which cannot be so readily controlled". [16]

Although multiple births form only a small proportion of all maternities, their number has risen steadily over the past ten years from over 6000 maternities in 1980 to over 8000 in 1990. While twin births are quite common at about 1% of all deliveries, the higher order births are much less so.

Fig. 1.15 shows the proportion of all maternities in England and Wales between 1939 and 1990 which resulted in a multiple birth. The proportion of all pregnancies which were a twin or higher order delivery rose from 1939 until the early 1950s and then fell until the late 1970s before starting to increase again. The increase in rate from the late 1970s is, at least in part, a consequence of the growing use of ovulation-stimulating drugs and, in the most recent years, various techniques of assisted conception such as IVF and GIFT.

A much more marked pattern is observed if triplets and higher order births are looked at separately. The ratio of higher order multiple births to twins has reduced in recent years such that by 1990 1 in 39 multiple births was a triplet or higher order delivery compared with 1 in 81 multiple births in 1970. Nevertheless, twins continue to dominate the pattern of all multiple births.

Fig. 1.16 shows the increase over time in the numbers of triplet and higher order births as a proportion of all pregnancies. The striking feature of this graph is the large increase in triplet and higher order births since 1970. These higher order birth rates showed a marked increase in 1986 when a record 123 sets of triplets were born in England and Wales. A similar level was recorded in 1987, but the rate then increased again in 1988 and 1989. In 1990 201 sets of triplets were registered. This was the most in any year since multiple birth

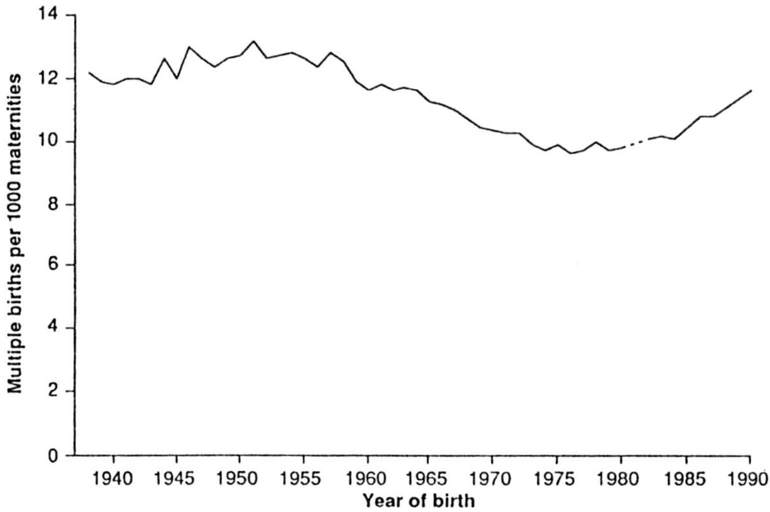

Fig. 1.15. Proportion of all maternities which resulted in a multiple birth, England and Wales, 1939–90.

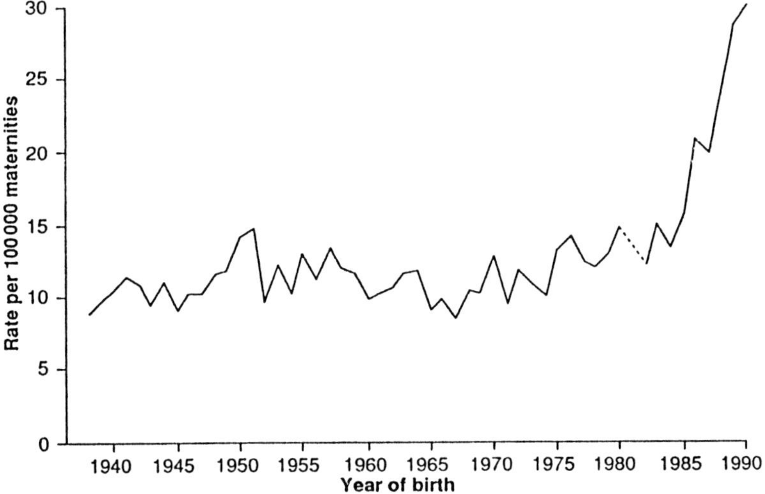

Fig. 1.16. Proportion of all maternities which resulted in a triplet or higher order birth, England and Wales, 1939–90.

statistics were first compiled, and represented a rate of 29 sets of triplets per 100 000 maternities (pregnancies leading to a registrable birth) compared with a rate of 14 in 1985.

The National Study of Triplet and Higher Order Births [17] had an overall aim to look for ways of improving the care and services provided for families with children from triplet and higher order births. It was planned by a Steering Group which included clinicians, parents and members of the Child Care and Development Group, the National Perinatal Epidemiology Unit and the Office of Population Censuses and Surveys. The Study was made up of a series of linked surveys of obstetricians, parents, general practitioners, community nursing departments and social services departments.

Data from the obstetric survey of the study of triplet and higher order births showed that 70% of mothers of quadruplet and higher order births in 1980 and 1982–85, and 36% of mothers of triplets in these years, had used drugs for ovulation induction. In contrast, only 6% of mothers of twins and 2% of mothers of singleton babies had used these drugs. These data are shown in Fig. 1.17. The impact of IVF and other assisted conception techniques was negligible in the survey period except in 1985.

Infertility Procedures

Some infertility is the result of a failure of ovulation and the woman can be prescribed drugs which stimulate ovulation. These drugs have the disadvantage that under some conditions more than one ovarian follicle ripens to produce multiple ova with a resulting higher risk of multiple births. It has been estimated

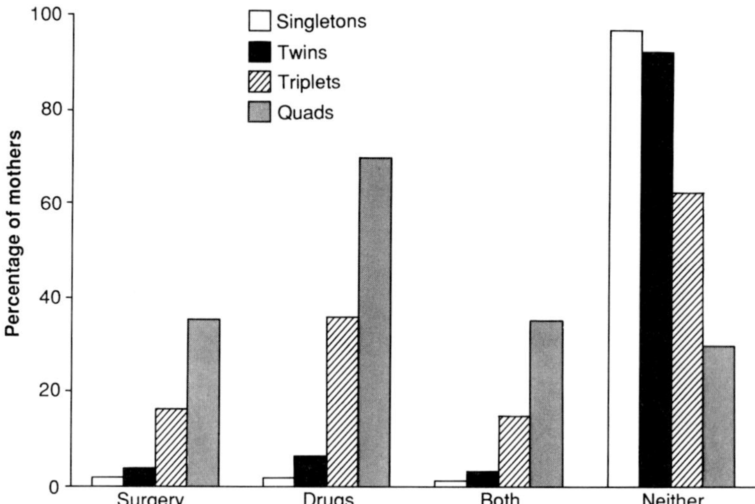

Fig. 1.17. Treatment for infertility, National Study of Triplet and Higher Order Births, 1980 and 1982–85.

that 6%–8% of pregnancies following treatment with the ovulation-induction drug clomiphene led to multiple births [18]. The risks with gonadotrophins were higher with estimates ranging from 18% to 54% [19]. The percentages varied according to the reasons for infertility and the dosage and timing of the drugs used. The majority of these multiple births will be twins, but a small proportion will be triplets and higher multiples.

More recently, multiple births have been the result of assisted reproduction techniques such as in vitro fertilisation (IVF) and gamete intrafallopian transfer (GIFT). The higher multiple pregnancy rate achieved with these assisted reproduction techniques is greater than expected [20].

The Sixth Report of the Interim Licensing Authority (ILA) [21] presents a statistical analysis of 1989 data reported to the ILA from centres licensed to perform IVF and/or GIFT. The overall multiple pregnancy rate for IVF in 1989 was 26.8% and for GIFT 20.4% (Fig. 1.18). The risk of multiple pregnancy increased with the number of eggs or pre-embryos transferred. When two or more pre-embryos were transferred during IVF and a pregnancy was established, the chance of a multiple birth was over 20%.

Mortality of Multiple Birth Babies

Babies born in a multiple birth have far higher mortality risk, partly due to being born preterm and of low birthweight. Fig. 1.19 shows that mortality risks for multiple births have not changed much since 1981 although this hides the increase in the number of triplet and higher order births which are more at risk than twins.

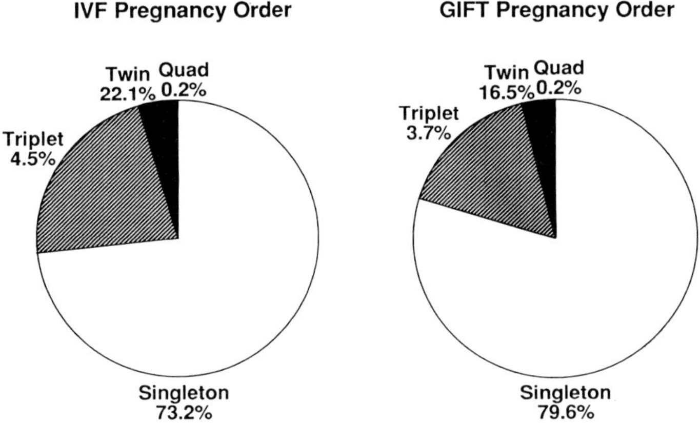

Fig. 1.18. Multiple pregnancies after IVF and GIFT, 1989. (Reproduced with permission of the Interim Licensing Authority.)

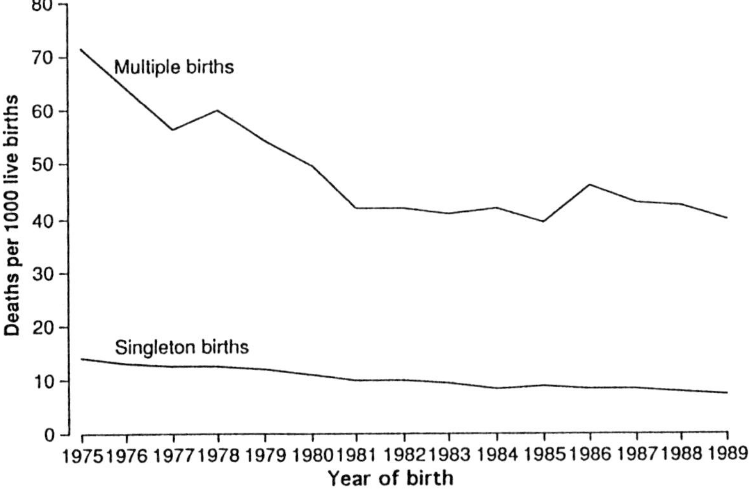

Fig. 1.19. Infant mortality rates for singleton and multiple births, England and Wales, 1975–89.

In 1989, as in previous years, over half of all live births in a multiple delivery were of low birthweight (less than 2500 g) compared with 6% of singleton live births. An even more marked pattern is seen within subgroups of this low birthweight category. Fig. 1.20 shows the distribution by birthweight of live births from singleton and multiple births in 1987–89. Of multiple births 8% were of very low birthweight (less than 1500 g) compared with 1% of singleton births.

Babies of very low birthweight (less than 1500 g) often die at very young ages, so they contribute disproportionately to very early deaths [22]. In 1987 54% of all singleton live-born babies that died during the first 30 min had

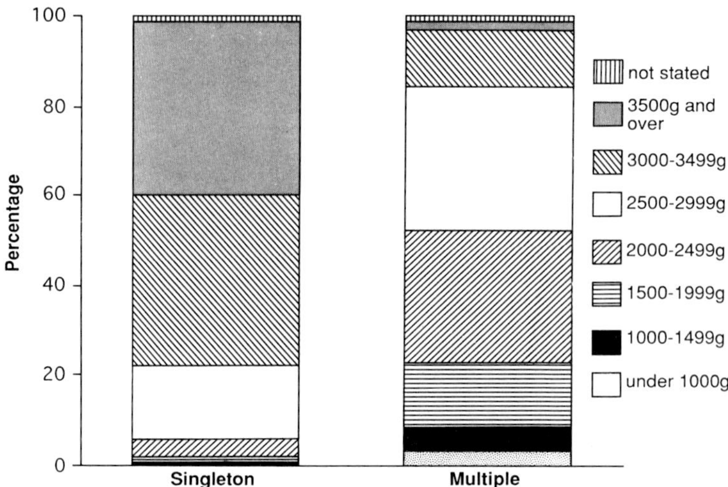

Fig. 1.20. Percentage distribution of singleton and multiple live births by birthweight, England and Wales, 1987–89.

weighed less than 1500 g compared with 92% of deaths among multiple births.

Low birthweight babies born in a multiple birth have higher mortality rates than heavier babies, whether from a singleton or multiple birth. For any given birthweight, however, babies from multiple births will, on average, be of an older gestational age and hence more mature. For babies born in a multiple birth in 1989, the overall stillbirth rate was three times higher than that for singletons, but for low birthweight stillbirths the rate was 45% lower. Again, the infant mortality rate for all multiple birth babies was five times higher than for singletons, but for low birthweight babies it was only 12% higher.

References

1. Craig J. Fertility trends within the United Kingdom. Population Trends, no. 67. London: HMSO, 1992.
2. OPCS. Birth statistics 1990. Series FM1; no. 19. London: HMSO, 1992.
3. Page H. The increasing demand for fertility treatment. Health Trends 1988; 115–18.
4. Greenhall E. Sub-fertility in the community. Thesis submitted for membership of the Faculty of Community Medicine. Oxford, 1987.
5. OPCS. General Household Survey 1988. Series GHS; no. 18. London: HMSO, 1989.
6. OPCS. Mortality statistics 1989 perinatal and infant: social and biological factors. Series DH3; no. 23. London: HMSO, 1992.
7. Department of Health. Confidential Enquiry into Maternal Deaths, United Kingdom 1985–87. London: HMSO, 1991.
8. Botting B. Trends in abortion. Population Trends; no. 64. London: HMSO, 1991.
9. Macfarlane A, Mugford M. Birth counts: statistics of pregnancy and childbirth. London: HMSO, 1984.
10. Bone M. Family planning services in England and Wales. OPCS Series SS; no. 467. London: HMSO, 1973.
11. Bone M. Family planning services: changes and effects. OPCS Series SS; no. 1055. London: HMSO, 1978.

12. Dunnell K. Family formation 1976. OPCS Series SS; no. 1080. London: HMSO, 1979.
13. OPCS. 1983 General Household Survey Report. Series GHS; no. 13. London: HMSO, 1985.
14. OPCS. 1986 General Household Survey Report. Series GHS; no. 16. London: HMSO, 1989.
15. Lin BH, Mahmood TA. Contraceptive practice among women seeking legal abortion in the Scottish Highlands. J Obstet Gynaecol 1989; 2:208–9.
16. General Register Office. Registrar General's statistical review for the year 1983. Part III, Civil. London: HMSO, 1944.
17. Botting BJ, Macfarlane AJ, Price FV, eds. Three, four and more: a study of triplet and higher order births. London: HMSO, 1990.
18. Merrell Pharmaceuticals Ltd. Clomid data sheet compendium. London: Data Pharm, 1981.
19. Schenker JG, Yarkoni S, Granat M. Multiple pregnancies following induction of ovulation. Fertil Steril 1981; 35:105–23.
20. MRC Working Party on children conceived by in-vitro fertilisation. Births in Great Britain resulting from assisted conception, 1978–87. Br Med J 1990; 300:1229–33.
21. Interim Licensing Authority. The sixth report of the Interim Licensing Authority for human in-vitro fertilisation and embryology. London: Interim Licensing Authority, 1991.
22. Alberman E, Botting B. Trends in prevalence and survival of very low birthweight infants, England and Wales: 1983–7. Arch Dis Child 1991; 66:1304–8.

Chapter 2

The Epidemiology of Infertility

A. Templeton

Introduction

Few countries collect information relevant to the study of the epidemiology of infertility. National data available in the UK are either indirect or incomplete. Furthermore, most published studies have methodological problems which mean they must be carefully interpreted. In particular there are difficulties studying the rates of resolved infertility and secondary infertility. Similarly, there are few data or studies in any European countries which provide information on secular trends in infertility, although such information may be available in the US. It is in the national interest to have this information both from the point of view of public health, and also in the planning, purchasing and provision of medical services. It is, however, clear that infertility is a common problem affecting a large proportion of otherwise healthy individuals, usually in their 20s and 30s. Infertility diminishes their sense of wellbeing and fulfilment and sometimes causes great personal distress. This chapter reviews the available information on the epidemiology of infertility and, even though all the current surveys are deficient in one way or another, conclusions are drawn where possible. It is hoped that the need for a robust system of national data collection will become apparent.

Historical Aspects

For the purpose of ascertaining the point by numerical data, I had a census taken of two villages of considerable size – near Grangemouth in Stirlingshire

and Bathgate in West Lothian – the one consisting principally of a seafaring population, and the other of persons engaged in agriculture and manufacture.
Fecundity, Fertility and Sterility. Matthews Duncan, 1872

Matthews Duncan (an Aberdeen graduate) was the first to publish a study of infertility rates in this country. His book *Fecundity, Fertility and Sterility* was written more than a hundred years ago and is all the more remarkable for its report of the first, and until recently only, population-based study of infertility [1].

Since that time surprisingly little attention has been paid to the prevalence of infertility, although a number of studies have employed census or parish record data in an attempt to define overall sterility rates [2]. Similarly, census data in the US indicate fluctuations in the rates of childless women during the 20th century, with a rise in the 1970s. An important, and indeterminate, factor in reviewing census data is the rate of voluntary infertility. This has apparently shown marked changes during the 20th century, particularly in the US where the prevalence has been reviewed by Poston and Kramer [3]. During the 1920s and up until the time of the Second World War, voluntary childlessness was extremely common, perhaps accounting for 25%–40% of cases of apparent infertility. After the War, there appeared to be a strong aversion to childlessness. Voluntary childlessness was extremely unusual and there was an overall reduction in infertility. In the 1960s, however, there was a very gradual increase in the overall incidence of childlessness, both voluntary and involuntary, with several studies highlighting the contribution by the voluntary category in the 1960s and 1970s [4–6]. Indeed some have concluded that the apparent increase in infertility in the 1960s and 1970s was due to voluntary rather than involuntary factors [3].

That having been said, there is no doubt that the major factor affecting the incidence of involuntary infertility is pelvic infection. There is a close association between sexually transmitted disease and fluctuations in infertility rates and the importance of preventing infection is well established [8,9]. Since 1984 when the Warnock Committee observed "even today there is very little factual information about the prevalence of infertility" [10], there have been a number of reports studying the problem. It is these studies, published mainly during the last decade, which form the basis of this review.

Methodology

The scarcity of useful data is not really surprising when the problems inherent in measuring the prevalence of infertility are considered [11].

The major problem with census data, as previously indicated, is that they cannot assess the impact of voluntary childlessness, nor do they give any indication of the extent of secondary infertility. A number of alternative approaches have been attempted, which have been reviewed by Greenhall and Vessey [11] and Thonneau and Spira [12]. Surprisingly few countries have adopted large-scale population surveys, the most notable exception being the

National Survey of Family Growth (NSFG) in the US which over the last 25 years has yielded important information on prevalence rates and trends [13–15]. In a population-based survey a stratified random sample of women in Denmark was interviewed in 1979 [16]. Such surveys carried out at intervals, using the same techniques in geographically defined populations, could yield important information on prevalence and trends, as well as assessing individuals' attitudes and the uptake of medical services. However, such information is available only from the US surveys. Such national surveys, on the other hand, do not lend themselves easily to interregional comparisons, nor can they give any indication of the aetiology of infertility. Other methods of sampling total populations have been attempted and include survey of general practitioner records [11,17] or random sampling from the Age–Sex Register [18] or Community Health Index [19,20].

Yet another approach has been the study of infertility among groups of women assembled for other purposes. These include screening for cervical cancer [21], a case–control study of breast cancer [11] and women stopping contraception presumably with the intention of becoming pregnant [22,23].

Thonneau and Spira [12] described two regionally based studies, one based in three French regions [24] and one based in Bristol, UK [25]. However, both these surveys included only those couples who had consulted a doctor for infertility and in the second case only those who were seen in a specialist clinic, and it is now well established that not all infertile couples will seek medical advice. However, these and other clinic-based approaches will give some indication of the relative rates of primary and secondary infertility as well as information on the several different causes of infertility. Clearly this information will have value in planning services.

A further problem relates to the duration of infertility defined in the various studies. Many of the studies described above use a one-year duration, most notably the NSFG studies in the US [14,15], the Oxford study [11], the French study [24], the Bristol study [25], and the Danish study [16]. A two-year definition was used in the Aberdeen studies [19,20] and has been supported by Trussell and Wilson [2]. This definition may be more in keeping with clinical practice and referral patterns.

Greenhall and Vessey [11], and Mosher and Pratt [15] have drawn attention to further problems in assessing the prevalence of subfertility. Infertile groups will include those who are sterile and those who have impaired fecundity. These authors have suggested therefore that infertility be categorised into "resolved" and "unresolved". There is a high number of pregnancies among apparently infertile women, independent of treatment [26,27] as well as following treatment, and this difference is important in attempting to understand the condition and in the planning and provision of services.

Prevalence of Infertility

The prevalence rates for all relevant studies published during the last ten years are given in Table 2.1. There is remarkable agreement among the studies, given the different methodologies and definitions used. In particular the region-

Table 2.1. Studies published in the last decade suggesting prevalence rates of infertility

Clinic based studies	Hull et al. [25]	17%
	Thonneau et al. [24]	14%
Population based studies	Rachootin and Olsen [16]	18%
	Hirsch and Mosher [14]	14%
	Johnson et al. [17]	14%
	Page [18]	13%
	Greenhall and Vessey [11]	24%
	Templeton et al. [19,20]	15%

or clinic-based studies could be expected to underrate the prevalence, as the estimates are based on only those couples who sought medical help. Similarly studies which employ shorter durations of infertility could be expected to report higher prevalence rates. Furthermore the reported studies span a decade, during which there may have been changes in the prevalence of both voluntary and involuntary infertility, as well as changes in referral patterns associated with greater awareness and understanding of the problem.

Overall, about 14% of individuals surveyed appear to have had a problem with infertility. Most studies, although not all, differentiate between primary and secondary infertility. In contrast to clinic-based studies, which indicate that the majority of couples will present with primary infertility [24,25,28–30], population-based studies indicate an equal or greater proportion of women with secondary infertility [11,14–16,19,20]. This difference reflects the uptake of medical services, and highlights the difficulties that clinic-based studies have in studying prevalence and change.

As indicated above, it has been suggested that infertility should be divided into resolved and unresolved categories [11]. Resolved infertility refers to those who became pregnant following a subfertile episode. The studies described in Table 2.1 indicate that the resolved rates for primary infertility vary from 5% to 14% and for secondary infertility from 3% to 12%, reflecting the differences in methodology and particularly the duration of infertility used. However, for unresolved infertility there appears to be more agreement, the range for primary infertility being 3%–6% and for secondary infertility 3%–7%.

These figures highlight the difficulties associated with oversimplification in describing infertility rates. Clearly the standard textbook figure suggesting that 10% of couples are infertile requires modification. From the studies presented above, which pertain mainly to Western society, it can be seen that around 14% of couples will be troubled by infertility, although rates as high as 25% have been reported when shorter durations of infertility are used [11,18]. At least half of these couples will present with secondary infertility, and the infertility will remain unresolved in a further half of each group. Thus overall only 3%–4% of couples will remain involuntarily sterile.

Changes in Prevalence

With the exception of the US and possibly France [12], national data that would allow the study of trends in the rates of infertility are unavailable.

Table 2.2. Prevalence of infertility among three age cohorts of women from a defined geographical area

Infertility	Aged 26–30 years (n = 465)	Aged 36–40 years (n = 1064)	Aged 46–50 years (n = 766)
Primary	6.7	7.1	7.3
Primary or secondary	0.6	2.3	1.6
Secondary	5.4	5.3	5.2
Total	12.7%	14.7%	14.1%

Census data, with all their limitations particularly with respect to voluntary infertility and secondary infertility, have nonetheless indicated an overall increase in the prevalence of infertility during the last three decades compared to the five decades before that [31,32].

Between 1965 and 1976 there was no overall increase of childlessness in the US [13]. However, this apparent stability obscured highly significant increases among young women, particularly younger black women in whom an increase from 3% to 13% was observed. During the same period there was a considerable increase in the rate of sexually transmitted disease [33,34] and the relationship between pelvic infection and infertility has been well documented [35]. During the same and subsequent decades, there has been a clearly documented increase in the associated condition of ectopic pregnancy [36–38].

However, despite these trends, any significant increase in the rate of infertility in Western Europe has been difficult to demonstrate. In fact, over the last 10–20 years the evidence points to a fairly stable background prevalence [12]. Two studies have attempted to address the problem by surveying different age cohorts in the same population. Johnson et al. [17] examined the general practitioner records of all women on their lists born in 1950 and 1935. The rates of involuntary childlessness were respectively 3.3% and 4.5%. Templeton et al. [20] surveyed women randomly selected from the Community Health Index for a geographically defined region. They have now compared three age cohorts, namely 26–30, 36–40 and 46–50 years. There was no overall increase in the prevalence of infertility whether primary or secondary, resolved or unresolved (Table 2.2). Further study (as yet unpublished) indicates no difference between rural and urban communities. Thus, despite an increase in sexually transmitted disease and a substantial increase in the rate of ectopic pregnancy in the same community [38] there has been no significant increase in the prevalence of infertility that can be determined using this methodology. However, this is perhaps not too surprising. Overall Mosher and Aral [33] have estimated that the incidence of infertility caused by sexually transmitted disease among white women is of the order of 0.7%–1%, a small proportion of the overall rate.

Use of Medical Services

It seems unlikely that the apparent increase in the prevalence of infertility in Western countries is associated mainly with an increased use of medical services, and this was first highlighted about a decade ago [39,40]. However, the available studies show very considerable differences in the uptake of services, and the evidence is worth reviewing.

Rachootin and Olsen [41] have demonstrated very clearly in Denmark, albeit ten years ago, that the majority of women with reduced fecundity did not seek medical care, with perhaps only a quarter of those with secondary infertility being referred to hospital. Similarly Hirsch and Mosher [14] have reported a greater use of services among US women with primary infertility and there were marked social differences (mainly higher income and education), among women with secondary infertility who made use of the services. Overall, however, only 50% of infertile women sought medical services. This is in stark contrast to the studies by Hull et al. [25] and Thonneau et al. [24] where the assumption was made that almost all infertile women would seek medical advice.

An increasing demand for specialist referral was documented in the above mentioned general practitioner study [17]. Nearly twice as many women born in 1950 had consulted a specialist compared to women born in 1935. Similarly, Templeton et al. [20] documented an increased use of both general practitioner and hospital services among younger women (Table 2.3). In fact, among young women with primary infertility only 5% had not consulted their doctor compared to 28% in the older age group. There were further differences in the uptake of medical services depending on the social class and urban or rural domicile, but these differences were less marked.

Thus there is variation in the uptake of medical services in different countries, although it is unlikely that there has been a considerable increase in all western countries in the past decade. It is important to know the present level of uptake in each region, so that future needs can be anticipated and services planned appropriately.

Table 2.3. Proportion of infertile women seeking medical advice. Two age cohorts from a defined geographical area

	36–40 years (n = 1064) (%)	46–50 years (n = 766) (%)
Attended hospital	76	62
Primary care only	13	7
Did not seek advice	11	31

From Templeton et al. [20].

Spontaneous Abortion in Infertile Women

A number of population-based studies have suggested a higher rate of spontaneous abortion among infertile women compared to fertile women in the same age groups [16,20]. In our study, among women in the 36–40-year age group the proportion of pregnancies ending in spontaneous abortion was as follows: women with primary infertility 15%, secondary infertility 19%, and no infertility 7%. These figures are striking, although it has to be remembered that the response is based on patient recall and there may be an element of under recording in some groups [42]. Nonetheless, surveys of this sort have been evaluated and the quality of the recall data appears to be acceptable [43].

Higher abortion rates have also been reported among infertile women attending infertility clinics, particularly those with secondary infertility [44–46]. Subgroups of the infertile population among whom spontaneous abortion rates appear to be high include those treated with gonadotrophins [44,47,48] or LHRH [49] and those with polycystic ovarian disease or even elevated luteinising hormone levels [50]. Among other common groups, including those with endometriosis [44,45] and tubal disease [45,51] there does not appear to be any increased risk. Other subgroups with a higher spontaneous abortion rate have been reported, but would be unlikely to contribute to an overall increase. These include women with uterine malformations [52,53] and autoimmune disease [54]. Higher abortion rates after assisted-conception techniques, particularly in older women, are well documented [55,56].

Voluntary Infertility

Although high rates of voluntary infertility have been reported in the recent past [3,57,58] there are few reliable contemporary data. Johnson et al. [17] reported an increase in voluntary childlessness among women born in 1950 (9.2%) compared to women born in 1935 (1.9%). On the other hand Templeton et al. [20] could detect no significant difference in the rate of voluntary infertility among two cohorts of women aged 36–40 years and 46–50 years. The overall rates of voluntary infertility were 8.6% and 7.3% respectively. The reason given by most women was that they had not married, whereas among married women the rate of voluntary infertility was extremely low, around 2%. This rate was even lower in a rural community within the same region. These figures are much lower than previous estimates [3,58] but clearly apply to only one region and an assumption should not be made about general applicability. Similar surveys in other parts of the country would be of interest.

Conclusions

All studies relating to the prevalence of infertility published in the last decade have been reviewed. The methodological difficulties in assessing prevalence rates are considerable, and different approaches are needed to answer different questions. Changes in prevalence rates can be discovered only from population-based studies and except for the US NSFG no such studies exist. Overall in Western countries, 14% of couples would appear to have a problem with infertility, with half or more of these having a problem with secondary infertility. In half of each group again, the infertility will remain unresolved, so that

3%–4% of couples will remain involuntarily childless. Uptake of medical services varies considerably from country to country and probably from region to region within countries, but there is evidence of a considerable increase in the last decade. In some communities almost all women with primary infertility will seek medical help, although there are social and other differences in the level of uptake. The spontaneous abortion rate among infertile women is increased. Voluntary infertility is probably much less frequent than previously estimated, although further studies are needed to assess this. This review has highlighted the need in the UK for a national survey of infertility, that can be repeated at fixed time intervals, to provide data on secular trends.

References

1. Duncan JM. Fecundity, fertility and sterility. Edinburgh: A and C. Black, 1866; 194.
2. Trussel J, Wilson C. Sterility in a population with natural fertility. Popul Stud 1985; 39:269–86.
3. Poston SL Jr, Kramer KB. Voluntary and involuntary childlessness in the United States. Soc Biol 1983; 30:291–306.
4. Veevers JE. Factors in the incidence of childlessness in Canada: an analysis of census data. Soc Biol 1972; 19:266–74.
5. Waller HH, Rao BR, Li CC. Heterogeneity of childless families. Soc Res 1973; 20:133–42.
6. Poston DL Jr. Characteristics of voluntarily and involuntarily childless wives. Soc Biol 1976; 23:198–209.
7. Sherris JD, Fox G. Infertility and sexually transmitted disease. A Public Health Challenge. Popul Rep [L] 1983; 4:113–51.
8. Belsey MA. The epidemiology of infertility: a review with particular reference to sub-Saharan Africa. Bull World Health Organ 1976; 54:319–41.
9. Meyer L, Job-Spira N, Bouyer J, Bouvet E, Spira A. Prevention of sexually transmitted diseases: a randomised community trial. J Epidemiol Community Health 1991; 45:152–8.
10. Department of Health and Social Security. Report of the Committee of inquiry into Human Fertilisation and Embryology. London: HMSO, 1984. (Warnock Report).
11. Greenhall E, Vessey M. The prevalence of subfertility: a review of the current confusion and are part of two new studies. Fertil Steril 1990; 54:978–83.
12. Thonneau P, Spira A. Prevalence of infertility; international data and problems of measurement. Eur J Obstet Gynecol Reprod Biol 1990; 38:43–52.
13. Mosher WD. Infertility trends among US couples: 1965–1976. Fam Plann Perspect 1982; 14:22–7.
14. Hirsch MB, Mosher WD. Characteristics of infertile women in the United States and their use of infertility services. Fertil Steril 1987; 47: 75–9.
15. Mosher WD, Pratt WF. Fecundity and infertility in the United States: incidence and trends. Fertil Steril 1991; 56:192–3.
16. Rachootin P, Olsen J. Prevalence and socioeconomic correlates of subfecundity and spontaneous abortion in Denmark. Int J Epidemiol 1982; 11:245–9.
17. Johnson G, Rolberts D, Brown R et al. Infertile or childless by choice? A multipractice survey of women. Br Med J 1987; 294:804–7.
18. Page H. Estimation of the prevalence and incidence of infertility in a population: a pilot study. Fertil Steril 1989; 51:571–7.
19. Templeton A, Fraser C, Thompson B. The epidemiology of infertility in Aberdeen. Br Med J 1990; 301:148–52.
20. Templeton A, Fraser C, Thompson B. Infertility – epidemiology and referral practice. Hum Reprod 1991; 6:1391–4.
21. Rantala M, Koskimies AI. Infertility in women participating in a screening program for cervical cancer in Helsinki. Acta Obstet Gynecol Scand 1986; 65:823–
22. Vessey M, Wright NH, McPherson K, Wiggins P. Fertility after stopping different methods of contraception. Br Med J 1978; i:265–7.

23. Spira N, Spira A, Schwartz D. Fertility of couples following cessation of contraception. J Biosoc Sci 1985; 17: 281–90.
24. Thonneau P, Marchand S, Tallec A et al. Incidence and main causes of infertility in a resident population (1 850 000) of three French regions (1988–1989). Hum Reprod 1991; 6:811–16.
25. Hull MGR, Glazener CMA, Kelly NJ et al. Population study of causes, treatment and outcome of infertility. Br Med J 1985; 291:1693–7.
26. Bernstein D, Levin S, Amsterdam E, Insler V. Is conception in infertile couples treatment-related? A survey of 309 pregnancies. Int J Fertil 1979; 24:65–7.
27. Collins JA, Wrixon W, Janes L, Wilson EH. Treatment-independent pregnancy among infertile couples. N Engl J Med 1983; 309:1201–6.
28. Templeton AA, Penney GC. The incidence, characteristics and prognosis of patients whose infertility is unexplained. Fertil Steril 1982; 37:175–82.
29. Haxton MJ, Black WP. The aetiology of infertility in 1162 investigated couples. Clin Exp Obstet Gynecol 1987; 14:75–9.
30. Randall JM, Templeton AA. Infertility: The experience of a tertiary referral centre. Health Bull 1991; 49:48–53.
31. Jacobson CK, Heaton TB, Taylor KM. Childlessness among American women. Soc Biol 1988; 35:187–97.
32. Hastings DW, Robinson JG. Incidence of childlessness for United States women, cohorts born 1891–1945. Soc Biol 1974; 21:178–84.
33. Mosher WD, Aral SO. Factors related to infertility in the United States, 1965–1976. Sex Transm Dis 1985; 12:117–23.
34. Aral SO, Holmes KK. Epidemiology of sexually transmitted diseases. In: Holmes KK, Mardh PA, Sparling PF, Wiesner PH, eds. Sexually transmitted diseases. New York: McGraw Hill, 1984; 127–41.
35. Westrom L. Incidence, prevalence and trends of acute pelvic inflammatory disease and its consequences in industrialized countries. Am J Obstet Gynecol 1980; 138:880–92.
36. Westrom L, Bengtsson L, Mardh PA. Incidence, trends and risk of ectopic pregnancy in a population of women. Br Med J 1981; 282:15–18.
37. Rubin GL, Peterson HB, Dorfman SF et al. Ectopic pregnancy in the United States 1970 through 1978. JAMA 1978; 239:1715–29.
38. Flett GMM, Urquhart DR, Fraser C, Terry PB, Fleming JC. Ectopic pregnancy in Aberdeen 1950–1985. Br J Obstet Gynaecol 1988; 95:740–6.
39. Aral SO, Cates W. The increasing concern with infertility. JAMA 1983; 250:2327–31.
40. Menken J, Trussell J, Larsen U. Age and infertility. Science 1986; 233:1389–94.
41. Rachootin P, Olsen J. Social selection in seeking medical care for reduced fecundity among women in Denmark. J Epidemiol Community Health 1981; 35:262–4.
42. Miller JF, Williamson E, Glue J. Fetal loss after implantation: a prospective study. Lancet 1980; i:554–6.
43. Joffe M. Feasibility of studying subfertility using retrospective self reports. J Epidemiol Community Health 1989; 43:268–74.
44. Jansen RPS. Spontaneous abortion incidence in the treatment of infertility. Am J Obstet Gynecol 1982; 143:451–73.
45. Pittaway DE, Vernon C, Fayez JA. Spontaneous abortions in women with endometriosis. Fertil Steril 1988; 50:711–15.
46. Chong AP, Keene ME, Forte CC, Dileo PE, Brown PC, McGBarry JG. Incidence of spontaneous abortion among treated infertility patients. Int J Fertil 1991; 36:219–21.
47. Radwanska E, Maclin V, Rana N, Henig I, Rawlins R, Dmowski WP. Early endocrine events in induced pregnancies. Int J Fertil 1988; 33:162–7.
48. Corsan GH, Kemmann E. Risk of a second consecutive first-trimester spontaneous abortion in women who conceive with menotropins. Fertil Steril 1990; 53:817–21.
49. Homburg R, Eshel A, Armar NA et al. One hundred pregnancies after treatment with pulsatile luteinising hormone releasing hormone to induce ovulation. Br Med J 1989; 298:809–12.
50. Regan L, Owen EJ, Jacobs HS. Hypersection of luteinising hormone, infertility, and miscarriage. Lancet, 1991; 337:119–20.
51. Singhal V, Li TC, Cooke ID. An analysis of factors influencing the outcome of 232 consecutive tubal microsurgery cases. Br J Obstet Gynaecol 1991; 98:628–36.
52. Hannoun A, Khalil A, Karam K. Uterine unification procedures: postoperative obstetrical outcome. Int J Gynaecol Obstet 1989; 30:161–4.
53. Michalas SP. Outcome of pregnancy in women with uterine malformation: evaluation of 62 cases. Int J Gynaecol Obstet 1991; 35:215–19.

54. Silman AJ, Black C. Increased incidence of spontaneous abortion and infertility in women with scleroderma before disease onset; a controlled study. Ann Rheum Dis 1988; 47:441–4.
55. Medical Research International, SART, AFS. In vitro fertilization–embryo transfer (IVF-ET) in the United States: 1990 results from the IVF–ET Registry. Fertil Steril 1992; 57:15–24.
56. Crosignani PG, Walters DE, Soliani A. Addendum to the ESHRE multicentre trial: a summary of the abortion and birth statistics. Hum Reprod 1992; 7:286–7.
57. Veevers JE. Voluntary childlessness: a review of issues and evidence. Marriage Fam Rev 1979; 2:1–26.
58. Porter M. Infertility: the extent of the problem. Biol Soc 1984; 1:128–35.

Chapter 3

The Causes of Infertility and Relative Effectiveness of Treatment

M. G. R. Hull

Introduction

There has been greatly increased demand for treatment of infertility during the last decade, not due to any increase in its incidence but because of the perception that effective treatment is now available. And the demands are worldwide. Infertility is as deeply painful to the affected individuals in undeveloped overpopulated countries as in developed countries. It also seems to be equally common (though accurate demographic studies are only available in developed countries, as described in Chapter 2), and the causes appear to be the same except in Africa, where tubal damage is relatively common due to sexually transmitted pelvic infection [1]. Therefore the needs of treatment are virtually the same the world over.

Assisted conception methods have given great encouragement to couples particularly with tubal disease, endometriosis and prolonged unexplained infertility. Yet effectiveness and public and personal affordability have been questioned [2]. Furthermore, properly controlled study of some conventional treatments (for example hormonal treatment of minor endometriosis or artificial insemination using husband's semen) has shown them to be ineffective. This review aims to assess the actual and comparative effectiveness of treatments in so far as there are well-defined and strictly comparable published data available. It will concentrate on studies that are confined to couples with a single reliably defined condition and which report pregnancy rates that are time-specific or cycle-specific.

Use of time-specific or cycle-specific conception rates is essential. Crude pregnancy rates per couple are almost meaningless. Pregnancy rates per cycle can also be misleading if limited to the first cycle or two because the rate may fall in subsequent cycles. Therefore cumulative rates will be described in this review whenever possible.

Time-specific prognostication is also of key practical importance to the patients. Their expectations need to be measured in terms of months or at most 1–2 years. A chance of success amounting to 10%, 20% or 30% in the course of the next two years offers no realistic hope to a couple who have been trying to conceive for some years already. Even a greater chance may seem hopeless to a woman already in her late 30s.

Because most infertility is not absolute and there is a chance of conceiving naturally without treatment, carefully controlled prospective studies are needed to assess the true benefit of treatment. This review will focus whenever possible on such studies. When controlled trials are lacking, results will be compared with pregnancy rates observed in similarly defined couples without treatment, preferably from the same centres to standardise diagnostic criteria. Comparison between centres can be very unreliable because of marked variation in entry criteria and assessment of results.

In some cases controlled trials of treatment are inappropriate, except of one treatment against another, for example ovulation induction therapy in women with amenorrhoea. In such cases a useful reference for comparison is the conception rate in a normal poopulation of proven fertility. The highest reported rates will be used for reference in this review [3,4]. It should be noted that some reports of treatment describe over-optimistic results by comparing them with "normal" rates that are too low.

Peak normal fertility observed per cycle is only about 33%, in the first month of trying, but falls quickly settling to about 5%. The average is only 20%–25% per cycle [5] and expectations of any fertility treatment must be judged against that. But less effective treatments may be acceptable. In couples with subfertility the choice of treatment depends on a balance of factors: the chance of pregnancy without treatment; or the chance with simple but only modestly effective treatment; or with more successful but more complex and costly treatment; and other factors must be taken into account like the duration of infertility or age. Comparable data to enable such choices will be presented wherever possible in this review.

There is only sufficient space here to consider the more common causes of infertility. Results will be selected for illustrative purposes when they seem to be representative and uncontroversial. In other cases all the useful data found will be considered and pooled and corrected for weight of numbers to give a representative overall picture. Discussion must be brief. The data will be presented so they can speak for themselves! The section on treatment is essentially drawn from a review published elsewhere [6].

Causes of Infertility

The true distribution of causes of infertility is difficult to determine accurately because there are numerous factors that bias every study and which are virtually unavoidable. Nevertheless, there is a fairly consistent pattern in most reports.

The causes can only be determined when couples have been investigated, therefore the only sources of information can be specialist clinics. Patterns of causes will be affected by the couples seeking referral, by referral policy, resources and special interests of the clinic. Secondary infertility is more common than primary and the patterns of causes are somewhat different, but more couples with primary infertility seek help. The pattern of causes will also be affected by how soon couples seek help. For instance unexplained infertility is less common with longer duration because many such couples are fairly normal and conceive without treatment. Age is not a relevant factor with respect to the causes of infertility, but affects the chance of pregnancy occurring either naturally or in response to treatment.

The distribution of causes is also affected by the diagnostic criteria employed. Less use of laparoscopy will reduce the incidence of endometriosis or peritubal adhesions. Use of only seminal microscopy without testing sperm function to assess male infertility will lead to inaccurate diagnosis (and treatment), though it may not affect the apparent frequency of male infertility because men both with normal and abnormal sperm function will be misdiagnosed and therefore interposed. Some spurious diagnoses may be considered, however, such as "hyperprolactinaemia" in women with ovulatory cycles. Such diagnoses have been excluded in the simplified list of main causes given in Table 3.1 [1,7–11]. This table describes the distributions of causes reported during the last decade from clinics in developed countries, along with some of the factors affecting

Table 3.1. Percentage frequency distribution of main causes of infertility in developed countries, and clinic population characteristics of age, duration of infertility and proportion with primary infertility

	Collins et al. 1983 [7]	Cates et al. 1985 [1]	Hull et al. 1985 [8]	Haxton and Black 1987 [9]	Randall and Templeton 1991 [10]	Thonneau et al. 1991 [11]
Country of origin	Canada	15 countries	UK	UK	UK	France
Number couples	1145	3904	708	1162	633	1318
Woman's age (mean yr)	–	–	28	27	30	31
Duration infertility (mean yr)	>2	>2.5	2.5	>2	4	3.5
Primary infertility (%)	71	71	59	71	62	67
Cause of infertility (%)						
Ovulatory disorder	30	33	21	38	19	32
Tubal/peritubal damage	16	36	14	13	22	26
Endometriosis	4	6	6	2	11	4
Cervical mucus defect	5	–	3	–	–	4
Sperm defects/dysfunction	25	22	24	11	14	52
Azoospermia	6		2	3		9
Coital impairment	–	–	6	–	–	–
Unexplained	13	14	28	31	23	8
Others	–	–	11	12	–	3
Totals	100	110	115	110	100	138

the distributions, namely the proportions with primary infertility, mean duration of infertility and the woman's age. Some reports gave only the single most important cause in each couple (therefore adding to totals of exactly 100%), but others showed that 10%–15% of couples had more than one cause. One study, from three regions in France [11], appeared to find several abnormalities more frequently than other authors and 38% of couples had more than one cause. Some causes like cervical mucus abnormality or coital impairment were not recorded by many authors.

In general, the main causes or categories of infertility appear to be: ovulatory disorder (mainly amenorrhoea or oligomenorrhoea) about 30%; sperm defects or disorder and unexplained infertility each about 25%; and tubal/pelvic damage about 20%. Much less frequent were endometriosis (5%), cervical mucus defects or disorder (4%) and coital impairment (6% in one report). Seminal antisperm antibodies accounting for sperm dysfunction were reported in 8% of couples by Hull et al. [8] but less than 1% by Thonneau et al. [11]. Genital tuberculosis was found in only 0.5% by Haxton and Black [9]. Uterine abnormalities were never specifically mentioned and appear to be rare. In primary infertility endometriosis and seminal abnormalities were consistently more frequent and tubal damage consistently less frequent compared with secondary infertility.

Treatment of Infertility

Ovulation Failure and Induction

Ovulatory disorder in women with normal menstrual cycles rarely occurs persistently enough to cause prolonged infertility. Subtle impairment of early luteal progesterone secretion is of some prognostic significance for natural fertility [12] but there is no specific effective treatment.

Amenorrhoea and oligomenorrhoea are the only clear indications of ovulatory failure and after excluding primary ovarian failure the patients are generally very successfully treated as shown in Fig. 3.1 [8]. In women with amenorrhoea, given accurate diagnosis of the underlying disorder and appropriate selection of treatment every method is highly effective (Fig. 3.2) [13,14]. A minority of patients undergoing the simpler treatments are unsuccessful but all can then respond to gonadotrophin therapy as shown. Conception rates with gonadotrophin therapy are virtually constant in successive cycles up to at least 10 or 12, and cumulative rates are apparently above normal [13,15].

Results are slightly below normal in women with oligomenorrhoea, mainly it seems because of polycystic ovarian disease which accounts for most cases.

Polycystic Ovarian Disease Unresponsive to Clomiphene

However, women with PCO presenting with either oligomenorrhoea or amenorrhoea who fail to respond adequately to clomiphene are a particularly problematic subgroup. They make up about 16% of women with oligo- or amenorrhoea [16] and respond variably to gonadotrophin therapy with relatively

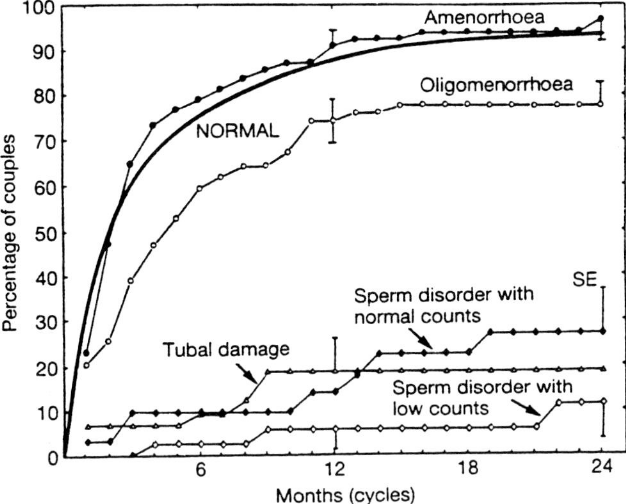

Fig. 3.1. Cumulative conception rates of some of the commonest single causes of infertility in a complete population of infertile couples treated by conventional methods, compared with normal [3,4]. Donor insemination, assisted conception methods, and reversal of sterilisation not included. SE, standard error. (From Hull et al. [8].)

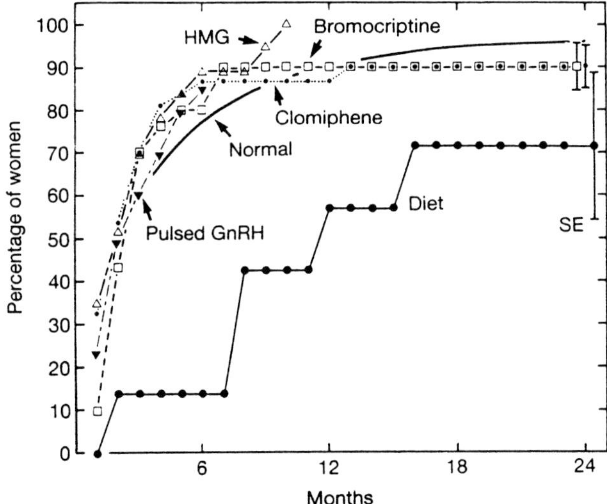

Fig. 3.2. Cumulative conception rates resulting from individual treatment methods as appropriate in women with amenorrhoea, compared with normal. GnRH, gonadotrophin releasing hormone; hMG, human menopausal gonadotrophin. (From Hull et al. [13]; Mason et al. [14].)

poor pregnancy rates and high miscarriage rates. Furthermore, a high proportion of treatment cycles have to be discontinued due to either inadequate or often excessive response. Cumulative pregnancy rates reported after six cycles of gonadotrophin therapy have been only 38%–63% [15–19], rates that have been matched by laparoscopic ovarian electrocautery in one controlled study [17]. Ovarian electrocautery treatment has been reviewed by Vaughan Williams [20].

There have been numerous recent optimistic reports of the results of gonadotrophin therapy but these must be judged with caution because they are per cycle and not cumulative. The heterogeneity of responsiveness of these relatively difficult cases of PCO is likely to result in diminishing cycle fecundity in successive cycles. Table 3.2 summarises an analysis [21] of the reports, comparing standard-dose [22–27] with more recent low-dose therapy aimed at achieving uni-ovulation [17,19,24,25,28,29]. Low-dose therapy succeeded in more than halving the risk of multiple pregnancy but at the price of a proportional reduction in the chance of achieving a pregnancy, which presents a clinical dilemma in the choice of treatment method.

It is notable that the relatively high risk of miscarriage was not reduced by low-dose gonadotrophin therapy, nor does it appear to be reduced when combined with pituitary down-regulation, nor by pulsed gonadotrophin releasing-hormone therapy following pituitary down-regulation (see Chapter 15). The relative effectiveness of the latter treatments involving pituitary down-regulation has not yet been adequately assessed.

Table 3.2. Summarised dose-related analysis of reported results of gonadotrophin therapy (without pituitary down-regulation) to induce ovulation in women with oligomenorrhoea or amenorrhoea and polycystic ovaries resistant to clomiphene. Rates for multiple pregnancy, miscarriage and on-going pregnancies

	Standard dose	Low dose
Patients	111	243
Cycles	210	786
Completed ovulatory cycles	163(78%)	570(73%)
Pregnancies	49	86
Per cycle	23%	11%
Per ovulatory cycle	30%	15%
Multiple pregnancy rate	23%	9%
Miscarriage rate	17%	37%
On-going pregnancy rates		
Per cycle	20%	7%
Per ovulatory cycle	25%	9%

From Hull [21], with permission.

Tubal/Pelvic Infective Disease and Surgery

There is growing disappointment with tubal surgery despite the advance of microscopic methods. It is difficult to assess the true benefits of surgery because of selectivity of cases and reporting, and lack of controlled studies. Fig. 3.1 illustrates the overall outcome in a complete population of women with tubal disease, some of whom had surgery and others did not because their condition was either inoperable or seemingly too minor to justify operation. The overall

2-year cumulative pregnancy rate was 19% [8], similar to the overall rate of 23% in the population reported by Wu and Gocial [30], who achieved a rate of 33% with surgery compared with 16% in those not operated upon.

Selectivity for surgery is essential. Table 3.3 gives the time-specific intrauterine pregnancy rates following microsurgery for different types of tubal/pelvic inflammatory disease reported by several leading centres [31–34]. Salpingostomy for distal tubal occlusion carries the worst prognosis: a 2–3-year pregnancy rate of 23%–27%. That can now be expected in a single cycle of IVF (in vitro fertilisation) treatment. Clearly the most limiting factor for surgical success is the irreversible damage to tubal mucosal and fimbrial function.

Table 3.3 also suggests that the severity of disease is important for the outcome of surgery and the specific nature of the tubal damage and extent of pelvic adhesions should be considered. In the specific case of distal tubal occlusion both hysterosalpingography and laparoscopy to assess the tubal mucosa are valuable prognostic indicators [35,36]. Wu and Gocial [30] developed a general system to classify tubal/pelvic disease into four grades of severity. Some patients offered surgery declined and, though not strictly controlled, the outcome can be compared with or without surgery (Figs. 3.3 and 3.4). Figs. 3.3 and 3.4 show that even the most minor degrees of infective damage (such as only adhesions with healthy-looking tubes) are associated with severe subfertility unless treated, and surgery appears to be of substantial benefit. In the second grade of severity, cumulative conception rates after surgery approach 50%.

It appears from the report by Wu and Gocial [30] that only 25%–50% at most are suitable for surgery, though others have estimated the proportion to be as low as 10% [37,38]. It also follows that most appropriate surgery should be possible laparoscopically. Even if suitable for surgery, IVF must be considered a year, or two at most, after surgery if still no pregnancy occurs. In the more severe cases surgery of any sort is wasted and IVF should be the immediate choice.

Table 3.3. Time-specific intrauterine pregnancy rates[a] after tubal microsurgery

Operative procedure	No. of patients	1 year (%)	2 years (%)	3 years (%)	Total actual (%)
Salpingolysis [31]	42	43	62	–	64
Fimbrioplasty [31] (patent phimosis)	132	37	57	60	60
Salpingostomy					
Tubes not distended [31]	27	15	40	45	48
Hydrosalpinges [31]	56	5	11	20	23
Mixed [33]	93	9	13	16	13
Mixed [34]	230	17	28	31	33
Tubocornual anastomosis [32]	26	48	56	74	58
Overall	606	22	34	37	39

[a]By life-table analysis where necessary to allow for incomplete follow-up to later time points. Hence total actual pregnancy rates are sometimes lower.

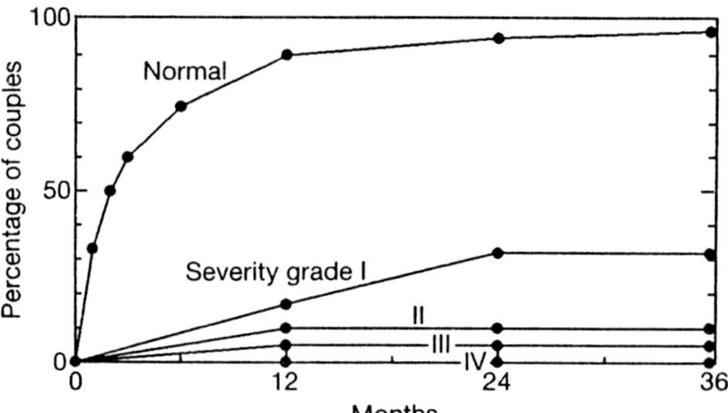

Fig. 3.3. Cumulative conception rates with untreated tubal/pelvic disease related to disease grading, compared with normal [30]. Ectopic pregnancies included, up to 10% of pregnancies (Wu, personal communication).

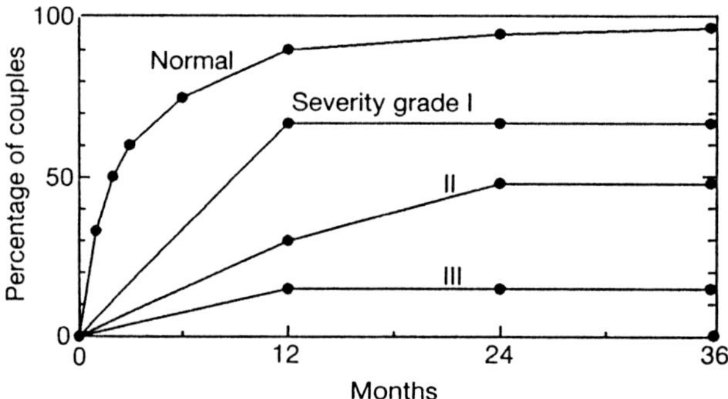

Fig. 3.4. Cumulative conception rates after surgery for tubal/pelvic disease related to disease grading as in Fig. 3.3. The most severe cases (grade IV) were not operated on.

Endometriosis

Endometriosis even in minor degree is associated with marked subfertility as shown in Fig. 3.5 [39–44]. Severe disease is usually treated but the few data available show very severe subfertility without treatment. It is notable, however, that relatively minor involvement of the ovaries such as limited adhesions on their undersides can be associated with severe subfertility and may require treatment.

Hormonal Treatment

Severe disease with extensive damage is an obvious cause of infertility, and relief of associated pain by hormonal suppression of endometriotic activity led to the assumption that subfertility associated with minor disease would also benefit from such treatment. However, controlled studies have shown no benefit of danazol or progestogen in minor endometriosis [40,45–47]. On the

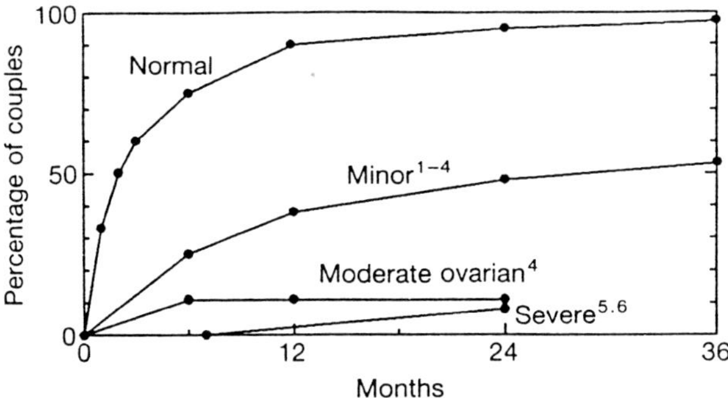

Fig. 3.5. Cumulative conception rates with untreated endometriosis related to disease grading, compared with normal. (From 1, [39]; 2, [40]; 3, [41]; 4, [42]; 5, [43]; 6, [44].)

contrary, the chance of pregnancy is delayed by the duration of treatment. Treatment by pituitary down-regulation appears to lead to reduced crude pregnancy rates [48].

Surgery

There remains no controlled study of ablative therapy for minor endometriosis but time-specific conception rates after laparoscopic laser therapy have been only slightly higher [49] or similar [50] to those reported without treatment (Fig. 3.5).

Fig. 3.6 shows that in severe endometriosis surgical therapy to ablate disease and lyse adhesions, particularly by microsurgery or laparoscopic laser, appears to be of substantial benefit compared with the chance of conceiving without treatment (shown in Fig. 3.5). These results may be over-optimistic and must be interpreted with caution, because the studies are uncontrolled and even lack information on how matched patients perform without surgery in the same centres [41,49–52]. Nevertheless they are encouraging, and seem to be notable

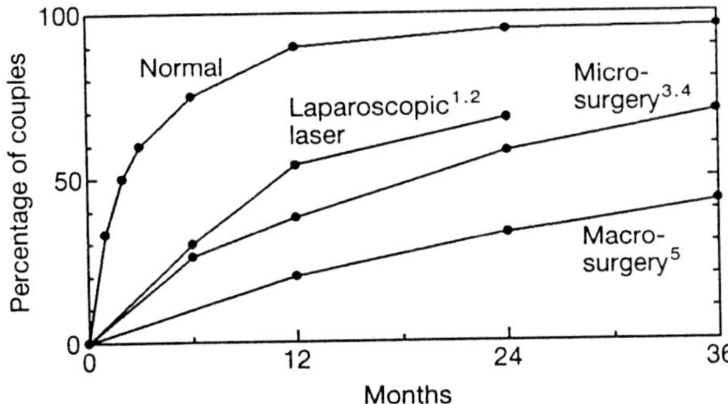

Fig. 3.6. Cumulative conception rates in severe endometriosis related to type of surgical treatment, compared with normal. (From 1, [50]; 2, [49]; 3, [51]; 4, [41]; 5, [52].)

in leading to a continued possibility of conception after one or two years, unlike tubal surgery for infective damage. Presumably women with endometriosis retain some degree of tubal function. Pregnancy rates are still markedly subnormal, however, and IVF or GIFT (gamete intrafallopian transfer) should be considered after no more than one or two years. Superovulation/IUI (intrauterine insemination) is probably not appropriate in view of likely remaining structural damage or adhesions.

Assisted Conception Methods

Whatever the mechanism of subfertility associated with endometriosis, fertilising ability is normal or only slightly impaired [53] and the results of treatment are relatively favourable.

In cases of severe endometriosis, however, there have been reports of relatively poor results with IVF due to reduced availability of eggs and impaired implantation [54,55], but favourable with GIFT [56]. The effect of prior surgical ablation in these cases is unknown.

By contrast, in minor endometriosis the results of both IVF and GIFT are particularly favourable, with pregnancy rates per cycle around 30% [54,56–58]. But should IVF or GIFT be applied in cases of minor endometriosis? The chance of natural conception is only 3% per cycle on average (Fig. 3.5). It is also likely, as with unexplained infertility (see below), that the chance of natural conception will be inversely related to the duration of infertility. After more than 3 years there will be an appreciable reduction in fecundity. Prognostic information related to duration of infertility and age in cases of minor endometriosis is lacking. However, after more than 2 years from the time of diagnosis the balance of choice seems to be clearly in favour of IVF and GIFT, and to a lesser extent superovulation/IUI. The effectiveness of these treatments appears to be maintained in successive cycles giving rise to favourable cumulative pregnancy rates as illustrated later.

Cervical Factors

The picture is generally unclear, partly because it is difficult to define disorders of cervical mucus production and function reliably (timing and repetition in different cycles are critical but often not done adequately), and partly because when well defined they are very uncommon.

In a report by Check et al. [59], treatment aimed specifically at improving mucus quality using mucolytic agents or oestrogen led to a 6-month cumulative pregnancy rate of 22%, but that was uncontrolled and may be no better than without treatment. A small subgroup of patients with evidence of underlying ovulatory disorder treated mainly with gonadotrophins appeared to do better, with a 6-month pregnancy rate of 53%.

In the usual absence of any underlying ovulatory disorder IUI is an obvious choice to simply by-pass the cervix, and one controlled study has shown significant benefit compared with natural intercourse [60], but reported results are very variable ranging from 1.5% to 16% per cycle (see references [59,60] and reviews by Irianni et al. [61] and Dodson and Haney [62]).

Results with combined IUI and superovulation using gonadotrophins are equally scanty and unreliable, though more encouraging (see reviews [61,62]). It seems likely that assisted conception methods like IUI or IVF and GIFT would succeed as well as they do for unexplained infertility (see below), and in the absence of specific proven therapy can be offered optimistically. In most cases there is no reasonable alternative.

Sperm Disorders

This subject is too complex and important to deal with briefly. Sperm disorders are the single most common cause of infertility but are heterogeneous in origin. Proper assessment of treatments is elusive because of fundamental differences and inaccuracies in the definition of sperm disorders in the first place. As an introductory summary, Fig. 3.1 illustrates three points: sperm disorders can occur in men with normal sperm "counts" on standard semen microscopy, cause severe subfertility, and there is virtually no treatment of proven benefit to restore natural fertility. That is why donor insemination (DI) is so often required to by-pass the problem.

Numerous prospective studies have shown that semen microscopy is at best only a weak predictor of fertility, and tests of sperm function show that there are men with low seminal sperm counts who have normal fertility, and vice versa. In prolonged "unexplained" infertility in which the male is defined only by normal seminal microscopy, one-third have been shown to have severely defective fertilising ability by testing hamster egg penetration [63]. That also correlated with simpler evidence of sperm dysfunction as tested by penetration of normal cervical mucus in vitro [64]. Ability to fertilise human eggs has also been correlated in several studies with mucus penetration, sperm migration into culture medium ("swim-up") and more complex measurements of motility. Defining men as infertile because of moderately reduced seminal sperm counts (as so often reported) when there is normal sperm behaviour in mucus or culture medium leads to claims of successful therapy which are spurious. It is of key importance to distinguish between properly defined sperm dysfunction and the commonly reported "male factor" infertility based on semen microscopy alone (Chapter 5). Assessment of reported treatments in the following review is based on defined sperm dysfunction as far as possible.

In summary, controlled trials have demonstrated no benefit of treatment either by hormonal stimulation (see review by Hewitt et al. [65] and references [66–69]) or by artificial insemination, using whole semen [70] or prepared sperm by IUI [60]. The classic report indicating benefit of IUI by Kerin et al. [71] is now recognised to be not of sperm dysfunction, but the negative report by te Velde et al. [60] is also not well defined. Uncontrolled reports of IUI have shown very poor results (about 1% pregnancy rate per cycle) when sperm disorder is well diagnosed [61].

Seminal antisperm antibodies are the only specific cause of sperm disorder to have been shown to benefit from treatment, specifically aimed at suppressing antibody production, but only in one study and the benefit was small (see below). Treatment of varicocele remains unproven and unencouraging. In general the main hope in cases of sperm dysfunction is from IVF treatment as discussed below.

Seminal Antisperm Antibodies

Optimistic reports of glucocorticoid therapy have been questioned [72]. Table 3.4 summarises the only properly controlled studies reported, all employing high-dose therapy [73–75]. Treatment was cyclical to minimise risks. Two of the three studies found no benefit. Hendry et al. [74] found statistically significant benefit only after more than six cycles of treatment and therefore emphasised the need for protacted therapy. A 9-month cumulative pregnancy rate of 27% seems small reward, however, for unpleasant and potentially dangerous treatment. Hendry et al. [74] noted that pregnancy was usually only achieved by men whose seminal antibody levels were completely suppressed. Perhaps a selective approach to continuation of therapy should be explored according to such effect and/or improvement in sperm-mucus penetration.

Assisted conception methods aim to by-pass the problem of sperm progression through cervical mucus and minimise binding of the seminal antibodies to the sperm by early treatment and recovery into serum-rich medium [76]. Encouraging results with such methods have been reported and are summarised in Table 3.5 [77–79]. Pregnancy rates are surprisingly good given the marked reduction in fertilisation rates, presumably as a result of a plentiful excess of eggs. For those reasons IVF should be preferred to GIFT or IUI.

Sperm preparation as just described is more favourable than by standard delayed "swim-up" [76] but still relatively inefficient at preventing or reducing sperm–antibody binding. Better separation of bound from unbound sperm by immunobead attachment has led to encouraging preliminary success [79].

Table 3.4. Results of placebo-controlled studies of high-dose glucocorticoid therapy for infertility due to seminal antisperm antibodies

Author and treatment	Patients	Cycles per patient	Pregnancy rates (%)	
			Cumulative	Monthly (estimated)
Haas and Manganiello [73]				
Methyl prednisolone	24	3	13*	4
Placebo	19	3	5	1.5
Hendry et al. [74]				
Prednisolone	33	9	27**	3.5
Placebo	27	9	7	1
Bals-Pratsch et al. [75]				
Prednisolone	17	3	0	–
Placebo	17	3	0	–

*Not significant; $^{**}P < 0.05$.

Table 3.5. Results of IVG, GIFT and ZIFT treatment for infertility due to seminal antisperm antibodies

Reference	Cycles	Fertilisation per oocyte (%)	Pregnancy per cycle (%)
Palermo et al. [77]	38	48	34
Van der Merwe et al. [78]	29	24	24
Elder et al. [76]	54	44	24
Overall	121	44	27

Assisted Conception Methods for Non-Immune Sperm Disorders

In cases of well-defined sperm disorder IVF and related treatments have been disappointing, but substantial improvements seem to have been made recently. Clearly what matters is sperm function, not simply number.

IUI with or without superovulation is completely useless according to Irianni et al. [61] if "male factor infertility" is well defined (pregnancy rate 1% per cycle). Not only is there sperm dysfunction but for IUI much larger numbers are needed than for IVF or GIFT.

When limits are placed on the number of embryos or eggs transferred by IVF or GIFT respectively it is clear that IVF should be preferred in cases of sperm dysfunction, to be able to adjust appropriately to the reduced fertilisation rate. Pregnancy and implantation rates are lower with GIFT than IVF when there is sperm dysfunction [80] but higher when sperm function is favourable [53].

IVF treatment may be able to take numerical advantage of increased superovulation or concentration of sperm in microdroplet culture. Stimulation of sperm motility using, for example, pentoxifylline has been reported optimistically [81] but a properly controlled trial is awaited. The use of Percoll density gradient separation of the favourable sperm to minimise their adverse exposure to free oxygen radicals released by damaged sperm or macrophages [82] has led to significant improvement in fertilisation and consequent pregnancy rates in a controlled study [83]. It seems that clinical pregnancy rates of about 17% per cycle can now be achieved using such simple methods [58]. However, very variable results are reported with IVF for sperm dysfunction, which should not be surprising given conditions varying and methods performing at their functional limits. In cases of sperm disorder IVF must always be undertaken speculatively and with diagnostic as much as therapeutic intent. Cumulative pregnancy data are not available, and would anyway be biased by couples who choose not to continue IVF treatment after initial failure or poor fertilisation rate.

Micromanipulation

Micromanipulation methods involve extreme attempts to facilitate entry or inject sperm through the zona pellucida. There have been few well-controlled

Table 3.6. Results of reported comparisons in the same treatment cycles of micromanipulation methods and standard IVF, expressed as monospermic fertilisation rates per oocyte

Method and authors	Micromanipulation		Standard IVF	
	Oocytes	Fertilised (%)	Oocytes	Fertilised (%)
Zona drilling				
Gordon et al. [84]	45	11	16	13
Partial zona dissection				
Cohen et al. [85]	50	56	45	33
Cohen et al. [86]	220[a]	23[a]	128	21
Hill et al. [87]	88	18	65	23
Zona cutting				
Payne et al. [88]	69	36	68	28
Subzonal injection				
Fishel et al. [89]	311	15[b]	250	8[b]

[a]Stated selection bias against micromanipulation method.
[b]Includes polyspermic fertilisations but most apparently monospermic.

comparisons with standard insemination of eggs for IVF. The results of reported comparisons in the same treatment cycles are summarised in Table 3.6 [84–89]. They show little or no benefit and point to an important change in basic policy.

The benefit shown by Cohen et al. in 1989 [85] is of the same order as that achieved with simple Percoll preparation in cases of only moderately severe sperm disorder [83]. Micromanipulation methods need to be focused on only the most severe cases. In a separate study Cohen et al. in 1991 [86] found similar monospermic fertilisation rates by partial zona dissection (15%) and subzonal injection (16%) but concluded that the two methods should be applied to different types of sperm disorder. Only Fishel [89] and Cohen have achieved implantation rates when embryos have been obtained. Pregnancy rates per treatment cycle have been 0–5%, not all from micromanipulated oocytes.

Azoospermia

Azoospermia, or virtual azoospermia, is an infrequent cause of infertility (excluding cases of regretted vasectomy) and there is insufficient space here to deal with the several aspects of the subject.

Primary spermatogenic failure is untreatable but not always complete and IVF may be considered. Hypothalamic–pituitary failure is treated by specific endocrine methods. In cases of obstructive azoospermia the results of microsurgical anastomosis are generally poor and current interest is focused on surgical recovery of epididymal sperm, which have proved surprisingly motile, for IVF. Only one centre reports optimistic results, however, placing emphasis on a combination of microsurgical skills to collect the spermatozoa, production of at least 10 mature oocytes, and willingness to transfer relatively large numbers of embryos to the fallopian tube [90].

Donor Insemination

Donor insemination (DI) treatment offers the most realistic option for most couples whose infertility is due to sperm dysfunction or obstructive azoospermia. Cryopreservation of semen leads to substantial functional damage to sperm and pregnancy rates are lower than with fresh semen, but use of cryopreserved semen is now obligatory therefore only results with such semen will be considered here. The large experience in France [91,92] seems representative and the results are shown in Fig. 3.7.

Most couples can achieve pregnancy if they are prepared to continue treatment long enough, which is feasible because of its simplicity. Women over 35 years old have a substantially reduced chance of success and may wish to consider assisted conception treatment without much delay. There are no reliable data about IUI using donor sperm but widespread (though little published) evidence of normal fertilising ability of cryopreserved donor sperm in vitro. It is not usually until after the age of 40 years that any substantial reduction in success by IVF treatment has been noted (see later).

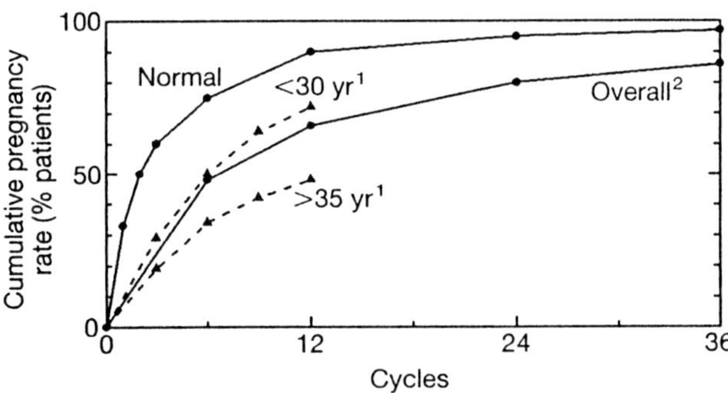

Fig. 3.7. Cumulative conception rates resulting from donor insemination treatment using frozen semen. Results are shown overall and related to the woman's age, compared with normal. (From 1, [91], and 2, [92].)

Unexplained Infertility

The literature on unexplained infertility is difficult to interpret because of varying, often poor, definition of the condition. There is general agreement that investigations should be included to demonstrate normal ovulatory cycles, normal tubal/pelvic state by laparoscopy, and normal semen microscopy. There is variation in the minimum duration of infertility required but general recognition of the importance of that factor. There are claims that an abnormality of some sort can be found to "explain" the infertility in most cases if enough investigations are done, but controlled studies have demonstrated no significant relevance to any other abnormalities suggested, except sperm dysfunction. The need to test sperm function is as important in the definition of unexplained infertility as it is conversely in the diagnosis of sperm disorder; it is the other side of the same argument discussed earlier. This review will focus as far as possible only on reports of unexplained infertility which include tests indicating favourable sperm function, usually including a normal postcoital test.

Given that definition, the main factors determining the chance of conceiving naturally are the woman's age and, more importantly, the duration of infertility so far [8] (Fig. 3.8). Those findings demonstrate the heterogeneous nature of unexplained infertility. In particular:

1. Couples with unexplained infertility of less than 3 years duration are mostly normal and have simply been unlucky so far. Most will conceive within 2 years. All they need (apart from diagnostic investigations) is advice and encouragement. There is no evidence of any benefit from simple treatment such as clomiphene [93] and the balance of choice does not yet justify assisted conception methods except in women aged in their late 30s.
2. After more than 3 years duration the chance of natural conception offers unrealistic hope. The monthly chance is down to 1%–3% (from Fig. 3.8); Crosignani et al. [94] estimated about 1% and at most 2%. Treatment is needed.

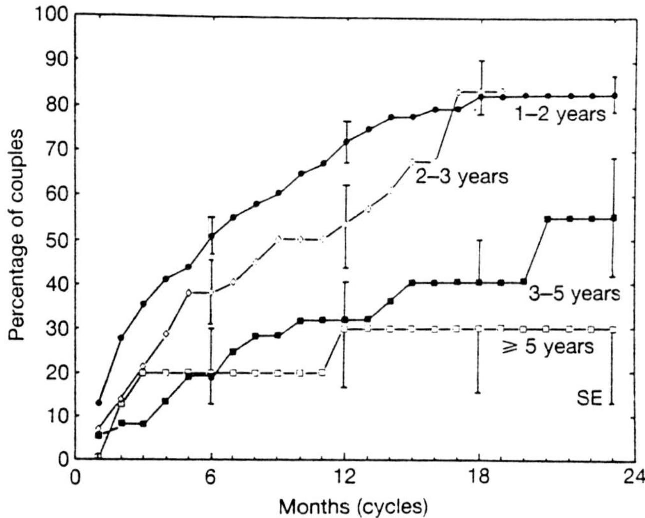

Fig. 3.8. Cumulative conception rates in unexplained infertility without treatment related to duration of infertility when investigated. (From [8].)

Treatment of Prolonged Unexplained Infertility

Controlled studies of clomiphene treatment have shown significant but only slight benefit and have not been extended in duration: pregnancy rates of

3%–5% per cycle for 3–4 cycles [93,95]. A controlled study of IUI (alone) [71] (which was described to be of male infertility though sperm function is now recognised to have been favourable), found pregnancy rates per cycle of 21% versus 2% in untreated controls. It was described as a preliminary report but the study has never been repeated, and such results have never been achieved by others. The only other controlled studies of treatments of unexplained infertility have been of IVF compared with tubal infertility, showing slightly reduced fertilisation rates per oocyte in unexplained infertility but roughly similar pregnancy rates [53,96,97]; and of IVF compared with GIFT, showing advantage in favour of GIFT [53].

Reported results are summarised in Table 3.7 [57,58,60–62,71,80,93–103], expressed as pregnancy rates per cycle. Cumulative rates over an extended series of cycles have not been reported specifically for unexplained infertility but are available pooled with other conditions that are favourable for assisted conception methods and are described later in that section of this review. It is clear that simple treatment with clomiphene is worth pursuing initially, perhaps for 6 months, but then only assisted conception methods combined with superovulation offer any substantial hope of success.

Assisted Conception Methods of Treatment

Assisted conception methods are here defined as involving a combination of superovulation therapy and delivery of prepared sperm to the eggs. The basic range is IVF, GIFT and superovulation/IUI. They are considered jointly here because they are the common solution to a variety of infertility problems and results as reported are often inseparable. Others are considered later as variants of the basic methods (zygote or embryo tubal transfer after IVF, or direct intraperitoneal insemination instead of IUI, for example). From the foregoing discussion it is clear that the main indications for assisted conception treatments are tubal disease, unexplained infertility, and endometriosis, when they have reached a stage that the chance of conceiving by any other means is less than 1%–2% per cycle or 20%–30% after 2 years, or the woman's age leaves insufficient time for speculative treatment.

Assisted conception treatment, particularly IVF, may also be undertaken speculatively as in cases of sperm disorder partly for diagnostic purpose; also in women over about 40 years old, in whom ovarian responsiveness declines uncertainly. Those are the commonest factors adversely affecting the success of treatment. It is important to classify results accordingly in order to compare the effectiveness of different methods, not only in terms of results per cycle but as valid cumulative rates. The validity of life-table analysis depends on the reasons being unbiased for couples not continuing in treatment as long as others. Cases of sperm disorder and women over 40 years old are likely not to continue.

For those reasons cumulative pregnancy rate data are lacking for treatment of (well defined) sperm disorders and older women. Tan et al. [104] have reported substantial reduction in pregnancy rates in women aged 35–38 years and further reduction aged 39 years or more, after three and six cycles of IVF treatment but also in the first cycle. Other authors, reporting simply rates per cycle have not noted such a marked effect, if at all, until after about 40 years of age.

Table 3.7. Reported results of treatments for prolonged unexplained infertility

Treatments, authors	Conception rates per cycle
Clomiphene	
Fisch et al. [95]	3%[a]
Glazener et al. [93]	5%[b]
HMG	
Serhal et al. [98]	6%
Crosignani et al. [94]	8%
IUI	
Kerin et al. [71]	21%[c]
Serhal et al. [98]	3%
te Velde et al. [60]	3%
Irianni et al. [61]	6%
Clomiphene and/or IUI	
Martinez et al. [99]	9%[d]
HMG and IUI (or IPI)	
Serhal et al. [98]	26%
Dodson and Haney [62]	15%
Crosignani et al. [94]	23%
Hull et al. [58]	18%
Turhan et al. (IPI) [100]	23%
IVF	
Navot et al. [96]	32%
Audibert et al. [97]	14%
Crosignani et al. [94]	28%
Hull et al. [58]	27%
1990/91 treatments:	32%
GIFT	
Braeckmans et al. [101]	26%
Borrero et al. [57]	31%
Wong et al. [102]	29%
Crosignani et al. [94]	29%
Hull et al. [58]	36%
ZIFT	
Devroey et al. [103]	48%
Tubal embryo transfer	
Tanbo et al. [80]	38%

a, b, c, d Controlled studies – results per cycle: [a], 1%, [b], 1%, [c], 2%, [d], zero.

The cumulative pregnancy rates illustrated in the following figures refer to mainly favourable conditions for treatment, though variously defined, including women under 35–40 years old. They appear to be valid and representative. Results using donated eggs are not included.

In Vitro Fertilisation (IVF)

Fig. 3.9 illustrates cumulative pregnancy rates reported from six centres or groups in Australia [105], France [106], Holland [107], UK [58,104] and USA [108]. In most cases the women were under 40 years old but the report by Tan et al. [103] from the UK was limited to women under 35. Pregnancy rates in the first cycle were 13%–28%, and after three cycles 35%–51%, after six cycles 54%–66%, and after nine cycles 71%–79%.

The lowest results relate to some of the earliest years of IVF practice (1980–85). There was a steady improvement in later years and marked improvement in pregnancy rates since about 1990. Pregnancy rates per cycle of 28%–30% seem to be the normal expectation now [58,94], despite strict limitation imposed on the number of embryos transferred for example in the UK [58].

Cumulative "take-home baby" rates are what really matter. The only useful report is of 60% after four cycles of IVF treatment [58]. Fig. 3.10 shows the rates not only for IVF but for combinations of assisted conception methods, for instance GIFT or IUI if appropriate following an initial cycle of IVF. The cumulative "take-home baby" rate was 70% after six cycles and 90% after nine cycles in all women under 40 years old and men with normal sperm [58].

It is important to note that all the foregoing reports of cumulative pregnancy rates where reliable data exist show little or no reduction in the chance of pregnancy in successive cycles, usually up to at least six cycles and perhaps more. By contrast, one report from the USA describes marked reduction of

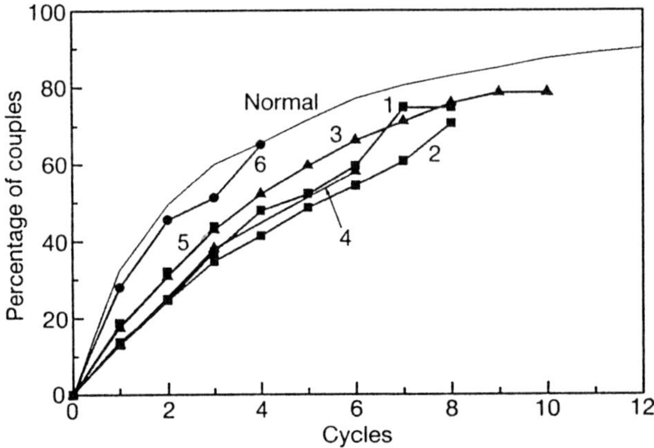

Fig. 3.9. Cumulative conception rates by IVF treatment per attempted egg recovery, done for mostly favourable indications in women aged less than 35–40 years old, compared with normal. (From 1, [108]; 2, [105]; 3, [106]; 4, [104]; 5, [107]; 6, [58].)

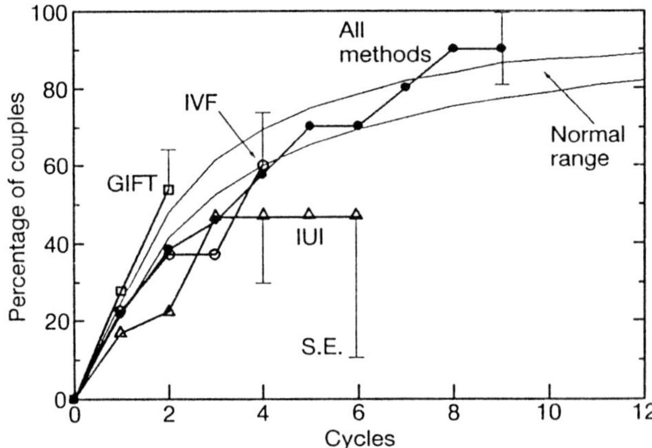

Fig. 3.10. Cumulative successful birth ("take-home baby") rates by IVF (mostly) or by GIFT or superovulation/IUI treatment, individually or in serial combinations as applied in practice, in women under 40 years old and men with normal sperm [58], compared with a normal population range (parous and nulligravid) [113].

cycle fecundity in successive cycles and draws attention to heterogeneity of the patient population treated [109]. That serves to emphasise the crucial importance of classifying results by specific diagnostic categories and other critical factors like age. In practice it is the only way to avoid bias in life-table analysis and to be able to give reliable advice to patients.

Gamete Intra-fallopian transfer (GIFT)

GIFT was introduced about 5 years after IVF, and reports are mostly limited to relatively small series so that useful cumulative pregnancy rates are not available. Reported rates per cycle are 26%–36% (Table 3.7).

Superovulation with IUI or IPI

The results reported by Chaffkin et al. [110] shown in Fig. 3.11 indicate clearly that neither IUI nor superovulation alone is effective and the two must be combined. The average pregnancy rate per cycle was nearly 20% but fell sharply after three or four cycles and the cumulative rate was about 68% after six cycles. Dodson and Haney [62], in a large study, reported an average pregnancy rate per cycle of 14% which did not decline significantly and the cumulative rate was 56% after six cycles.

Other Methods and Relative Effectiveness

There have been optimistic isolated reports of ZIFT (zygote intrafallopian transfer) [103] and tubal embryo transfer [80] as extended variations of IVF, and substantial reports of direct intraperitoneal insemination (IPI) as an alternative to IUI [100,111], but they have not yet been fully assessed.

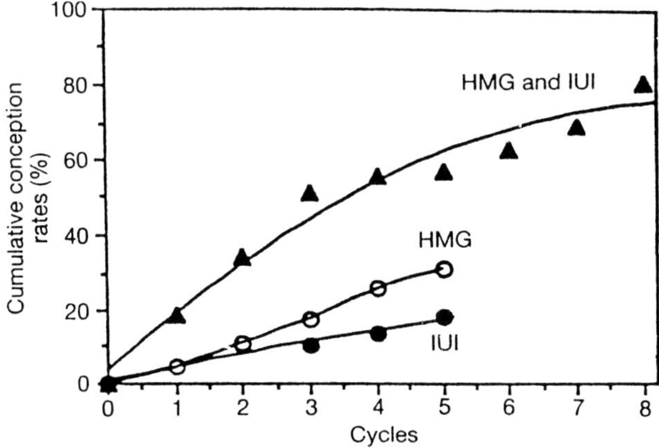

Fig. 3.11. Cumulative conception rates by IUI, HMG or combined superovulation/IUI, done for various favourable indications. (From [110].)

Proper comparison of reported results with different assisted conception methods is difficult, for several reasons: pooling of diagnostic indications; unspecified and unrestricted numbers of eggs and embryos transferred, or of follicular stimulation for IUI and IPI; and inclusion of early "biochemical" pregnancies in some reported "success" rates. Few fertility reports give birth rates though clinically confirmed pregnancies would be acceptable for comparison, given miscarriage and ectopic pregnancy rates.

The few controlled comparisons of treatment show that IVF is about equally effective in endometriosis [53,54,56] and unexplained infertility [53,96,97] compared with tubal infertility.

Comparison of GIFT with IVF for non-tubal infertility has revealed variable findings. Leeton et al. [112] and Crosignani et al. [94] found no difference but did not specify numbers of eggs and embryos transferred. Tanbo et al. [80] found reduced success with GIFT but later recognised sperm dysfunction contributing to failure in many of the GIFT treated couples. The only strictly comparable study showed significant advantage in favour of GIFT [53].

Comparison of superovulation/IUI with GIFT cannot be made strictly because of variation in follicular stimulation for IUI. Crosignani et al. [94] found similar rates but gave no information about the limits of ovarian stimulation if any. Mills et al. [53] applied cautious limits and achieved half the pregnancy rate with IUI compared with GIFT (20% versus 40% per cycle). Nevertheless IUI may seem to be a cost-effective choice particularly in relatively young women, and encouraging cumulative pregnancy rates have been reported, as described ealier [62,110]. The concern about multiple birth rates remains, however.

References

1. Cates W, Farley TMM, Rowe PJ. Worldwide patterns of infertility: is Africa different? Lancet 1985; ii:596–8.
2. World Health Organisation. Consultation on the place of in vitro fertilization in infertility care. Report EUR/ICP/MCH 122(S), WHO, Geneva, 1990.

3. Tietze C. Statistical contributions to the study of human fertility. Fertil Steril 1956; 7:88–95.
4. Tietze C. Fertility after the discontinuation of intrauterine and oral contraception. Int J Fertil 1968; 13:385–9.
5. Spira A. Epidemiology of human reproduction. Hum Reprod 1986; 1:111–15.
6. Hull MGR. Infertility treatment: relative effectiveness of conventional and assisted conception methods. Hum Reprod 1992; (in press).
7. Collins JA, Wrixon W, Janes LB, Wilson EH. Treatment-independent pregnancy among infertile couples. N Engl J Med 1983; 309:1201–6.
8. Hull MGR, Glazener CMA, Kelly NJ et al. Population study of causes, treatment, and outcome of infertility. Br Med J 1985; 291:1693–7.
9. Haxton MJ, Black WP. The aetiology of infertility in 1162 investigated couples. Clin Exp Obstet Gynecol 1987; XIV:75–9.
10. Randall JM, Templeton AA. Infertility: the experience of a tertiary referral centre. Health Bull 1991; 49/1:48–53.
11. Thonneau P, Marchand S, Tallec A et al. Incidence and main causes of infertility in a resident population (1 850 000) of three French regions (1988–1989). Hum Reprod 1991; 6:811–16.
12. Dunphy BC, Barratt CLR, Li TC, Lenton EA, Macleod IC, Cooke ID. The interaction of parameters of male and female fertility in couples with previously unexplained infertility. Fertil Steril 1990; 54:824–7.
13. Hull MGR, Savage PE, Jacobs HS. Investigation and treatment of amenorrhoea resulting in normal fertility. Br Med J 1979; i:1257–61.
14. Mason P, Adams J, Morris DV et al. Induction of ovulation with pulsatile luteinising hormone releasing hormone. Br Med J 1984; 288:181–5.
15. Dor J, Itzkowic DJ, Mashiach S, Lunenfeld B, Serr DM. Cumulative conception rates following gonadotrophin therapy. Am J Obstet Gynecol 1980; 136:102–5.
16. Hull MGR. Epidemiology of infertility and polycystic ovarian disease: endocrinological and demographic studies. Gynecol Endocrinol 1987; 1:235–45.
17. Abdel Gadir A, Mowafi RS, Alnaser HMI, Alrashid AH, Alonezi OM, Shaw R. Ovarian electrocautery versus human menopausal gonadotrophins and pure follicle stimulating hormone therapy in the treatment of patients with polycystic ovarian disease. Clin Endocrinol 1990; 33:585–92.
18. Ginsburg J, Hardiman P. Ovulation induction with human menopausal gonadotropins – a changing scene. Gynecol Endocrinol 1991; 5:57–58.
19. Hamilton-Fairley D, Kiddy D, Watson H, Sagle M, Franks S. Low-dose gonadotrophin therapy for induction of ovulation in 100 women with polycystic ovary syndrome. Hum Reprod 1991; 6:1095–9.
20. Vaughan Williams C. Ovarian electrocautery or hormone therapy in the treatment of polycystic ovary syndrome. Clin Endocrinol 1990; 33:569–72.
21. Hull MGR. Gonadotrophin therapy in anovulatory infertility. In: Howles CM, ed. Gonadotrophins, gonadotrophin-releasing hormone analogues and growth factors in infertility: future perspectives. Hove, East Sussex: Medi-Fax International, 1992; (in press).
22. Flamigni C, Venturoli S, Paradisi R, Fabbri R, Porcu E, Magrini O. Use of human urinary follicle-stimulating hormone in infertile women with polycystic ovaries. J Reprod Med 1985; 30:184–8.
23. Garcea N, Campo S, Panetta V, Venneri M, Siccardi P, Dargenio R, De Tomasi F. Induction of ovulation witth purified urinary follicle-stimulating hormone in patients with polycystic ovarian syndrome. Am J Obstet Gynecol 1985; 151:635–40.
24. Seibel MM, McArdle C, Smith D, Taymor ML. Ovulation induction in polycystic ovary syndrome with urinary follice-stimulating hormone or human menopausal gonadotrophin. Fertil Steril 1985; 43:703–8.
25. Buvat J, Dehaene JL, Buvat-Herbaut M, Verbecq P, Marcolin G, Renouard O. Purified follice-stimulating hormone in polycystic ovary syndrome: slow administration is safer and more effective. Fertil Steril 1989; 52:553–9.
26. McFaul PB, Traub AI, Sheridan B, Leslie H. Daily or alternate-day FSH therapy in patients with polycystic ovarian disease resistant to clomiphene citrate treatment. Int J Fertil 1989; 34:194–8.
27. Neyro JL, Barrenetxea G, Montoya F, Rodriguez-Escudero FJ. Pure FSH for ovulation induction in patients with polycystic ovary syndrome and resistant to clomiphene citrate therapy. Hum Reprod 1991; 6:218–21.
28. Homburg R, Eshel A, Kilborn J, Adams J, Jacobs HS. Combined luteinizing hormone releasing hormone analogue and exogenous gonadotrophins for the treatment of infertility associated

with polycystic ovaries. Hum Reprod 1990; 5:32–5.
29. Larsen T, Bostofte E, Larsen JF, Felding C, Schioler V. Comparison of urinary human follicle-stimulating hormone and human menopausal gonadotropin for ovarian stimulation in polycystic ovarian syndrome. Fertil Steril 1990; 53:426–31.
30. Wu CH, Gocial B. A pelvic scoring system for infertility surgery. Int J Fertil 1988; 33:341–6.
31. Donnez J, Casanas-Roux F. Prognostic factors of fimbrial microsurgery. Fertil Steril 1986; 46:200–4.
32. McComb P. Microsurgical tubocornual anastomosis for occlusive cornual disease: reproducible results without the need for tubouterine implantation. Fertil Steril 1986; 46:571–7.
33. Laatikainen TJ, Tenhunan AK, Venesmaa PK, Apter DL. Factors influencing the success of microsurgery for distal tubal occlusion. Arch Gynecol Obstet 1988; 243:101.
34. Winston RML, Margara RA. Microsurgical salpingostomy is not an obsolete procedure. Br J Obstet Gynaecol 1991; 98:637–42.
35. Boer-Meisel ME, te Velde ER, Habbema JDF, Kardaun JWPF. Predicting the pregnancy outcome in patients treated for hydrosalpinx: a prospective study. Fertil Steril 1986; 45:23–9.
36. Mage G, Pouly JL, de Joliniere JB, Chabrand S, Riouallon A, Bruhat MA. A preoperative classification to predict the intrauterine and ectopic pregnancy rates after distal tubal microsurgery. Fertil Steril 1986; 46:807–10.
37. Lilford RJ, Watson AJ. Has in-vitro fertilisation made salpingostomy obsolete? Br J Obstet Gynaecol 1990; 97:557–60.
38. Watson AJS, Gupta JK, O'Donovan P, Dalton ME, Lilford RJ. The results of tubal surgery in the treatment of infertility in two non-specialist hospitals. Br J Obstet Gynaecol 1990; 97:561–8.
39. Portuondo JA, Echanojauregui AD, Herran C, Alijarte I. Early conception in patients with untreated mild endometriosis. Fertil Steril 1988; 39:22–4.
40. Hull MEO, Moghissi KS, Magyar DF, Hayes MF. Comparison of different treatment modalities of endometriosis in infertile women. Fertil Steril 1987; 47:40–4.
41. Badawy SZA, Elbakry MM, Samuel F, Dizer M. Cumulative pregnancy rates in infertile women with endometriosis. J Reprod Med 1988; 33:757–60.
42. Hull MGR. Indications for assisted conception. Br Med Bull 1990; 46:580–95.
43. Garcia C, David SS. Pelvic endometriosis: infertility and pelvic pain. Am J Obstet Gynecol 1977; 129:740–7.
44. Olive DL, Stohs GF, Metzger DA, Franklin RR. Expectant management and hydrotubations in the treatment of endometriosis-associated infertility. Fertil Steril 1985; 44:35–41.
45. Thomas EJ, Cooke ID. Successful treatment of asymptomatic endometriosis: does it benefit infertile women? Br Med J 1987; 294:1117–19.
46. Bayer SR, Seibel MM, Saffan DS, Berger MJ, Taymor ML. Efficacy of danazol treatment for minimal endometriosis in infertile women. J Reprod Med 1988; 33:179–83.
47. Telimaa S. Danazol and medroxyprogesterone acetate inefficacious in the treatment of infertility in endometriosis. Fertil Steril 1988; 50:872–5.
48. Mahmood TA, Templeton A. Pathophysiology of mild endometriosis: review of literature. Hum Reprod 1990; 5:765–84.
49. Nezhat C, Crowgey S, Nezat F. Videolaseroscopy for the treatment of endometriosis associated with infertility. Fertil Steril 1989; 51:237–40.
50. Olive DL, Martin DC. Treatment of endometriosis – associated infertility with CO_2 laser laparoscopy: the use of one- and two-parameter exponential models. Fertil Steril 1987; 48:18–23.
51. Donnez J, Nisolle-Pochet M, Lemaire-Rubbers M, Casanas-Roux F, Karaman Y. Combined (hormonal and microsurgical) therapy in infertile women with endometriosis. Fertil Steril 1987; 48:239–42.
52. Guzick DS, Bross DS, Rock JA. Assessing the efficacy of The American Fertility Society's classification of endometriosis: application of a dose–response methodology. Fertil Steril 1982; 38:171–6.
53. Mills MS, Eddowes HA, Cahill DJ, Fahy U, Abuzeid MIM, McDermott A, Hull MGR. A prospective controlled study of in-vitro fertilisation (IVF), gamete intrafallopian transfer (GIFT) and intrauterine insemination (IUI) combined with superovulation. Hum Reprod 1992; 7:490–4.
54. Chillick C, Rosenwaks Z. Endometriosis and in vitro fertilisation. Semin Reprod Endocrinol 1985; 3:4:377–80.
55. Yovich JL, Matson PL, Richardson PA, Hilliard C. Hormonal profiles and embryo quality in women with severe endometriosis treated by in vitro fertilization and embryo transfer. Fertil

Steril 1988; 50:308–13.
56. Yovich JL, Matson PL. The influence of infertility etiology on the outcome of IVF-ET and GIFT treatments. Int J Fertil 1990; 35:26–33.
57. Borrero C, Ord T, Balmaceda JP, Rojas FJ, Asch RH. The GIFT experience: an evaluation of the outcome of 115 cases. Hum Reprod 1988; 3:227–30.
58. Hull MGR, Eddowes HA, Fahy U et al. Expectations of assisted conception treatments for infertility. A complete account of current practice. Br Med J 1992; 304:1465–9.
59. Check JH, Dietterich C, Lauer C, Liss J. Ovulation-inducing drugs versus specific mucus therapy for cervical factor. Int J Fertil 1991; 36:108–12.
60. te Velde ER, Kooy RJ, Waterreus JJH. Intrauterine insemination of washed husband's spermatozoa: a controlled study. Fertil Steril 1989; 51:182–5.
61. Irianni FM, Acosta AA, Oehninger S, Acosta MR. Therapeutic intrauterine insemination (TII) – controversial treatment for infertility. Arch Androll 1990; 25:147–67.
62. Dodson WC, Haney AF. Controlled ovarian hyperstimulation and intrauterine insemination for treatment of infertility. Fertil Steril 1991; 55:457–67.
63. Aitken RJ, Best FSM, Richardson DW, Djahanbakhch O, Mortimer D, Templeton AA, Lees MM. An analysis of sperm function in cases of unexplained infertility: conventional criteria, movement characteristics, and fertilising capacity. Fertil Steril 1982; 38:212–21.
64. Schats R, Aitken RJ, Templeton AA, Djahanbakhch O. The role of cervical mucus–semen interaction in infertility of unknown aetiology. Br J Obstet Gynaecol 1984; 91:371–6.
65. Hewitt J, Cohen J, Steptoe P. Male infertility and in vitro fertilisation. In: Studd J, ed. Progress in obstetrics and gynaecology Vol. 6, Edinburgh: Churchill Livingstone, 1987; 253–75.
66. Wang C, Chan C, Wong K, Yeung K. Comparison of the effectiveness of placebo, clomiphene citrate, mesterolone, pentoxifylline, and testosterone rebound therapy for the treatment of idiopathic oligospermia. Fertil Steril 1983; 40:358–65.
67. Sokol RZ, Petersen G, Steiner BS, Swerdloff RS, Bustillo M. A controlled comparison of the efficacy of clomiphene citrate in male infertility. Fertil Steril 1988; 49:865–70.
68. Clark RV, Sherins RJ. Treatment of men with idiopathic oligozoospermic infertility using the aromatase inhibitor, testolactone results of a double-blinded, randomized, placebo-controlled trial with crossover. J Androl 1989; 10:240–7.
69. Gerris J, Peeters K, Comhaire F, Schoonjans F, Hellemans P. Placebo-controlled trial of high-dose mesterolone treatment of idiopathic male infertility. Fertil Steril 1991; 55:603–7.
70. Glazener CMA, Coulson C, Lambert P, Watt EM, Hinton RA, Kelly NJ, Hull MGR. The value of artificial insemination with husband's semen in infertility due to failure of postcoital sperm–mucus penetration – controlled trial of treatment. Br J Obstet Gynaecol 1987; 94:774–8.
71. Kerin JFP, Kirby C, Peek J, Jeffrey R, Warnes GM, Matthews CD. Improved conception rate after intrauterine insemination of washed spermatozoa from men with poor quality semen. Lancet 1984; i:533–5.
72. Smarr SC, Wing R, Hammond MG. Effect of therapy on infertile couples with antisperm antibodies. Am J Obstet Gynecol 1988; 158:969–73.
73. Haas GG, Manganiello P. A double-blind, placebo-controlled study of the use of methylprednisolone in infertile men with sperm-associated immunoglobulins. Fertil Steril 1987; 47:295–301.
74. Hendry WF, Hughes L, Scammell G, Pryor JP, Hargreave TB. Comparison of prednisolone and placebo in subfertile men with antibodies to spermatozoa. Lancet 1990; 335:85–8.
75. Bals-Pratsch M, Doren M, Karbowski B, Schneider HPG, Nieschlag E. Cyclic corticosteroid immunosuppression is unsuccessful in the treatment of sperm antibody-related male infertility: a controlled study. Hum Reprod 1992; 7:99–104.
76. Elder KT, Wick KL, Edwards RG. Seminal plasma anti-sperm antibodies and IVF: the effect of semen sample collection into 50% serum. Hum Reprod 1990; 5:179–84.
77. Palermo G, Khan I, Devroey P, Wisanto A, Camus M, Van Steirteghem AC. Assisted procreation in the presence of a positive direct mixed antiglobulin reaction test. Fertil Steril 1989; 52:645–9.
78. van der Merwe JP, Hulme VA, Kruger TF, Menkveld R, Windt M. Treatment of male sperm autoimmunity by using the gamete intrafallopian transfer procedure with washed spermatozoa. Fertil Steril 1990; 53:682–7.
79. Grundy CE, Robinson J, Guthrie KA, Gordon AG, Hay DM. Establishment of pregnancy after removal of sperm antibodies in vitro. Br Med J 1992; 304:292–3.
80. Tanbo T, Dale PO, Abyholm T. Assisted fertilization in fertile women with patent Fallopian tubes. A comparison of in-vitro fertilization, gamete intra-Fallopian transfer and tubal embryo stage transfer. Hum Reprod 1990; 5:266–70.
81. Yovich JM, Edirisinghe WR, Cummins JM, Yovich JL. Influence of pentoxifylline in severe

male factor infertility. Fertil Steril 1990; 53:715–22.
82. Aitken RJ, West KM. Analysis of the relationship between reactive oxygen species production and leucocyte infiltration in fractions of human semen separated on Percoll gradients. Int J Androl 1990; 13:433–51.
83. Nice L, Ray B, Grant S, Williams J, McDermott A, Hull MGR. Use of Percoll in IVF: a comparison between sperm dysfunction and tubal patients (in press).
84. Gordon JW, Talansky BE, Grunfeld L, Richards C, Garrisi GJ, Laufer N. Fertilization of human oocytes by sperm from infertile males after zona pellucida drilling. Fertil Steril 1988; 50:68–73.
85. Cohen J, Malter H, Wright G, Kort H, Massey J, Mitchell D. Partial zona dissection of human oocytes when failure of zona pellucida penetration is anticipated. Hum Reprod 1989; 4:435–42.
86. Cohen J, Malter H, Alikari et al. Microsurgical fertilization and teratozoospermia. Hum Reprod 1991; 6:118–23.
87. Hill DL, Surrey M, Adler D, Danzer H, Rothman C, Friedman S. Micromanipulation in a center for reproductive medicine. Fertil Steril 1991; 55:36–8.
88. Payne O, McLaughlin KJ, Depypere HT, Kirby CA, Warnes GM, Matthews CD. Experience with zona drilling and zona cutting to improve fertilization rates of human oocytes in vitro. Hum Reprod 1991; 6:423–31.
89. Fishel S, Antinori S, Jackson P, Johnson J, Rinaldi L. Presentation of six pregnancies established by sub-zonal insemination (SUZI). Hum Reprod 1991; 6:124–30.
90. Silber SJ, Ord T, Balmaceda J, Patrizio P, Asch R. Congenital absence of the vas deferens. The fertilizating capacity of human epididymal sperm. N Engl J Med 1990; 323:1788–92.
91. Federation CECOS, Schwartz D, Mayaux MJ. Female fecundity as a function of age. N Engl J Med 1982; 306:404–6.
92. Federation CECOS, Lannou DL, Lansac J. Artificial procreation with frozen donor semen: experience of the French Federation CECOS. Hum Reprod 1989; 4:757–61.
93. Glazener CMA, Coulson C, Lambert PA, Watt EM, Hinton RA, Kelly NG, Hull MGR. Clomiphene treatment for women with unexplained infertility: placebo-controlled study of hormonal responses and conception rates. Gynecol Endocrinol 1990; 4:75–83.
94. Crosignani PG, Walters DE, Soliani A. The ESHRE multicentre trial on the treatment of unexplained infertility: a preliminary report. Hum Reprod 1991; 6:953–8.
95. Fisch P, Collins JA, Casper RF, Reid RL, Brown SE, Simpson C, Wrixon W. Unexplained infertility: evaluation of treatment with clomiphene citrate and human chorionic gonadotrophin. Fertil Steril 1989; 51:828–33.
96. Navot D, Veeck LL, Muasher SJ, Kreiner D, Oehninger S, Rosenwaks Z, Liu HC. The value of in vitro fertilization for the treatment of unexplained infertility. Fertil Steril 1988; 49:854–7.
97. Audibert F, Hedon B, Arnal F et al. Results of IVF attempts in patients with unexplained infertility. Hum Reprod 1989; 4:766–71.
98. Serhal PF, Katz M, Little V, Woronowski H. Unexplained infertility – the value of Pergonal superovulation combined with intrauterine insemination. Fertil Steril 1988; 49:602–6.
99. Martinez AR, Vermeiden JPW, Bernardus RE, Schoemaker J, Voorhorst FJ. Intrauterine insemination does and clomiphene citrate does not improve fecundity in couples with infertility due to male or idiopathic factors: a prospective, randomized, controlled study. Fertil Steril 1990; 53:847–53.
100. Turhan NO, Artini PG, D'Ambrogio G, Droghini F, Volpe A, Genazzani AR. Studies on direct intraperitoneal insemination in the management of male factor, cervical factor, unexplained and immunological infertility. Hum Reprod 1992; 7:66–71.
101. Braeckmans P, Devroey P, Camus M. Gamete intra-Fallopian transfer: evaluation of 100 consecutive attempts. Hum Reprod 1987; 2:201–5.
102. Wong PC, Ng SC, Hamilton MPR, Anandakumar C, Wong YC, Ratnam SS. Eighty consecutive cases of gamete intra-Fallopian transfer. Hum Reprod 1988; 3:231–3.
103. Devroey P, De Grauwe E, Staessen C, Wisanto A, Camus M, Van Steirteghem AC. Zygote intrafallopian transfer as a successful treatment for unexplained infertility. Fertil Steril 1989; 52:246–9.
104. Tan SL, Steer C, Royston P, Rizk P, Mason BA, Campbell S. Conception rates and in-vitro fertilisation. Lancet 1990; 335:299.
105. Kovacs GT, Rogers P, Leeton JF, Trounson AO, Wood C, Baker HW. In-vitro fertilization and embryo transfer. Med J Aust 1986; 144:682–3.
106. de Mouzon J, Bachelot A, Gagnepain A, Pessione F. Analyse des resultats 1989 et 1986–1989. Contraception Fertilite Sexualite 1990; 18:589–91.
107. Haan G, Bernardus RE, Hollanders HMG, Leerentveld BO, Prak FM, Naaktgeboren N.

Selective drop-out in successive in-vitro fertilization attempts: the pendulum danger. Hum Reprod 1991; 6:939–43.
108. Guzick DS, Wilkes C, Jones HW. Cumulative pregnancy rates for in vitro fertilization. Fertil Steril 1986; 46:663–7.
109. Hershlag A, DeCherney AH, Kaplan EH, Lavy G, Loy RA. Heterogeneity in patient populations explains differences in in vitro fertilization programs. Fertil Steril 1991; 56:913–17.
110. Chaffkin LM, Nulsen JC, Luciano AA, Metzger DA. A comparative analysis of the cycle fecundity rates associated with combined human menopausal gonadotrophin (hMG) and intrauterine insemination (IUI) versus either hMG or IUI alone. Fertil Steril 1991; 55:252–7.
111. Seracchioli R, Melega C, Maccolini A, Cattoli M, Bulletti C, Bovicelli L, Flamigni C. Pregnancy after direct intraperitoneal insemination. Hum Reprod 1991; 6:533–6.
112. Leeton J, Healey D, Rogers P, Yates C, Caro C. A controlled study between the use of gamete intrafallopian transfer (GIFT) and in vitro fertilization and embryo transfer in the management of idiopathic and male infertility. Fertil Steril 1987; 48:605–7.
113. Vessey MP, Wright NH, McPherson K, Wiggins P. Fertility after stopping different methods of contraception. Br Med J 1978; ii:265–7.

Discussion

Beral: Professor Templeton showed data that 3% to 6% of women who want to have children are permanently childless, and yet many of the figures show that about 15% of women are what are called infertile. The difference between those figures is huge, and in terms of population, of services needed, and so forth, on a national scale is quite massive. If 5% of women who wanted children were nonetheless childless and 15% were called infertile, are those 10% infertile, or are they women who take a long time to get pregnant? Is there something in between or is it the end of the distribution?

Templeton: Between 3% and 4% remain childless, and another 3%–4% remain without having completed their families. Overall there are between 6% and 8% who have primary infertility and a further 6%–8% with secondary infertility. Thus half of each group have unresolved infertility, and half eventually achieve a pregnancy.

As far as services are concerned, the important figure is 15% because this is the proportion of women who are likely to seek advice.

Beral: There is a big difference between those numbers, huge if it is multiplied back to the number of women of childbearing age and related to the services required. Is that difference just people who take a long time to conceive, or people with real pathological conditions?

Hull: One can easily calculate what the cumulative rates will be at fixed rates. Even at 1% per cycle, after a decade the majority, certainly more than half, will have conceived. The issue in practice from the clinical point of view and from the patient's standpoint is this. Is it appropriate to take a 10- or 20-year view of their chances of eventually conceiving and perhaps conceiving only once? For them their perspective is in a much shorter time frame, and that is the time frame in which we need to work. They want an accelerated chance of conceiving not just once, but probably more than once within that sort of time frame, and that is a very different perspective from looking retrospectively at ultimate fertility or infertility in the population.

Beral: The implication is that infertility is a perception.

Hull: It is not a perception at all. There is a degree of subfertility, which we can define, and work out, from cumulative conception rates, the monthly fecundity. In some cases, for instance in short unexplained infertility, it averages 8% per cycle, which is slightly below average but well within the normal range. Then there are others with prolonged infertility and for example endometriosis, with clearly demonstrable marked subfertility; one can define it at perhaps 3% per cycle, or less.

Barlow: The Aberdeen population study described those with both voluntary and involuntary infertility. Within the voluntary group are we including women who are not in relationships and therefore they have no opportunity to become pregnant, or are these a separate group?

Templeton: The total voluntary infertility figure I first showed, about 8%, includes all women, including single women. Among the married women, 2%–3% deliberately chose not to have children.

Shaw: In your chapter were you describing age-related infertility rates?

Templeton: No, the figures represented the cumulative experience. We chose the 46–50-year age group first because they are at the end of their reproductive life. It was the total experience of infertility in that age group. We then looked at the 36–40-year-old group, the vast majority of whom had decided that their reproductive life was over for all effective purposes and so it was a cumulative experience for them as well.

Beral: It seems to be quite important in this sort of discussion to know what the prevalence is. It is quite extraordinary that we do not know whether prevalence is increasing over time. With the increase in sexually transmitted diseases, for example, one might expect that the prevalence is increasing, but the evidence is not there.

Templeton: There is some evidence of an increase in subsections of the population from the US surveys, but there are methodological problems because the surveys were done in slightly different ways for each of the five years and the results are therefore open to question.

Baird: I should like to pursue this further. There was a question about the difference between the 5% who remained childless and the 15% who sought infertility services. Really the relevant question is not are these 15% infertile, the relevant question to ask is what is a reasonable time for somebody to try and get pregnant without seeking medical help, and hence having resource implications. Those are the data they collected in Aberdeen and are based on a two-year wait, which most people would say, subject to variation in age, in other things, is reasonable.

The other relevant question to be borne in mind is how do we identify those for whom treatment may make a difference. Quite a large percentage of women do become pregnant relatively quickly without treatment, and those of us who run infertility clinics know this. But perhaps more importantly, there is a significant proportion of couples who attend infertility clinics for whom treatment will not make any difference and in the triage system we should avoid pouring resources into those types and causes of infertility.

It is unproductive to spend a long time asking if this is a disease, or is it a perception, or something. It is really what is acceptable.

Templeton: A question on the methodology of doing cumulative conception rates. Professor Hull showed one chart for normal fertility and one for mild endometriosis. The starting points are different: mild endometriosis starts from the point of diagnosis and the normal from the point of trying to achieve conception. If we look at mild endometriosis from the point of attempting to achieve conception, conception rates are superimposed on the normal rates. The reason that they appear lower than the normal rates is that women with mild endometriosis who go to infertility clinics only do so because they are infertile and it then happens that the mild endometriosis is diagnosed. It would be wrong to draw the conclusion that mild endometriosis causes infertility.

Hull: I accept the implication. Those results beg very important questions about an association with subfertility, a causal relationship, whether it exists, what the mechanism might be if there is a cause, and treatment methods. I have said it begs all those suggestions and I am not suggesting there is a causal relationship.

Templeton: I wanted to make the general point that very frequently in reporting fertility rates, starting points are different, and so results are not comparable.

Hull: Could I make another plea? I do not think we are to have the opportunity to evaluate particular treatments and assisted conception methods. The way the data are presented is largely uninterpretable, from the literature around the world, and even indeed from the licensing authority. One of the problems from around the world is that there are often no restrictions on numbers of eggs and embryos transferred by GIFT and IVF, and that makes interpretation difficult. And even in the UK where we now have strict regulation, and in the last few years we have only been allowed to transfer four in exceptional circumstances, there are still problems in the interpretation of the data as presented. They are not classified by, for example, the woman's age, they are not classified by diagnostic background. Last year for the first time the licensing authority published data by diagnostic class, and extraordinarily they included in the male infertility group those who were treated with donor sperm [1]. We simply cannot unravel that. This plea is directed at the new licensing authority that if it is getting into data publication, then please do it in a sensible classified way.

It is of practical importance because when we come to attempting to give couples prognostic information about their expectations, on a cumulative basis one can only apply treatment repetitively if we know that the situation is favourable, or at least defined. Professor Templeton raised an issue about life-table analysis. It is only valid if the assumption is made that those who did not continue in a study would have done as well as those who did, and therefore we must exclude bias from our calculation of cumulative rates. That can only be excluded if the diagnostic classification is accurate and specific at the start.

Templeton: Dr Botting, from the OPCS perspective, how easy would it be to ask specific questions about infertility on the back of the General Household Survey.

Botting: The General Household Survey seems to me an ideal vehicle for that kind of survey because the resources involved in mounting a GP-based survey are huge. Every two years there are questions in the General Household Survey on contraception and if a woman answers that she is not using any method of contraception, I believe there is a follow-up question which asks why she is not currently using contraception and so identifies cases where the woman is not currently in a sexual partnership. It, therefore, seems that there is the potential for adding in there additional questions. This seems to be an ideal facility and it would give results every other year, and by age, by marital status, by geography. It would give a wide range of data.

Beral: So the message is that OPCS would not mind a request that some information on infertility be collected as part of the General Household survey and it could easily be done, although there would be some resource implications.

Asch: There was a reference to multiple pregnancies. What percentage of all multiple pregnancies results from assisted reproduction techniques? And do we have the numbers in women aged 40 years and over?

Botting: I do not have the figures in detail. It is a small proportion. But the problem is that it includes not only the IVF and GIFT data, which we know, but also a large number of cases resulting from ovulation induction, and that is not measurable. There are no data at all.

With IVF and GIFT, in excess of one-third of triplets and above are results of assisted conception techniques. It is a smaller proportion of twins because there are large numbers in the general population.

Whittall: About the data that HFEA is collecting, there are a variety of ways in which the information can be analysed and what was suggested should be possible.

In terms of worldwide data, an international group meets on an *ad hoc* basis and somebody from the authority has started to attend that group. They are looking at ways of getting some sort of conformity in data collection and presentation.

Beral: I have been involved in this international group in looking at statistics from IVF and assisted conception and there really is a need, particularly from the HFEA, to have data that are relevant, that can be used for scientific purposes and for the evaluation of the efficacy of treatment; and also, something that has not been mentioned, looking at the health of the child. The treatment of infertility is not only to produce a pregnancy but to produce a child that is healthy and it is extremely important to have a broad view of the long-term outcome and to ensure that the children are healthy as a result.

Cooke: To take up the point about triage. Professor Baird was referring to resources but the other element of triage is the economic cost. In any appraisal of health economics there is the economic cost and there is also the assessment of the utility of any exercise, and unless both of those are addressed, the technical developments do not sit in perspective.

Beral: A good note to end on and a good beginning to these deliberations.

Reference

1. The Sixth Report of the Interim Licensing Authority for Human In Vitro Fertilisation and Embryology 1991. ILA: London, p. 28.

Section II

Investigation and Treatment of the Male

Chapter 4

Male Infertility Due to Testicular Dysfunction

E. Nieschlag and H. M. Behre

Herod: After all, you are barren.
Herodias: I barren? ...It is absurd to say that.
I have borne a child.
You have never had a child,
not even by one of your slaves.
It is you who are sterile not I that am barren.
Herod: Be quiet. I tell you that you are barren.
You have not borne me a child.

Oscar Wilde, *Salomé* (1892)

Fertility disturbances are among the most common disorders of young and middle-aged men. Neither the public nor the medical profession is aware of this dimension of the problem, since the field is still surrounded by taboos. The dialogue Oscar Wilde ascribed to Herod and Herodias a hundred years ago in a plot taking place 2000 years ago is still valid today. Men have difficulties in accepting the possibility of infertility – and in obtaining medical advice. Of men consulting for infertility 60% have let more than two years pass before deciding to consult a doctor [1]. A further obstacle to overcoming taboos is in the fact that in many cases medical assistance would be possible but is not life preserving, as in other medical disciplines, so that the patient has no physical pressure to seek help. Although two decades of intensive research have increased considerably our knowledge about male reproductive functions, andrology as the science and clinical practice of male fertility has been slow to gain academic status. However, the number of patients consulting for infertility is increasing, probably not because of an increasing incidence of male infertility, but because the taboos surrounding male potency and fertility are gradually diminishing.

Interdependence of Male and Female Reproductive Functions

Disturbances of male fertility may go unrecognised for a long time and become evident only when marriage remains without issue despite unprotected intercourse and the wish for a child. Fertility disorders must be assumed when, after 6–12 months of unprotected intercourse, no pregnancy occurs, since in the general population 70% (by 6 months) to 90% (by 12 months) of pregnancies occur within this time frame [2]. As illustrated by Fig. 4.1, fertility disturbances of one partner may become evident only through the other partner's problems, whereas optimal reproductive function of one partner may compensate for impaired function of the other and may prevent the couple from attending a clinic. Thus, infertility is mostly a disorder of the couple that should be investigated and treated as a single entity or at least in parallel. A worldwide survey indicated that these couples in whom fertility problems exist in both partners make up about a quarter of infertility patients in general [3], whereas they constitute the majority of cases in specialised centres. This means that unless azoospermia has been diagnosed, the significance of male fertility disorders has to be considered relative to the female reproductive status.

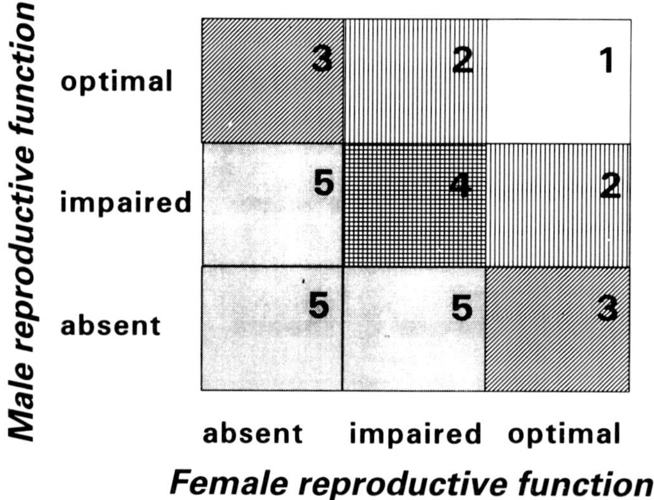

Fig. 4.1. Interdependence of male and female reproductive functions. Couples of group 1 and also couples of group 2, where one is compensating the other's problems, will rarely seek medical assistance. In group 3 diagnosis and treatment will concentrate on one partner, and groups 4 and 5 represent the "hard core" of infertile patients and require concerted action by the andrologist and gynaecologist.

Classification of Male Fertility

Male fertility disturbances may be associated with other symptoms which may become clinically relevant independent of the wish for offspring. In particular, endocrine disorders may have their origin in the dual functions of the testes, i.e. sperm production and testosterone synthesis. Whereas disturbances of spermatogenesis and spermiogenesis may occur without impaired testosterone synthesis, endocrine insufficiency of the testes always leads to infertility. Infertility may also have its origin in disorders of the outflow ducts (e.g. infections, traumata, agenesis/aplasia), the blood vessels (varicocele) or the immune system (sperm antibodies); infertility may also be the consequence of other diseases (e.g. renal insufficiency, diabetes mellitus, neoplasia, myotonic dystrophy) or exposure to pharmaceuticals and toxins. Usually the term "infertility" is used to describe disorders without clinically overt endocrine insufficiency of the testes, and the term "hypogonadism" describes infertility associated with symptoms of testosterone deficiency.

In order to account for the manifold causative mechanisms of male fertility disturbance we use localisation of the cause or the level of origin of the disturbance as a basis for a systematic classification, as shown in Table 4.1. As a complete review of male infertility is beyond the scope of this chapter, we will concentrate here on disturbances of spermatogenesis and spermiogenesis which are involved in the disorders most frequently encountered, as shown by a survey of more than 5000 cases from our Institute (Table 4.2). Therefore, endocrine hypogonadism, maldescended testes, varicocele and idiopathic infertility will be the object of this chapter preceded by general diagnostic considerations.

Table 4.1. Classification of male fertility disorders based on the localisation of the cause

Localisation of the cause	Disorder	Cause	Androgen-deficiency	Infertility
Hypothalamus/ Pituitary	Idiopathic hypogonadotropic hypogonadism and Kallmann syndrome	Congenital disturbances of GnRH secretion	+	+
	Prader–Labhart–Willi syndrome	Congenital disturbances of GnRH secretion	+	+
	Laurence–Moon–Biedl syndrome	Congenital disturbances of GnRH secretion	+	+
	Familiar cerebellar ataxia	Congenital disturbances of GnRH secretion	+	+
	Constitutionally delayed puberty	"Delayed biological clock"	+	+
	Secondary disturbance of GnRH secretion	Tumours, infiltrations, traumata, irradiation, malnutrition, disturbed blood circulation, general disease	+	+
	Panhypopituitarismus	Tumours, infiltration, traumata, irradiation, ischaemia, surgery	+	+
	Pasqualini syndrome	Congenital disturbance of LH secretion	+	(+)
	Isolated FSH deficiency	Congenital disturbance of FSH secretion	–	+
	Hyperprolactinaemia	Adenoma, pharmaceuticals	+	+

Continued overleaf

Table 4.1. *Continued*

Localisation of the cause	Disorder	Cause	Androgen-deficiency	Infertility
	Exogenous disturbances	Pharmaceuticals, drugs	+	+
Testes	Congenital anorchism	Fetal loss of testes	+	+
	Acquired anorchism	Trauma, torsion, tumour, infections, surgery	+	+
	Gonadal dysgenesis	Defective Y-chromosome	+	+
	Persistent oviduct syndrome	Lack of antiMüllerian hormone	−	(−)
	Germ cell aplasia	Congenital or acquired	−	+
	Leydig cell aplasia	Congenital	+	+
	Male pseudohermaphroditism	Enzymatic defect in testosterone synthesis	+	+
	Klinefelter syndrome	Meiotic non-dysjunction	+	+
	XXY syndrome	Numerical chromosome abnormality	(+)	(+)
	XX male	Translocation of the Y-chromosome	+	+
	Male Turner syndrome	Translocation of the Y-chromosome	+	+
	Maldescended testes	Congenital, lack of testosterone	(+)	+
	Testicular tumours	?	+	+
	Varicocele	Incompetent venous valves	(−)	+
	Orchitis	Infectious destruction of tubular epithelium	(−)	+
	Globozoospermia	Disturbed spermiogenesis	−	+
	Immotile cilia syndrome	Disturbed spermiogenesis	−	+
	General diseases and exogenously caused disturbances	e.g. Renal insufficiency, liver cirrhosis occupational hazards, irradiation	+	+
Seminal ducts and accessory glands	Infections	Bacteria, viruses	−	+
	Obstructions	Congenital abnormalities, infection, vasectomy	−	+
	Cystic fibrosis	Mutation of the CF gene	−	+
	Young syndrome	?	−	+
	Disturbed liquefication	?	−	+
	Immunologic infertility	Autoimmunity	−	+
Semen deposition	Penis deformities	Congenital or acquired	−	+
	Hypo-/epispadias	Congenital	−	+
	Phimosis	Congenital	−	(+)
	Retrograde ejaculation	Disturbed innervation	−	+
	Erectile dysfunction	Vascular, neurological, endocrine, psychological	(+)	(−)
Androgen target organs	Testicular feminisation	Complete androgen receptor deficiency	+	+
	Reifenstein syndrome	Incomplete androgen receptor deficiency	+	+

Continued

Table 4.1. *Continued*

Localisation of the cause	Disorder	Cause	Androgen-deficiency	Infertility
	Prepenile scrotum syndrome	Incomplete androgen receptor deficiency	+	+
	Androgen resistance in infertility	Mild androgen receptor deficiency	–	+
	Receptor positive androgen resistance	Post-androgen receptor disturbance	(+)	(+)
	Perineoscrotal hypospadias with pseudovagina	5 α-Reductase deficiency	+	+

Table 4.2. Percentage distribution of diagnoses from 5061 consecutive infertile patients attending the Institute of Reproductive Medicine at the University of Münster

Idiopathic infertility	30.2
Varicocele	15.4
Endocrine hypogonadism	9.7
Infections	8.5
Maldescended testes	8.0
Ejaculatory/erectile dysfunction	6.7
General diseases	5.2
Sperm antibodies	3.8
Testicular tumours	2.1
Obstruction	1.5
Remainder	8.9

Diagnosis of Male Infertility

Although of essential importance, semen analysis should not be the only diagnostic procedure. In particular, in the era of in vitro fertilisation the andrological work-up is often reduced to a glass vial with a semen sample, disregarding the diagnostic clues a complete investigation provides. At first a thorough history has to be taken from the couple, including sexuality and professional or environmental exposure to toxins. General physical examination is followed by investigation of the genitalia and the secondary sex characteristics. Determination of the consistency and size of the testes by palpatory comparison with an orchidometer or, more exactly, by ultrasonography [4] provides a first diagnostic clue to spermatogenesis, since testis size correlates with sperm count (Fig. 4.2). In addition, sonography reveals a number of abnormalities which would otherwise be overlooked (Table 4.3) [5].

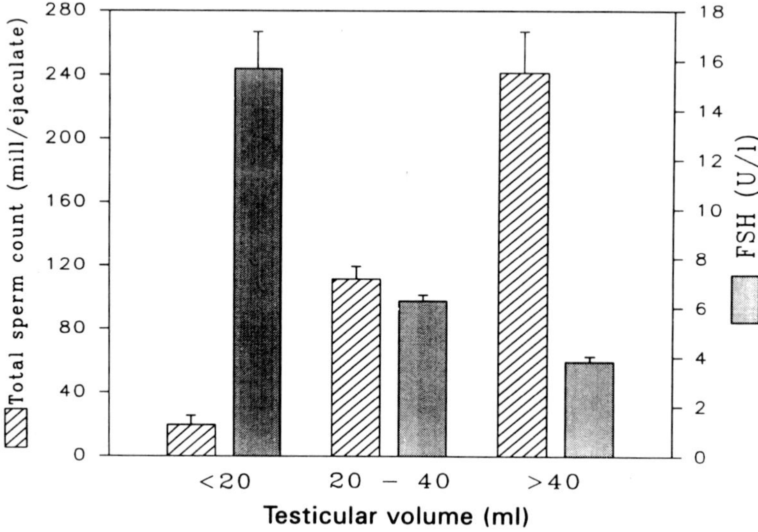

Fig. 4.2. Testicular volumes (determined by sonography and combined from both sides) of 478 infertile patients show a positive correlation with the total sperm count and a negative correlation with serum FSH.

Table 4.3. Results of scrotal sonography from 650 consecutive infertile patients attending the Institute of Reproductive Medicine of the University of Münster [5]

	n	%
Normal	360	55
Varicocele	141	22
Hydrocele	43	7
Epididymal pathology	43	7
Spermatocele	40	6
Inhomogeneity	26	4
Cyst	7	1
Tumour	4	0.6

Semen Analysis

Although semen analysis is the most important laboratory test for male infertility, it is the last area of the medical laboratory to be subjected to quality control. Reasons for this lack of quality control are the difficulty in working with live gametes and the absence of standardised methodology. However, since 1980 the World Health Organization (WHO) has provided guidelines for semen evaluation. A third edition of these guidelines appears in 1992 [6] and should be used by all andrological laboratories. Application of the recommended techniques is a prerequisite for standardisation and quality control. Variability

Table 4.4. Criteria of normality for semen samples analysed according to WHO guidelines [6]

Volume	>2.0 ml
pH	7.2–8.0
Sperm concentration	$>20 \times 10^6$/ml
Total sperm count	$>40 \times 10^6$ per ejaculate
Sperm motility	>50% a+b or <25% a
Sperm morphology	>30% normal
Sperm vitality	>75% live
White blood cells	$<1 \times 10^6$/ml
Immunobead test	<20% sperm with particles
MAR-test	<10% sperm with particles
α-Glucosidase	>20 mU/ejaculate
Zinc	>2.4 μmol/ejaculate
Fructose (total)	>13 μmol/ejaculate

a = rapid, b = slow progressive motility, MAR = mixed agglutination reaction.

of results can thus be reduced to an acceptable level [7], but the comparability of results generated in different laboratories still leaves much to be desired [8].

If properly conducted, semen analysis will provide values for sperm number, motility and morphology as well as for seminal markers and sperm antibodies. These values can be classified as "normal" and "subnormal" according to criteria established by WHO (Table 4.4). It is, however, important to bear in mind that these criteria do not discriminate between fertile and infertile men, as there is a wide overlap of seminal parameters from "spontaneously" fertile men with those from patients attending an infertility clinic who eventually become fathers and those who remain infertile [9,10] (Fig. 4.3). Therefore, as long as sperm are present, subnormal values should not induce despair but should encourage andrologists and gynaecologists to collaborate closely in an effort to improve the couple's chances for pregnancy.

However, there are disturbances of spermiogenesis which cannot be treated. These disorders are globozoospermia, where sperm lack the acrosome and therefore cannot interact with the egg [11] and axonemata defects of the sperm tail such as the "9+0" axoneme resulting in sperm immotility [12] (Fig. 4.4). For these cases microinjection of sperm into the oocyte may provide a solution in the future.

In addition to semen analysis, sperm cervical mucus interaction and other sperm function tests as described in the next chapter are important tools for evaluating male fertility further.

Serum FSH

The most important endocrine parameter in the infertility work-up is serum follicle stimulating hormone (FSH), since it reflects intratesticular events well.

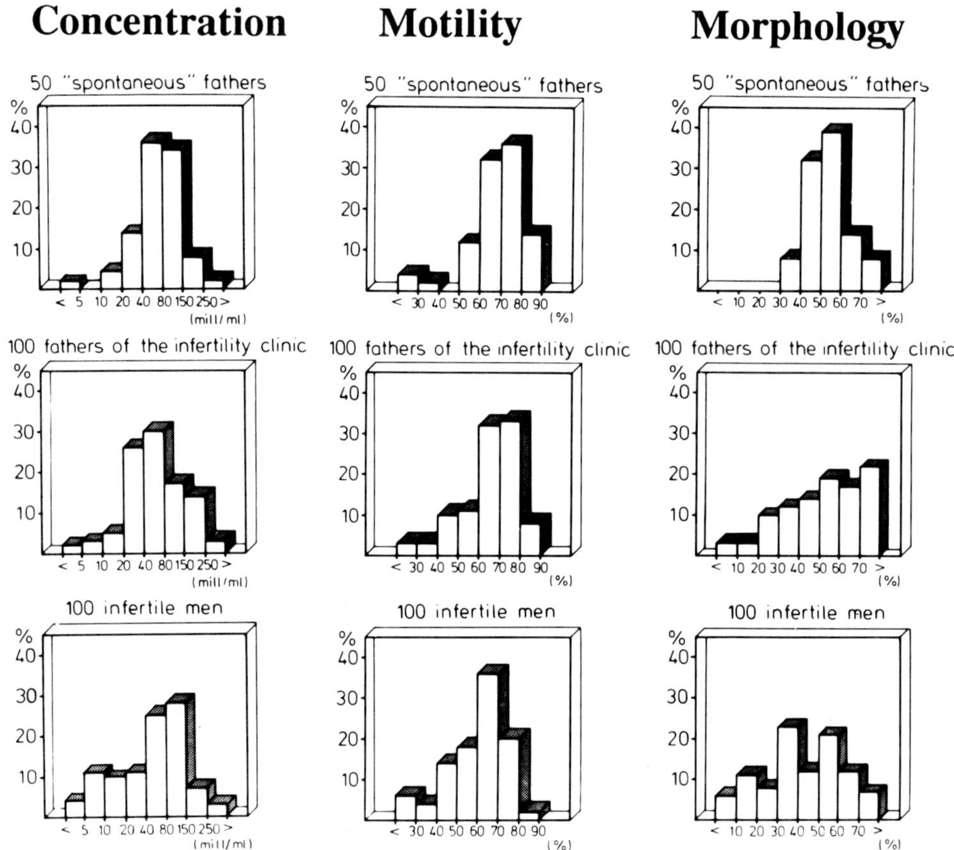

Fig. 4.3. Distribution (%) of seminal parameters from three groups of men: "spontaneous fathers" ($n = 50$), patients who fathered a child after attending our infertility clinic ($n = 100$) and patients who remained infertile ($n = 100$). Data indicate that subnormal semen parameters do not exclude paternity. (Modified from reference [9].)

Highest FSH levels are found in orchidectomised and Klinefelter patients, i.e. in the absence of spermatogenetic tissue. De Kretser [13] postulated that elevated FSH levels could be observed in patients with Sertoli cell only (SCO) syndrome, with spermatogenetic arrest before the appearance of spermatids and/or with hyalinisation of the basal tubular membrane. In a recent series of 97 cases we found that a relatively good correlation exists between serum FSH and testicular histology when the mass of SCO tubules is taken into consideration; the highest FSH values were found in azoospermic patients with bilateral complete SCO syndrome, followed by azoo- or oligozoospermic patients with bilateral focal SCO tubules and patients with unilateral focal SCO tubules, whereas oligozoospermic patients without SCO tubules showed FSH within the normal range [14] (Fig. 4.5). The Sertoli cells of the SCO tubules also showed a higher than normal immunoreactivity for FSH and inhibin, indicating the functionally diseased state of these cells.

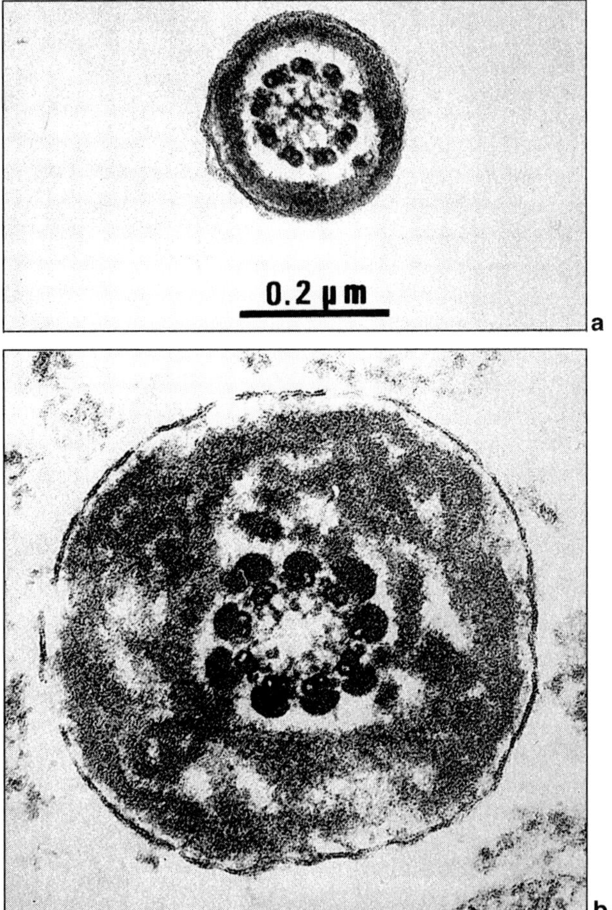

Fig. 4.4. Cross-section through human sperm tails. **a** Normal man: two single microtubules and the circle of nine doublets form the axoneme which is surrounded by a fibrous sheath and the plasma membrane. **b** Sperm tail from a patient with "9+0 syndrome". The central microtubules and their central sheath are missing. All other structures of the axoneme (microtubular doublets with dynein, nexin, spokes, and spoke heads) are present. The fibrous sheath is much thicker than in the normal sperm (see reference [12]).

These findings underline the quantitative aspect of the relationship between FSH and testicular histology. A certain number of seminiferous tubules needs to be affected before a rise in FSH occurs. In animal experiments we could demonstrate that 30% of testicular tissue needs to be destroyed in order to result in an FSH elevation [15]. Whether such a strict correlation also exists in humans under clinical conditions is difficult to assess, since biopsies may not be fully representative of the remaining testis and entire testes cannot be obtained for this purpose.

However FSH has remained the single most important parameter in the evaluation of spermatogenetic activity. Elevated FSH levels make testicular

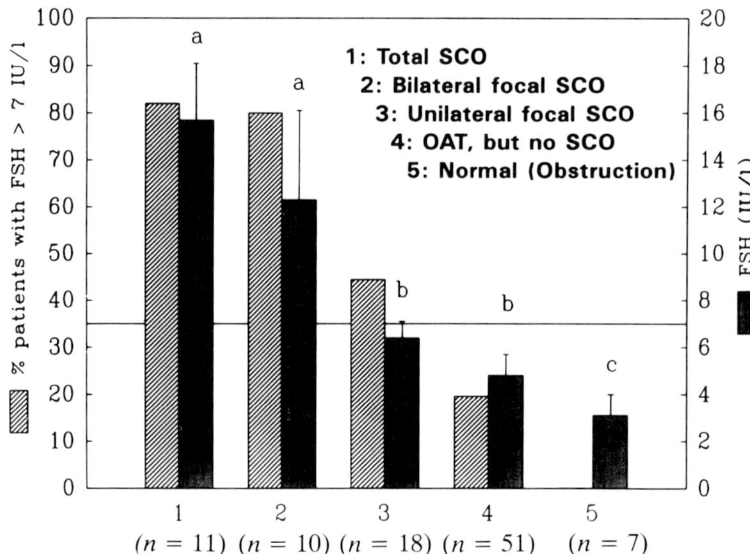

Fig. 4.5. Testicular histology of 97 infertile men in relation to serum FSH. Solid bars: mean values ± SE of serum FSH levels (different superscripts indicate significant differences). Hatched bars: percentage of patients in one group with serum FSH above the upper normal value of 7 IU/l. SCO = Sertoli cell only; OAT = oligoasthenoteratozoospermia. (Modified from reference [14].)

biopsy superfluous since a rational treatment of the underlying condition does not exist. Biopsies are mainly reserved for the differential diagnosis of azoospermia with normal FSH due to obstruction or to spermatogenetic disorders prior to eventual reconstructive surgery. Testicular biopsies are further warranted in the diagnosis of neoplasia including carcinoma in situ.

It should be mentioned here that the SCO syndrome is not a pathological entity but may result from different causes. The SCO syndrome may be congenital in origin, or may be the final stage of various influences on the testes, e.g. viral infections, chemotherapy, radiation, maldescent, varicocele.

FSH is usually measured by immunoassay but immunoactivity may not necessarily reflect bioactivity of the hormone [16, 17]. However, in cases of male infertility a good correlation exists between FSH immuno- and bioactivity [18] so that for diagnostic purposes it is sufficient to rely on (reliable!) immunoassays. Inhibin, so far, has not shown any correlation to spermatogenetic activity [19] and has thus not contributed to infertility diagnosis.

Hypogonadism

The endocrine regulation of testicular function is the best elucidated aspect of male reproductive function. The feedback mechanisms between hypothalamus, pituitary and testes are well known and their disturbances result in rational diagnostic and therapeutic measures in the case of secondary hypogonadism. Although testosterone (in pharmacological doses) or FSH are sufficient to induce or maintain qualitatively normal spermatogenesis, the presence and interplay of both hormones appears to be necessary for qualitatively normal spermatogenesis (for review see reference [20]).

Therefore secondary hypogonadism has to be treated with a combination of hCG (= luteinising hormone (LH) which stimulates testosterone) and hMG (= FSH) to induce spermatogenesis or, alternatively, in the case of idiopathic hypogonadotropic hypogonadism (IHH) caused by hypothalamic dysfunction, GnRH may be administered in pulsatile fashion. Unless other confounding factors are encountered, such as maldescended testes, these therapies are effective in inducing spermatogenesis and pregnancies. However, the current schemes of hCG/hMG and GnRH application may not entirely reflect the physiological situation since normal testis sizes and sperm counts are not achieved in most cases [21–23]. This may, for example, be due to insufficient administration of FSH as current therapeutic schemes appear to provide FSH for too short periods [24]. Perhaps, in the course of introducing biosynthetic FSH [25], pharmacokinetic studies may be performed and result in improved therapeutic schemes. Nevertheless, even the subnormal number of sperm produced by the current regimen is sufficient to induce pregnancies, since these are probably healthy sperm.

In cases of primary hypogonadism, among which Klinefelter syndrome is most frequently diagnosed (1/500 men in the general population), spermatogenesis cannot be improved. Klinefelter syndrome as well as other severe disturbances of spermatogenesis may be caused by numerical chromosomal abnormalities. More recently, qualitative chromosomal abnormalities have also been identified as causes of azoospermia, such as deletions in the short arm of the Y-chromosome [26]. In the future, molecular genetics may further help to identify causes of primary hypogonadism and severe spermatogenetic disturbances.

So far, at least the lack of testosterone can be treated satisfactorily in patients with primary hypogonadism. Conventional substitution therapy consists of intramuscular injections of testosterone enanthate, subcutaneous implantation of crystalline testosterone or oral application of testosterone undecanoate (for review see reference [27]). New forms of substitution therapy are under development, such as transdermal testosterone [28] or the long-acting ester testosterone buciclate [29], which may provide serum testosterone closer to physiological levels thus providing the patient with improved well being and long-term effects.

Maldescended Testes

A surprisingly high percentage of patients attending our infertility clinic complain of disturbed testicular descent (Table 4.2). In most cases maldescended testes have already been recognised and treated, and it is rare that the testes are actually in an abnormal position at the time of presentation with infertility. This indicates that prior treatment did not lead to complete prevention of infertility. In the past it was the usual procedure to deal with maldescended testes in advanced childhood or in puberty. This opinion has changed and for the past 20 years it has become accepted that the position of the testes should be corrected by the end of the second year of life. Most recently the consensus was formed that treatment by hCG, GnRH or orchidopexy should occur before the end of the first year of life [30]. As the average age of our patients is 32 years, patients given such early treatment will begin to turn up only gradually in infertility clinics or will remain conspicuous by their absence, if the concept of early treatment as a preventive measure proves itself.

There is great variability in the histological appearance of maldescended testes: in the worst of cases this may represent SCO syndrome [31]. These testes not only cause fertility disturbances, but also become malignant about ten times as often as completely descended testes. However, the chance of this occurring (0.04%) is low and drops further after the age of 30. It is not yet clear whether the early treatment recommended today actually reduces this risk. However, treatment should be carried out to make screening easier as both palpation and sonography can be performed better when the testes are in a scrotal position.

There is no possibility of improving ejaculatory parameters in patients with this condition wishing fertility. As in the case of idiopathic infertility, only assisted fertilisation can be implemented, observing the minimal criteria required for these procedures.

Varicocele

Varicocele results from inadequate functioning of the valves of the testicular vein leading to a reflux of venus blood into the pampiniform plexus and the testes. As a consequence, spermatogenesis is disturbed, which may result in oligoasthenoteratozoospermia and hence in impaired fertility. The possible pathogenic mechanisms underlying varicocele are increased venous pressure with increasing scrotal temperature (which can be measured thermographically [32]) and reduced removal of toxic substances. The results are impaired spermatogenesis and a reduction of testicular volume on the affected side.

Varicoceles are observed in about 10% of men, but only a small percentage of these become infertile [33]. Treatment is generally considered in a patient wishing fertility when there is a lower testicular volume on the affected side and subnormal semen parameters are observed. Venous reflux is intercepted by surgical ligation or radiological embolisation of the spermatic vein. Semen parameters may improve within 3–12 months after such therapy and pregnancy rates of 20%–50% have been reported. In a randomised study involving 71 couples we found that treatment by surgical ligation is as effective in terms of pregnancy rates as embolisation (unpublished). Thus, both methods appear to be equivalent. Since the rate of complications using sclerosing agents of the testicular vein appears to be high and pregnancy rates have not been established, the sclerosing procedure cannot be recommended.

Even if several studies have observed a positive influence on semen parameters and pregnancy rates, it should not go unmentioned that not all investigators are convinced that a varicocele requires treatment. Well-designed studies should resolve this controversy.

Idiopathic Infertility

As shown in Table 4.2, the largest group of patients with disturbed fertility suffer from "idiopathic infertility". These patients show impaired spermatogenesis and subnormal semen parameters with or without elevated FSH, but the pathological cause of this disturbance remains unclear. This diagnosis probably harbours a multitude of different pathologies and represents a constant challenge to andrological research. New insights are expected from molecular biology, paracrinology and gamete biology. However, since the patients require treatment now, there is a constant temptation to use a variety of drugs and procedures without final proof of effectiveness.

As hormones are so effective in the treatment of hypogonadism it was tempting to try their application in idiopathic infertility. Ever since gonadotrophins became available for clinical use in the early 1960s, they were also applied in this condition. Only relatively recently was a controlled double-blind study performed by which we could demonstrate that in comparison to the placebo group, no improvement of seminal parameters or increase in pregnancy rates could be produced [34]. Similarly, pulsatile GnRH therapy recommended for infertile men with normal or elevated FSH levels did not lead to improved semen values and pregnancy rates in our experience [35].

Time and again therapy of idiopathic infertility with androgens has been attempted although rationally not justifiable. After many years of mesterolone use for infertility, WHO performed a randomised, controlled, double-blind study involving 256 couples. Compared to placebo, no significant increase in pregnancy rates could be demonstrated [36].

Currently antioestrogens (in particular tamoxifen) have taken over from mesterolone. Antioestrogens block oestrogen action in the pituitary and thereby increase gonadotropins in serum. The results obtained to date are controversial and a representative controlled study based on a number of patients large enough to allow statistically valid conclusions is still lacking. The same is true for kallikrein, which is used in some countries without final proof of its effectiveness.

Today, techniques of assisted fertilisation are used in many centres for idiopathic infertility. Details are given in later chapters but it should be emphasised here that techniques of assisted fertilisation have to be subjected to the same rigorous criteria as any other method in order to establish effectiveness.

Conclusion

As this brief review shows, a multitude of disorders may be hidden behind the diagnosis of male infertility, of which only a few are defined as pathogenic entities. Endocrine disorders are the best defined and can be diagnosed and treated rationally, and thus form the backbone of andrology. Disturbances which cannot be treated well, in particular idiopathic infertility, must provide continued impetus for intensive research, but must not mislead the doctor to use untested therapy or purposeless polypragmatic treatment. All procedures whose effectiveness has not been clearly proved in controlled studies should be applied only in the framework of clinical trials until their effectiveness has been established. At the same time it should not be forgotten that fertility disturbances affect a couple and that one partner with especially good reproductive function may, to a certain extent, compensate for the deficits of the other. For this reason, the optimisation of female reproductive function must be an essential part of any strategy for treatment of male infertility. To come back to the historic example mentioned at the beginning, it would not suffice to treat Herod alone. Herodias, her early proven fertility notwithstanding, would also have to be examined and treated, if necessary.

Acknowledgement. Our own work reported here was supported by the Max-Planck Society, the Deutsche Forschungsgemeinschaft, the Federal Ministry of Health and the World Health Organization (WHO).

References

1. Bruckert E. How frequent is unintentional childlessness in Germany? Andrologia 1991; 23:245–50.
2. Knuth UA, Mühlenstedt D. Kinderwunschdauer, kontrazeptives Verhalten und Rate vorausgegangener Infertilitätsbehandlung. Geburtshilfe Frauenheilkd 1991; 51:1–7.
3. Cates W, Farley TMM, Rowe PJ. Worldwide patterns of infertility: is Africa different? Lancet 1985; ii:596–8.
4. Behre HM, Nashan D, Nieschlag E. Objective measurement of testicular volume by ultrasonography: evaluation of the technique and comparison with orchidometer estimates. Int J Androl 1989; 12:395–403.
5. Nashan D, Behre HM, Grunert JH, Nieschlag E. Diagnostic value of scrotal sonography in infertile men: report on 658 cases. Andrologia 1990; 22:387–95.
6. WHO laboratory manual for the examination of human semen and semen–cervical mucus penetration. 3rd edn, Cambridge: Cambridge University Press, 1992.
7. Cooper TG, Neuwinger J, Bahrs S, Nieschlag E. Internal quality control of semen analysis. Fertil Steril (in press).
8. Neuwinger J, Behre HM, Nieschlag E. External quality control in the andrology laboratory: an experimental multicenter trial. Fertil Steril 1990; 45:308–14.
9. Freischem CW, Knuth UA, Langer K, Schneider HPG, Nieschlag E. The lack of discriminant seminal and endocrine variables in the partners of fertile and infertile women. Arch Gynecol 1984; 236:1–12.
10. Cooper TG, Jockenhövel F, Nieschlag E. Variations in semen parameters from fathers. Hum Reprod 1991; 6:859–66.
11. Flörke-Gerloff S, Töpfer-Petersen E, Müller-Esterlw et al. Biochemical and genetic investigation of round-headed spermatozoa in infertile men including two brothers and their father. Andrologia 1984; 16:187–202.
12. Neugebauer DCH, Neuwinger J, Jockenhövel F, Nieschlag E. '9+0' axoneme in spermatozoa and some nasal cilia of a patient with totally immotile spermatozoa associated with thickened sheath and short midpiece. Hum Reprod 1990; 5:981–6.
13. De Kretser DM. Testicular biopsy in the management of male infertility. Int J Androl 1982; 5:449–51.
14. Bergmann M, Behre HM, Nieschlag E. FSH and inhibin immunoreactivity in testicular biopsies from infertile men. Int J Androl (submitted).

15. Weinbauer GF, Drobnitzky F, Galhotra MM, Nieschlag E. Intratesticular injection of glycerol as a model for studying the quantitative relationship between spermatogenic damage and serum FSH. J Endocrinol 1987; 115:83–90.
16. Jockenhövel F, Khan SA, Nieschlag E. Varying dose–response characteristics of different immunoassays and in vitro bioassay for FSH are responsible for changing ratios of biologically active to immunologically active FSH. J Endocrinol 1990; 127:523–32.
17. Simoni M, Nieschlag E. In vitro bioassays of follicle stimulating hormone: methods and clinical applications. (Review). J Endocrinol Invest 1991; 14:938–97.
18. Jockenhövel F, Khan SA, Nieschlag E. Diagnostic value of bioactive FSH in male infertility. Acta Endocrinol 1989; 121:802–20.
19. De Kretser DM, MacLachlan RI, Robertson DM, Burger HG. Serum inhibin levels in normal men and men with testicular disorders. J Endocrinol 1989; 120:517–23.
20. Weinbauer GF, Nieschlag E. Peptide and steroid regulation of spermatogenesis in primates. Ann NY Acad Sci 1991; 637:107–21.
21. Liu L, Banks SM, Barnes KM, Sherins RJ. Two-year comparison of testicular responses to pulsatile gonadotropin-releasing hormone and exogenous gonadotropins from the inception of therapy in men with isolated hypogonadotropic hypogonadism. J Clin Endocrinol Metab 1988; 67:1140–5.
22. Saal W, Happ J, Cordes U, Baum RP, Schmidt M. Subcutaneous gonadotropin therapy in male patients with hypogonadotropic hypogonadism. Fertil Steril 1991; 56:319–24.
23. Schopohl J, Mehltretter G, von Zumbusch R, Eversman T, von Werder K. Comparison of gonadotropin-releasing hormone and gonadotropin therapy in male patients with idiopathic hypothalamic hypogonadism. Fertil Steril 1991; 56:1143–50.
24. Jockenhövel F, Fingscheidt U, Khan SA, Behre HM, Nieschlag E. Bio- and immunoactivity of FSH in serum after intramuscular injection of highly purified urinary human FSH in normal men. Clin Endocrinol 1990; 33:573–84.
25. Mannaerts B, De Leeuw R, Geelen J, Van Ravestein A. Comparative in vitro and in vivo studies on the biological characteristics of recombinant human follicle stimulating hormone. Endocrinology 1991; 129:2623–30.
26. Vogt P. Y chromosome function in spermatogenesis. In: Nieschlag E, Habenicht UF, eds. Spermatogenesis – fertilization – contraception: molecular cellular and endocrine events in male reproduction. Heidelberg: Springer, 1992; 225–66.
27. Nieschlag E, Behre HM. Pharmacology and clinical use of testosterone. In: Nieschlag E, Behre HM, eds. Testosterone: action, deficiency, substitution. Heidelberg: Springer, 1991; 92–114.
28. Bals-Pratsch M, Knuth UA, Yoon YD, Nieschlag E. Transdermal testosterone substitution therapy for male hypogonadism. Lancet 1986; ii:943–6.
29. Behre HM, Nieschlag E. Testosterone buciclate (20-Aet-1) in hypogonadal men: pharmacokinetics and pharmacodynamics of the new long-acting testosterone ester. J Clin Endocrinol Metab (in press).
30. Stellungnahme der Sektion Pädiatrische Endokrinologie in der Deutschen Gesellschaft für Endokrinologie (DGE). Zur Therapie des Hodenhochstandes. Endokrinologie-Informationen 1991; 15:20–2.
31. Hadziselimovic F. Cryptorchidism. Management and implications. Heidelberg: Springer, 1983.
32. Jockenhövel F, Gräwe A, Nieschlag E. A portable digital data recorder for long-term monitoring of scrotal temperatures. Fertil Steril 1990; 54:694–700.
33. Nagao RR, Plymate SR, Berger RE, Perin EB, Paulsen CA. Comparison of gonadal function between fertile and infertile men with varicoceles. Fertil Steril 1986; 46:930–3.
34. Knuth UA, Hönigl W, Bals-Pratsch M, Schleicher G, Nieschlag E. Treatment of severe oligozoospermia with hCG/hMG. A placebo-controlled double-blind trial. J Clin Endocrinol Metab 1987; 65:1081–7.
35. Bals-Pratsch M, Knuth UA, Hönigl W, Klein HM, Bergmann M, Nieschlag E. Pulsatile therapy in oligozoospermic men does not improve seminal parameters despite decreased FSH levels. Clin Endocrinol 1989; 30:549–60.
36. World Health Organisation Task Force on the Diagnosis and Treatment of Infertility: mesterolone and idiopathic male infertility: a double-blind study. Int J Androl 1989; 12:254–65.

Chapter 5

Diagnosis of Male Infertility

R. J. Aitken

Introduction

Recent evidence suggests that among patients attending infertility clinics for the first time, approximately 1 in 4 is characterised by a clear-cut defect in the functional competence of the spermatozoa [1]. Once the major causes of infertility, such as anovulation or tubal damage, have been successfully addressed, the male factor is by far the most important single cause of infertility in the residuum of patients, for whom there is little prospect of an accurate diagnosis or effective therapy. Moreover there is growing evidence that semen quality has declined in the population at large during the past 40 years [2]. Thus the male factor is not just of current interest as a major reproductive pathology but may be of growing importance in the aetiology of human infertility.

Development of accurate techniques to diagnose and treat male infertility must ultimately depend on improvements in our understanding of the cell biology of the human spermatozoon and definition, at the molecular level, of the defects responsible for the loss of fertilising potential. Progress has been made in recent years through the development of a range of bioassays with which to monitor the various attributes of sperm behaviour that contribute to the fertilising potential of these highly specialised cells. With the aid of these functional assays we are now beginning to develop an understanding of the cellular mechanisms responsible for the control of human sperm function and the specific nature of the defects responsible for male infertility. The following account is a brief summary of the status of the bioassays and biochemical tests that are currently used for evaluating the functional status of human spermatozoa.

Sperm Movement

One of the most unique and striking features of a spermatozoon is its capacity to exhibit specific patterns of movement that are exquisitely adapted to meet the different physical demands of penetrating cervical mucus, ascending the female reproductive tract and breaching the zona pellucida.

The advent of computerised image analysis systems for monitoring the movement characteristics of spermatozoa has enabled us to undertake objective assessments of sperm motility and relate these measurements to the functional competence of human spermatozoa. Such computerised systems are programmed with algorithms that permit the sperm head to be identified on the basis of its size, shape and brightness. Once identified, the trajectory followed by the sperm head can be traced and accurately measured. The major problem facing such cell identification systems is the presence of leukocytes, precursor germ cells and non-cellular debris in human ejaculates, that might be confused with immotile spermatozoa. Motility analysers such as the Hamilton Thorn 2030 system overcome this problem by using a background subtraction procedure in order to temporarily remove from the analysis all immotile objects in the field. The motile population of cells is then examined in order to confirm that the objects identified are of a size and optical intensity typical of spermatozoa, and are not being confused with other motile particles, such as bacteria or drifting debris. The mean size and optical intensity of the sperm head is then computed for this motile population, in order to derive values which can subsequently be applied to all static background objects, to identify and quantify the immotile sperm population.

The trajectory followed by the sperm head during movement gives the general appearance of a pseudosinusoidal wave, which can be characterised in terms of its amplitude, frequency and velocity. The latter can be described in terms of the total distance travelled by the sperm head in unit time (curvilinear velocity or VCL) or computed as a 5-point running average to give the average path travelled by the sperm head (average path velocity or VAP). The straight-line distance travelled by the sperm head (straight-line velocity or VSL) can also be measured and compared with the curvilinear velocity (VSL/VCL) to give a measure of the linearity (LIN) of sperm progression.

In order to standardise these measurements as much as possible, it is important that the conditions under which such assessments are performed are carefully controlled. Temperature, for example, has a dramatic effect on the movement characteristics of human spermatozoa and should set to 37°C. Even with a heated microscope stage, care should be taken to note any instability in temperature due, for example, to the pulsatile behaviour of the heating element, since we have observed that temperature oscillations within 1.0°C can have a dramatic effect on the parameters of sperm movement.

The thresholds of sperm head size and optical intensity also have to be chosen carefully and verified for accuracy using the playback facility (if available) and ultimately adjusted if errors are found in cell identification and tracking. Unfortunately, it is not possible to define a set of thresholds that will apply to every computerised sperm analysis system, because these values will vary according to the physical properties of the specimen being examined, the type of illumination used to visualise the specimen, the dimensions and type of chamber used to house the sample and the quality of the optical system used to generate an image for analysis.

The values obtained for any analysis of sperm movement will also depend on the rate at which the sperm track is sampled. In Europe, video equipment is designed to generate a complete field every 1/25th of a second, giving a video framing rate of 25 frames/second or 25 Hz. In the United States the equivalent figure is 30 Hz. The speed at which spermatozoa are sampled makes a considerable difference to the values obtained for the movement characteristics of these cells, particularly with respect to apparent amplitude and frequency of the pseudosinusoidal wave described by the sperm head [3]. In general, the faster the framing rate, the more accurate the analysis, and computerised systems are now being developed which operate at speeds of 60 Hz. With the existing range of hardware available, the rational approach is to analyse semen samples at the maximum frame rate possible. As for the number of frames collected for analysis, 20–30 frames is normal, the major objective being to obtain adequate detail on a given sperm track without creating problems of track identification due to the crossing over of sperm trajectories. Clearly, the frequency with which sperm tracks will cross over depends on the concentration of these cells in suspension. For this reason there is an optimal working concentration for most systems which, in the case of the Hamilton Thorn motility analyser is in the order of 5–50 million spermatozoa/ml; at concentrations over 150 million/ml a warning is automatically given on the data screen, to alert the operator to possible inaccuracies in the results.

In view of this dependency on sperm concentration, it is occasionally necessary to adjust the concentration of spermatozoa in a semen sample by adding homologous seminal plasma. If the analysis is being performed on a washed sperm suspension then it is a simple matter to adjust the concentration of spermatozoa to the optimal working range of 10–20 million/ml.

While sperm trajectories that are too long may create problems due to an excessive incidence of cross-over, sperm tracks that are too short may also give rise to inaccuracies because the algorithms designed to calculate ALH (Amplitude of Lateral Head Displacement), mean path velocity, linearity etc. will be meaningless on a cell that has moved insufficiently to give a pseudosinusoidal track. For this reason, it is usual to define a lower threshold of movement (VAP of >10 μm/s is the convention) which must be exceeded if a trajectory is to be analysed.

Another source of variation in the analysis of sperm movement derives from the type of chamber in which the spermatozoa are held. For the analysis of spermatozoa in semen, where the flagellar beats tend to possess a narrow amplitude, a 10 μm deep Makler chamber or a 20 μm microcell slide is probably optimal. These types of chambers are also sufficiently well engineered to be used for performing sperm counts. However, for washed, capacitated sperm suspensions associated with the high amplitude flagellar waves typical of hyperactivated cells, deeper chambers are preferred, such as 100 μm deep, oval cross-sectioned capillary tubes supplies by Vitrodynamics (Camlab, Cambridge UK).

Having established the optimal conditions for the analysis of sperm movement in terms of sperm concentration, frame number, framing rate, temperature and chamber size, the number of cells analysed has to be sufficient to account for the heterogeneity inherent in every human semen specimen. A formal analysis of this problem by Ginsburg et al. [4] found that most parameters of sperm movement stabilised after approximately 200 cells had been analysed in a total of 12 fields.

The analysis of sperm movement, in the kind of detail now possible with computerised image analysis systems, is only rational if we can be confident that the objective assessment of sperm movement conveys information of relevance to the fertilising potential of the ejaculate, that could not have been obtained by the assessment of percentage motility alone. This is a simple question of fundamental importance that has not yet been adequately answered. However, within donor insemination [5] and in vitro fertilization [6] programmes, data are gradually emerging to support the notion that the detailed analysis of sperm movement can yield information of prognostic value, that could not have been obtained from the conventional semen profile.

Cervical Mucus Penetration

Cervical mucus penetration is a good example of a biological function that spermatozoa must perform, that is heavily dependent on their capacity for movement. Hence multiple regression analyses employing the penetration of cervical mucus as the dependent variable have repeatedly shown that the outcome of such tests is closely correlated with the concentration and morphology of the spermatozoa and their capacity for movement [7–11]. All the various measures of sperm head velocity (curvilinear velocity, average path velocity and straight line velocity) appear to be positively correlated with cervical mucus penetration, but it is the path velocity which is repeatedly selected as the most informative variable in stepwise multiple regression analyses. Measures of the straightness of individual sperm trajectories (linearity and mean linear index) are also positively correlated with the ability of human spermatozoa to penetrate cervical mucus, as is the lateral displacement of the sperm head. The latter is such a functionally important aspect of sperm movement that cases of infertility have been identified in which the only defect in the semen is a reduced amplitude of lateral sperm head displacement, as a result of which the spermatozoa are unable to penetrate the cervical barrier [8–12]. This relationship is thought to exist because a small amplitude of lateral sperm head displacement reflects a low amplitude flagellar wave [13] and it is the latter that determines the propulsive force that can be generated by the spermatozoa as they arrive at the cervical mucus interface.

Clinically, the most difficult aspect of performing cervical mucus penetration assays is the amount of time and effort that has to go into timing the aspiration of cervical mucus from the female partner, which must be performed at mid-cycle if the results are to be meaningful. It would clearly be beneficial if an artificial substitute for cervical mucus could be identified, the penetration of which depended on the same characteristics of sperm movement as the native material. Recent independent studies suggest that hyaluronic acid polymers can serve just such a role. The penetration of human spermatozoa into hyaluronate polymers has been shown to correlate with their ability to penetrate into both human and bovine cervical mucus and to depend on the same attributes of semen quality, including sperm number, morphology and movement [7,11]. Of the parameters of sperm movement examined, penetration of both cervical mucus and hyaluronate were found to depend on a similar progressive, linear mode of motility, associated with a significant amplitude of lateral sperm head displacement. Stepwise regression analysis indicated that the most informative single variable was the percentage of cells exhibiting a mean path velocity of more than 25 μm/s. This variable could, together with data derived from the conventional semen profile (morphology, motility and sperm count), account for 70% of the variability in hyaluronate or cervical mucus penetration (Table 5.1).

Sperm penetration into hyaluronate polymers is so closely dependent on the movement characteristics of human spermatozoa that the outcome of such tests can be used to obtain an extremely accurate assessment of the overall quality of sperm motility. Such relationships raise the question of whether the diagnostic potential of cervical mucus penetration assays is simply a consequence of their close correlation with sperm movement or whether they are providing additional information of relevance to the fertilising potential of the spermatozoa. If it is the relationship with sperm movement that is the key to the clinical significance of such assays, then it would be simpler and more objective to assess the movement characteristics of human spermatozoa directly, rather than become engaged in the logistical and technical problems of carrying out a cervical mucus penetration assay. The one area where cervical mucus penetration tests might be said to be providing important additional data would be in cases of autoimmunity, characterised by the presence of antisperm IgA antibodies. One part of the IgA molecule, the Fc portion, is capable of binding with great tenacity to cervical mucin chains, so that spermatozoa coated with antisperm IgA become bound to the cervical mucus and display a characteristic "shaking phenomenon". In situations where there is limited access to computerised image analysis systems or the reagents necessary to carry out antisperm antibody tests, the cervical mucus penetration assay possesses merit in terms of low cost and a documented ability to predict the fertilising potential of human spermatozoa in vivo and in vitro [14].

Table 5.1. Stepwise regression of the relationship between the conventional semen profile, sperm movement and Theoretical Vanguard Distance achieved in Hyaluronate Polymer (Sperm Select®) of Bovine Cervical Mucus (Penetrak®) [7]

Medium	Criteria	*R*	Standardised β
Hyaluronate	√Rapid (VAP >25 μm/s)	0.807	0.380
	Motility	0.834	0.312
	√Total Count	0.856	0.270
Bovine cervical mucus	√Rapid (VAP >25 μm/s)	0.808	0.874
	√Sperm concentration	0.822	−0.235
	Morphology	0.838	0.193

R = Regression Coefficient.
Standardised β indicates the relative information content of each variable.

Hyperactivation

The penetration of cervical mucus is not the only attribute of human sperm function dependent on movement. The ability of spermatozoa to cross the zona pellucida also depends on the way in which spermatozoa move. Zona penetration presents a different kind of physical challenge to the spermatozoon, necessitating the evolution of a particular form of movement known as hyperactivation. The current thinking is that as spermatozoa capacitate in the female reproductive tract, their changing physiological status results in a modified flagellar beat pattern involving increases in amplitude and asymmetry. The increase in beat amplitude seems to occur first, resulting in high amplitude sperm trajectories that are still progressive and characteristic of spermatozoa that have entered the transitional phase of hyperactivation [15]. Further capacitation results in an increasing asymmetry of the flagellar wave, so that the swimming trajectories become less progressive and may adopt a number of different configurations, variously described as helical, starspin or thrashing. Such highly motile, non-progressive cells are fully hyperactivated and are regarded as having reached a terminal stage of capacitation (Fig. 5.1).

The high amplitude, thrashing movements of the sperm tail that characterise such hyperactivated cells are thought to be necessary for generating the propulsive forces needed to achieve penetration of the zona pellucida [16]. The expression of a hyperactivated form of movement by human spermatozoa has been a source of controversy for many years. It now appears that this

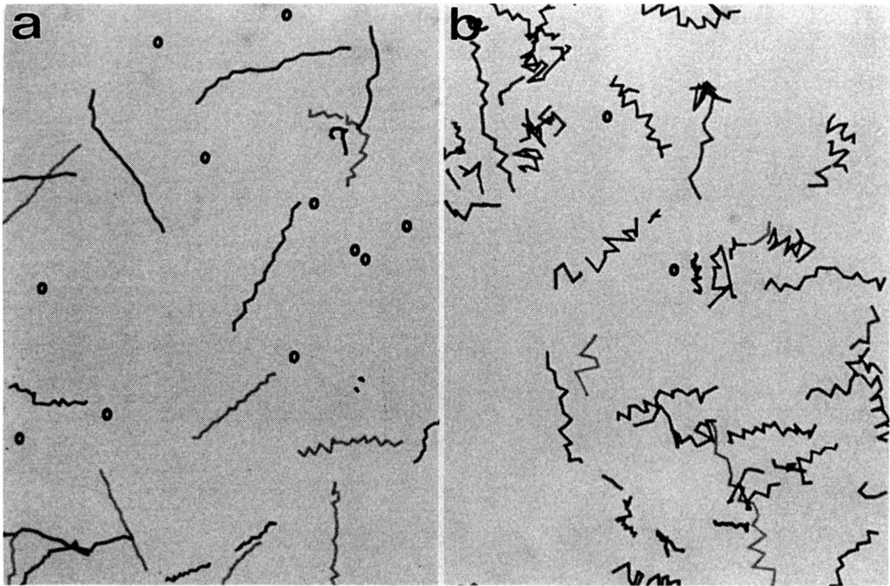

Fig. 5.1. Sperm movement analysed by a computerised image analysis system: **a** linear progressive tracks of a non-capacitated sperm population; **b** high amplitude, non-progressive tracks of a hyperactivated, capacitated sperm population.

discordance was largely due to a difference in the frequency with which human spermatozoa hyperactive in vitro, in relation to the spermatozoa of common laboratory species. Hence, whereas 70% of hamster spermatozoa consistently express hyperactivated motility after incubation in a simple culture medium [17], this figure may range from 3% to 50% for suspensions of human spermatozoa [15]. Kinetic studies also indicate that within a sample, the incidence of hyperactivation will vary with time, being maximal within 2–3 h of sperm preparation [18] and coinciding with a period of tight binding of the spermatozoa to the zona pellucida and an elevated incidence of spontaneous acrosome reactions.

Recently, the analysis of hyperactivated motility and assessment of its functional significance has been facilitated by the development of computerised image analysis systems that may be programmed with threshold values for velocity, linearity and lateral sperm head displacement that are typical of hyperactivated cells (Table 5.2). This facility therefore permits the automatic quantification of hyperactivated spermatozoa within a given sperm population. Such criteria are currently being used in many laboratories to provide information on the capacitation status of human sperm suspensions. The results obtained to date suggest that modifications to the incubation medium (elevated osmolarity, addition of fetal cord serum) that are thought to enhance sperm capacitation [15] also elevate the incidence of hyperactivation. However, the considerable differences between semen samples in their competence for capacitation is due to a multitude of factors, many of which are not reflected in the capacity of the cells to exhibit hyperactivated motility. For this reason, there does not appear to be a simple correlation between hyperactivation and other measures of capacitation, such as the outcome of the zona-free hamster oocyte penetration assay [19].

Sperm–Zona Interaction

Hyperactivation is only one aspect of the complex interactions between human spermatozoa and the zona pellucida during fertilisation. Before spermatozoa

Table 5.2. Sort criteria which may be used to automatically identify hyperactivated spermatozoa using the Hamilton Thorn Motility Analyzer

Criterion of movement	Lower limit	Upper limit
Path velocity (μm/s)	5	290
Progressive velocity (μm/s)	0	290
Curvilinear velocity (μm/s)	100	290
Linearity (%)	0	65
Amplitude of head displacement (μm)	7.5	30
Beat frequency	0	30

From the Hamilton Thorn Operations Manual, June 1989.

can physically penetrate the zona pellucida, they must first recognise and bind to it. The specificity of this interaction is extremely important because a spermatozoon may make contact with hundreds, or even thousands, of different cells on its journey through the female reproductive tract to the ovum. In order to accomplish this specificity, the spermatozoon has evolved surface receptors that will bind only to a unique protein component of the zona pellucida, known as ZP3. This particular protein is a highly conserved constituent of the zonae pellucidae of all mammals examined to date, the primary amino acid sequence exhibiting around 70% homology in species as disparate as the mouse and the human [20,21]. Despite the conserved nature of the peptide core of the ZP3 molecule, sperm binding to the zona pellucida exhibits great species specificity. For example, human spermatozoa will only bind to the zona pellucida of another hominoid ape, such as the gibbon [22]. This species specificity is thought to depend upon the unique configuration of the O-linked oligosaccharide side chains of the ZP3 molecule. Treatment of the ZP3 molecule with glycosidases to remove the O-linked side chains effectively destroys the capacity of this molecule to bind and activate spermatozoa [23].

Quantification of the ability of human spermatozoa to bind to the zona pellucida is thought to give an indication of their fertilising capacity [24]. The use of this end-point in a diagnostic context has been facilitated by the discovery that the ability of the human zona pellucida to bind human spermatozoa can be preserved indefinitely, if the ova are stored in high salt solutions containing 1.5M-$MgCl_2$ and 0.1% dextran. As a consequence of this finding, it is now feasible to preserve ova that have been rejected from IVF programmes, because of a failure to fertilise, and reuse them for diagnostic purposes to test the functional competence of patients' spermatozoa.

One of the major problems with this approach is that the sperm-binding capacity of the zona pellucida shows variation from patient to patient and even between ova from the same patient. At least part of the reason for this variation lies in the biological properties of the zona pellucida and derives from the fact that the sperm-binding capacity of this structure varies with the state of oocyte maturation. Hence, metaphase II human oocytes are surrounded by zonae pellucidae that bind significantly more spermatozoa than ova at earlier stages of development, at prophase I or metaphase I [25]. A rational solution to this problem of inter-zona variability has been to use each zona as its own control, as in the "hemi-zona" assay [24,26,27]. In this procedure, each zona pellucida is cut into halves using a micromanipulator. One half is incubated with control spermatozoa from a donor of proven fertility, while the remaining half is placed with a sperm sample from a patient of unknown fertility status. The spermatozoa are prepared at a concentration of 500 000/ml and incubated with the hemi-zonae for 4 h at 37°C. The number of spermatozoa tightly bound to each hemi-zona is then determined and a "hemi-zona index" calculated as a percentage of the control value (number of patient's spermatozoa bound/ number of control spermatozoa bound × 100). Preliminary data have been obtained indicating that the hemi-zona assay may be of some diagnostic significance, correlating with the fertility of the spermatozoa in vitro and their capacity for hyperactivation [26].

The future of this particular area lies in the use of recombinant ZP3 to study the capacity of human spermatozoa to recognise this molecule and become activated by it. This proposition stems from the fact that the human zona pellucida is more than just a recognition site for spermatozoa; it is also the site at which these cells become activated and initiate the cascade of intracellular events leading to the acrosome reaction. The gene encoding human ZP3 has recently been cloned and sequenced and it can only be a matter of time before biologically active recombinant human ZP3 is available [21].

The Acrosome Reaction

Spermatozoa acrosome-react as they bind to the surface of the zona pellucida and establish intimate contact with ZP3. The liberation of proteases from the acrosomal vesicle is thought to facilitate the passage of the spermatozoa through the zona pellucida. In addition, the acrosome reaction is, in a way which is not yet understood, associated with the sudden acquisition of the spermatozoon's ability to recognise and fuse with the vitelline membrane of the oocyte. In view of the functional significance of the acrosome reaction, analysis of the ability of human spermatozoa to undergo this change should be of some diagnostic value. Unfortunately, the acrosomal vesicle of the human spermatozoon is so small that the acrosome reaction cannot be resolved at the light microscope level. As a consequence, it has become necessary to develop reagents that can be used to probe this structure, so that its integrity can be easily monitored. Two classes of probe have been introduced, comprising monoclonal antibodies and plant lectins respectively (Fig. 5.2), usually conjugated to a label such as

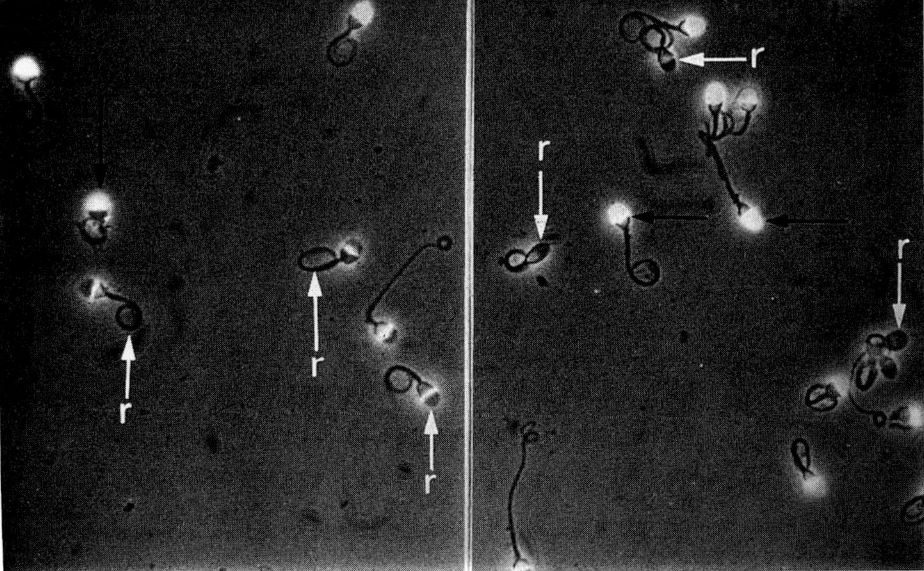

Fig. 5.2. Use of a fluorescein-conjugated lectin in combination with the hypo-osmotic sperm swelling test to monitor the acrosome reaction in populations of human spermatozoa. r, acrosome reacted cells in which the label is confined to the equatorial segment.

fluorescein. The monoclonal antibodies or lectins may be directed against the acrosomal contents or the outer acrosomal membrane. With such probes, the acrosomal region of acrosome-intact cells exhibits a uniform bright fluorescence. However, when the acrosome reaction occurs, the fluorescence over the acrosomal region gradually dissipates. In the case of the most commonly used label, fluorescein-conjugated peanut agglutinin (Fig. 5.2), the acrosome reaction is associated with the appearance of a punctate pattern of labelling over the acrosomal region followed by the restriction of the label to a band around the equatorial region of the sperm head [28,29].

Identifying appropriate labels to monitor the state of the acrosome is not the only problem to be addressed in developing a diagnostic test around this secretory event. A second issue concerns the viability of the spermatozoa, since acrosomes may be lost as a result of pathological cell senescence, as well as a biologically meaningful acrosome reaction. It is therefore necessary to include some means of monitoring the viability of the spermatozoa, so that cells undergoing a physiological acrosome reaction can be accurately identified. Discrimination between the physiological and pathological event can, for example, be accomplished with DNA-sensitive fluorochromes, such as H33258, which exhibit limited membrane permeability and stain only cells that have lost their membrane integrity and hence their viability [30]. Alternatively, the viability of the spermatozoa can be assessed using the hypo-osmotic swelling test, which identifies living cells with an intact, fluid plasma membrane, by virtue of the coiled configuration adopted by the sperm tail when the spermatozoa are forced to swell by immersion in a hypo-osmotic medium (Fig. 5.2).

A further problem with the acrosome reaction as a diagnostic test is that the spontaneous incidence of this event in vitro is extremely low, amounting to no more than 10% of the total sperm population, even after very prolonged periods of incubation [29]. As a consequence, the test has a limited dynamic range and the clinician is faced with the problem of differentiating between a normal fertile specimen exhibiting around a 5% acrosome reaction rate, and a subfertile specimen in which this figure might be reduced to 1% or 2%. In order to determine whether there is a real difference between fertile and potentially infertile specimens, when the dynamic range is so limited, very large numbers of cells would have to be counted. One solution to this problem, which has considerable potential, is to use flow cytometry to characterise the acrosomal status of large numbers of human spermatozoa [31].

An alternative approach to enhancing the discriminatory power of acrosome reaction tests, is to induce this process artificially. Under normal circumstances, the acrosome reaction would occur on the surface of the zona pellucida, as a consequence of the interaction between the sperm surface and the zona glycoprotein, ZP3. As a consequence of the inductive power of the zona pellucida, around 80% of human spermatozoa bound to the zona surface are acrosome reacted, compared with only 5%–10% in the ambient medium. Until recombinant ZP3 becomes available, the acrosome reaction will have to be induced by chemical means. To achieve this end, the divalent cation ionophore A23187 has been used to induce the intracellular changes in calcium and pH that precipitate the acrosome reaction. Since the physiological acrosome reaction is also induced by these second messengers, the A23187-induced event does have a rational biological basis and does appear to correlate with the fertilising capacity of human spermatozoa, at least in vitro [31,32].

The Hamster Oocyte Penetration Assay

Concomitant with the acrosome reaction, the human spermatozoon acquires a capacity to fuse with the oocyte. This change is limited to a narrow band of plasma membrane around the equatorial segment of the sperm head, which suddenly acquires the ability to recognise receptor sites on the surface of the oocyte and initiate fusion with the vitelline membrane. Recent studies in the guinea pig indicate that the sperm surface receptor responsible for initiating fusion with the oocyte is a dimeric protein complex that combines a recognition motif in one subunit and a fusion domain in the other. The portion of the molecule capable of initiating fusion shares many features with viral fusion peptides, suggesting that biological events as disparate as the fusion of an enveloped virus with its target cell and the fertilisation of the mammalian oocyte involve similar molecular mechanisms [33,34].

Monitoring the competence of human spermatozoa to fuse with the oocyte following the acrosome reaction is a key area of sperm function, which should be addressed in any diagnostic work up of a potentially infertile patient. Such assessments would pose severe logistical and ethical problems, were it not for the existence of an alternative to the human ovum for monitoring the ability of acrosome-reacted human spermatozoa to engage in sperm–oocyte fusion. In 1976, Yanagimachi et al. [35] made the serendipitous discovery that the oocyte of the golden hamster, once stripped of its zona pellucida, is susceptible to fusion with spermatozoa from a wide variety of different mammalian species, providing these cells have undergone the acrosome reaction. The condition that fusion depends upon the previous occurrence of the acrosome reaction suggests that the process of sperm–oocyte fusion in this heterologous model is physiologically meaningful. Moreover, the ultrastructural details of fusion in this system appear to reflect the biological situation, in that this process is initiated by the plasma membrane overlying the equatorial segment of the sperm head [36].

In order to bring an element of standardisation into the way in which this bioassay is performed, the World Health Organisation [37,38] has described a basic protocol for performing the test, employing an overnight incubation in order to capacitate the spermatozoa. With such a protocol, the levels of sperm–oocyte fusion are low, since they depend on the occurrence of spontaneous acrosome reactions within the sperm population. Ideally, such assays should be supplemented with recombinant ZP3 in order to induce a biologically relevant acrosome reaction. While we wait for such materials to become available, the divalent cation ionophore A23187 has been used for this purpose [39]. In the presence of this reagent, high rates of sperm activation are observed in the normal fertile population, giving penetration rates of 70%–100%. In contrast, spermatozoa from subfertile males, such as the oligozoospermic population, exhibit penetration values of less than 10% after stimulation with A23187, with about half of such patients' spermatozoa failing to fuse with any oocytes [39]. The diagnostic significance of the zona-free hamster oocyte penetration test carried out in the presence of ionophore has been demonstrated both in vivo [40] and in vitro [41] and probably represents as good an overall test of human sperm function as is currently available.

Biochemical Criteria of Semen Quality

The development of bioassays to assess the individual components of sperm function provides us with a means of classifying patients according to the defects that are present in their spermatozoa. With the aid of such functional assays, we can now start to address the molecular basis of defective sperm function and pave the way to the development of biochemical assays that should be much easier to perform and standardise than the more traditional bioassays. Moreover, a deeper understanding of the cellular mechanisms responsible for defective sperm function should provide us with a platform from which to develop rational therapeutic strategies. In the past five years there have been two major developments in terms of the identification of biochemical criteria for the diagnosis of defective sperm function: free radical-mediated lipid peroxidation and creatine phosphokinase.

Reactive Oxygen Species and Lipid Peroxidation

The fact that human spermatozoa can generate reactive oxygen species such as superoxide anion and hydrogen peroxide was originally suggested by MacLeod in 1943 [42] and confirmed independently in 1987 by Aitken and Clarkson [43] and Alvarez et al. [44]. The primary product of the spermatozoon's free radical-generating system appears to be superoxide anion, which dismutates to hydrogen peroxide under the influence of intracellular superoxide dismutase. The biological significance of this highly specialised, and potentially pernicious, free-radical generating system is currently unknown, although its role in the aetiology of defective sperm function appears to be significant [43–49]. In both retrospective and long-term prospective studies the excessive generation of reactive oxygen species has been shown to be associated with impaired male fertility [43,49]. The mechanism by which sperm function is impaired under such circumstances involves a loss of membrane function, as a consequence of the peroxidation of unsaturated fatty acids in sperm plasma membrane [48]. Given that this is the mechanism by which sperm function is frequently impaired, the therapeutic benefit of antioxidants is clearly worthy of investigation.

Creatinine Phosphokinase

Another biochemical marker of defective sperm function that has come to light in recent years is creatine phosphokinase (CPK), a key enzyme involved in the synthesis, transport and dephosphorylation of creatine phosphate [50,51]. These studies have highlighted the existence of a highly significant inverse relationship between sperm CPK activity and sperm concentrations in normospermic, moderately oligozoospermic and severely oligozoospermic samples. The increased CPK activity observed in oligozoospermic samples was shown to be related to higher cellular sperm concentrations of this enzyme and appears to be due to the retention of excess cytoplasm during the differentiation of these defective cells [50,51].

The diagnostic potential of sperm CPK measurements has been indicated in an analysis of the fertilising potential of human spermatozoa in an intrauterine insemination service [52]. In this study, the fertilising potential of oligozoospermic samples in vivo was correlated with sperm CPK concentrations, before and after isolation of the most motile spermatozoa using a swim-up procedure.

Conclusions

We can identify three sequential phases in our approach to male fertility diagnosis, which may be categorised as: (a) descriptive, (b) functional and (c) molecular. The original approach was descriptive, and was based on the weak premise that fertility could be predicted on the basis of the appearance of the ejaculate, in terms of the number, motility and morphology of the spermatozoa. The development and application of bioassays to determine the functional competence of human spermatozoa rapidly revealed that impaired sperm function could be detected in spermatozoa from apparently normal ejaculates. With the aid of such functional assays accurate predictions of male fertility have been obtained, which could not have been generated from the conventional semen profile on the same patients. Although such functional assays are clearly of prognostic value they are expensive and labour intensive to perform, with the result that potentially valuable procedures, such as the zona-free hamster oocyte penetration test, are run in only a handful of centres worldwide. The long-term future of this area of reproductive medicine must lie in the utilisation of such bioassays to determine, at a biochemical level, the nature of the defects that are responsible for the loss of fertilising potential in different groups of patients. With the aid of such information, it should be possible to develop biochemical assays that will be much better suited to routine diagnostic laboratories than the current bioassays. Moreover, understanding male infertility at the biochemical level should also lead to the development of rational forms of therapy that are specifically designed to counteract defined cellular lesions, rather than the empirical treatments that are frequently employed at the present time.

Acknowledgements. Grateful thanks are due to Victor Souaid for his helpful criticisms of the manuscript.

References

1. Hull MGR, Glazener CMA, Kelly NJ et al. Population study of causes, treatment and outcome of infertility. Br Med J 1985; 291:1693–7.

2. Carlsen E, Giwercman A, Keiding N, Skakkebaek NE. Evidence for decreasing semen quality during the last half-century. Br Med J (in press).
3. Mortimer D, Serres C, Mortimer ST, Jouannet P. Influence of image sampling frequency on the perceived movement characteristics of progressively motile human spermatozoa. Gamete Res 1988; 20:313–27.
4. Ginsburg KA, Moghissi KS, Abel EL. Computer assisted semen analysis: sampling errors and reproducibility. J Androl 1988; 9:82–90.
5. Irvine DS, Aitken RJ. Predictive value of in vitro sperm function tests in the context of an AID service. Hum Reprod 1987; 1:539–45.
6. Lui DY, Clarke GN, Baker HWG. Relationship between sperm motility assessed with the Hamilton Thorn motility analyzer and fertilization rates in vitro. J Androl 1991; 12:231–9.
7. Aitken RJ, Bowie H, Buckingham D, Harkiss D, Richardson DW, West KM. Sperm penetration into a hyaluronic acid polymer as a means of monitoring functional competence. J Androl 1992; 13:44–54.
8. Aitken RJ, Sutton M, Warner P, Richardson DW. Relationship between the movement characteristics of human spermatozoa and their ability to penetrate cervical mucus and zona-free hamster oocytes. J Reprod Fertil 1985; 73:441–9.
9. Aitken RJ, Warner PE, Reid C. Factors influencing the success of sperm–cervical mucus interaction in patients exhibiting unexplained infertility. J Androl 1986; 7:3–10.
10. Mortimer D, Mortimer ST, Shu MA, Swart R. A simplified approach to sperm-cervical mucus interaction testing using a hyaluronate migration test. Hum Reprod 1990; 5:835–41.
11. Mortimer D, Pandya IJ, Sawers RS. Relationship between human sperm motility characteristics and sperm penetration into cervical mucus in vitro. J Reprod Fertil 1986; 78:93–102.
12. Feneux D, Serres C, Jouannet P. Sliding spermatozoa: a dyskinesia responsible for human infertility? Fertil Steril 1985; 44:508–11.
13. David G, Serres C, Jouannet P. Kinematics of human spermatozoa. Gamete Res 1981; 4:83–6.
14. Hull MGR. Indications for assisted conception. Br Med Bull 1990; 46:580–95.
15. Burkman LJ. Hyperactivated motility of human spermatozoa during in vitro capacitation and implications for fertility. In: Gagnon C, ed. Controls of sperm motility: biological and clinical aspects. Boston: CRC Press, 1990; pp. 303–31.
16. Katz DF, Yanagimachi R. Movement characteristics of hamster spermatozoa within the oviduct. Biol Reprod 1990; 22:759–64.
17. White DR, Aitken RJ. Relationship between calcium, cAMP, ATP and intracellular pH and the capacity to express hyperactivated motility by hamster spermatozoa. Gamete Res 1989; 22:163–78.
18. Robertson L, Wolf DP, Tash JS. Temporal changes in motility patterns related to acrosomal status: identification and characterization of populations of hyperactivated human spermatozoa. Biol Reprod 1988; 39:797–805.
19. Wang C, Leung A, Tsoi W-L, Leung J, Ng V, Lee K-F, Chan SYW. Evaluation of human sperm hyperactivated motility and its relationship with the zona free hamster oocyte penetration assay. J Androl 1991; 12:253–8.
20. Chamberlin ME, Dean J. Genomic organization of a sex specific gene: the primary sperm receptor of the mouse zona pellucida. Dev Biol 1989; 131:207–14.
21. Chamberlin ME, Dean J. Human homolog of the mouse sperm receptor. Proc Natl Acad Sci USA 1990; 87:6014–18.
22. Bedford JM. Sperm–egg interaction: the specificity of human spermatozoa. Anat Rec 1977; 188:477–88.
23. Bleil JD, Wassarman PM. Mammalian sperm–egg interaction: identification of a glycoprotein in mouse egg zonae pellucidae possessing receptor activity for sperm. Cell 1980; 20:873–80.
24. Burkman LJ, Coddington CC, Kruger TF, Rosenwaks Z, Hodgen GD. The hemizona assay (HZA): development of a diagnostic test for the binding of human spermatozoa to the human hemizona pellucida to predict fertilization potential. Fertil Steril 1988; 49:688–97.
25. Oehninger S, Veeck L, Franken D, Kruger TF, Acosta AA, Hodgen GD. Human preovulatory oocytes have a higher sperm-binding ability than immature oocytes under hemizona assay conditions: evidence supporting the concept of zona maturation. Fertil Steril 1991; 55:1165–70.
26. Coddington CC, Franken DR, Burkman LJ, Oosthuizen WT, Kruger T, Hodgen GD. Functional aspects of human sperm binding to the zona pellucida using the hemizona assay. J Androl 1991; 12:1–8.
27. Franken DR, Oehninger SC, Burkman LJ et al. The hemizona assay (HZA) a predictor of human sperm fertilizing potential in in vitro fertilization (IVF) treatment. J In Vitro Fert

Embryo Transf 1989; 6:44–50.
28. Mortimer D, Curtis EF, Miller RG. Specific labelling by sperm agglutinin of the outer acrosomal membrane of the human spermatozoon. J Reprod Fertil 1988; 81:127–35.
29. Aitken RJ. Evaluation of human sperm function. Br Med Bull 1990; 46:654–74.
30. Cross N, Morales P, Overstreet JW, Hanson FW. Two simple methods for detecting acrosome reacted sperm. Gamete Res 1986; 15:213–26.
31. Fenichel P, Hsi BL, Farahifar D, Donzeau M, Barrier-Delpech D, Yeh CJG. Evaluation of the human sperm acrosome reaction using a monoclonal antibody, GB 24, and fluorescence-activated cell sorter. J Reprod Fertil 1989; 87:699–706.
32. Cummings JM, Pember SM, Jequier AM, Yovich JL, Hartmann PE. A test of the human sperm acrosome reaction following ionophore challenge (ARIC). J Androl 1991; 12:98–103.
33. Blobel CP, Wolfsberg TG, Turck CW, Myles DG, Primakoff P, White JM. A potential fusion peptide and an integrin ligand domain in a protein active in sperm oocyte fusion. Nature 1992; 356:248–52.
34. Aitken J. A family of fusion proteins. Nature 1992; 356:196–7.
35. Yanagimachi R, Yanagimachi H, Rogers BJ. The use of zona-free animal ova as a test system for the assessment of the fertilizing capacity of human spermatozoa. Biol Reprod 1976; 15:471–6.
36. Koehler JK, De Curtis I, Stenchever MA, Smith D. Interaction of human sperm with zona-free hamster eggs: a freeze fracture study. Gamete Res 1982; 6:371–86.
37. World Health Organization. WHO Laboratory manual for the examination of human semen and semen–cervical mucus interaction. Cambridge: Cambridge University Press, 1987.
38. Aitken RJ. The zona-free hamster oocyte penetration test and the diagnosis of male infertility. Int J Androl (Suppl) 1986; 6:1–199.
39. Aitken RJ, Ross A, Hargreave T, Richardson D, Best F. Analysis of human sperm function following exposure to the ionophore A23187: comparison of normospermic and oligozoospermic men. J Androl 1984; 5:321–9.
40. Aitken RJ, Irvine DS, Wu FC. Prospective analysis of sperm–oocyte fusion and reactive oxygen species generation as criteria for the diagnosis of infertility. Am J Obstet Gynecol 1991; 164:542–51.
41. Aitken RJ, Thatcher S, Glasier AF, Clarkson JS, Wu FC, Baird DT. Relative ability of modified versions of the hamster oocyte penetration test, incorporating hyperosmotic medium or the ionophore A23187, to predict IVF outcome. Hum Reprod 1989; 2:227–31.
42. MacLeod J. The role of oxygen in the metabolism and motility of human spermatozoa. Am J Physiol 1943; 138:512–18.
43. Aitken RJ, Clarkson JS. Cellular basis of defective sperm function and its association with the genesis of reactive oxygen species by human spermatozoa. J Reprod Fertil 1987; 81:459–69.
44. Alvarez JG, Touchstone JC, Blasco L, Storey BT. Spontaneous lipid peroxidation and production of hydrogen peroxide and superoxide in human spermatozoa. J Androl 1987; 8:338–48.
45. Aitken RJ, Clarkson JS. Generation of reactive oxygen species by human spermatozoa. In: Dormandy T, Rice-Evans C, eds. Free radicals: recent developments in lipid chemistry, experimental pathology and medicine. London: Richelieu Press, 1987, pp. 333–5.
46. Aitken RJ, Clarkson JS. Significance of reactive oxygen species and antioxidants in defining the efficacy of sperm preparation techniques. J Androl 1988; 9:367–76.
47. Aitken RJ, Clarkson JS, Fishel S. Generation of reactive oxygen species, lipid peroxidation and human sperm function. Biol Reprod 1989; 40:183–97.
48. Aitken RJ, Clarkson JS, Hargreave TB, Irvine DS, Wu FCW. Analysis of the relationship between defective sperm function and the generation of reactive oxygen species in cases of oligozoospermia. J Androl 1989; 10:214–20.
49. Aitken RJ, Buckingham D, West K, Wu FC, Zikopoulos K, Richardson DW. On the contribution of leucocytes and spermatozoa to the high levels of reactive oxygen species recorded in the ejaculates of oligozoospermic patients. J Reprod Fertil 1992 (in press).
50. Huszar G, Corrales M, Vigue L. Correlation between sperm creatine phosphokinase activity and sperm concentrations in normospermic and oligozoospermic men. Gamete Res 1988; 19:67–75.
51. Huszar G, Vigue L, Corrales M. Sperm creatine phosphokinase quality in normospermic, variable spermic and oligospermic men. Biol Reprod 1988; 38:1061–6.
52. Huszar G, Vigue L, Corrales M. Sperm creatine kinase activity in fertile and infertile men. J Androl 1990; 11:40–6.

Discussion

Hull: Were the WHO method standards established by critical prospective evaluation?

Nieschlag: A good question. The guidelines are based on the published literature and on common knowledge, but they are not the result of concentrated effort to verify whether each of these methods is correct.

Hull: That is what worries me. Standard methodology is being imposed upon us without, as far as I can see, proper and adequate evaluation.

Nieschlag: Yes. That is a very heuristic approach; but for practical purposes we have to start somewhere. These guidelines have triggered several studies which may amend the guidelines over time.

Templeton: Many clinics feel that the examination of a man prior to investigation is not helpful. I disagree and I agree with Dr Nieschlag's approach, which is that the man and woman should be examined together prior to investigation. But Professor Cooke and Bruce Dunphy have demonstrated in a study that it does not help in management or improve the prognosis to do that [1]. It is quite an important issue and I should like to hear Dr Nieschlag's view.

Nieschlag: The wide range of pathologies we find is the best defence. Furthermore if we have our diagnosis we can alert the gynaecologist to take more or less care of the female.

Cooke: It depends very much on the population one is dealing with. Someone who is running an Institute of Reproductive Medicine is likely to be referred large numbers of males with problems of the male urogenital system, whereas someone who is running exclusively an infertility clinic is likely to be sent couples where the male primarily has an infertility problem. That polarises the populations and therefore alters dramatically the percentage of patients that are likely to fall into the categories. I would not recognise a distribution of 10% of hypogonadism, for example, in my infertility clinic.

Aitken: Andrology is still a very young science and although there may not be much clinical utility in examining the male at the present time, obviously the more we understand of male reproductive pathologies and the better able we are to treat those pathologies, the more meaningful considering the male will become.

Winston: One of the issues affecting the Study Group is how we go forward in a practical sense. I detected a note of considerable caution about the interpretation of simple zona attachment tests and perhaps Dr Aitken would elaborate on their usefulness.

Aitken: Most of the data on the utility of sperm zona binding comes from Norfolk using their hemi-zona assay; it is all from one laboratory. I find it difficult to answer. Sperm zona binding is determined by many different factors.

It is not an entity to be considered in isolation, which they tend to do. For example, if the sperm cells exhibit poor motility, there will be a reduction in sperm zona binding and a very sophisticated test has been used to try to understand something that is relatively simple.

The technology in this area is developing rapidly. We already have biologically active recombinant ZP3. If we want to understand whether the binding sites for ZP3 are on the surface of the cells, in a very short time there will be simple ELISA assays which everybody will be able to use and will be used by laboratory technical staff without any need for access to human oocytes to understand that problem.

Abdalla: In assessing prognosis, we need to collect data in relation to count and motility. This is what we need to advise patients.

Aitken: That is an important point but we have ongoing studies doing just that. We have taken cohorts of men, whose female partners are normal which is obviously important, and they are left untreated for long periods of time, which is difficult to establish in many centres. If that can be achieved, then we can get meaningful studies on the relationship between a variety of different diagnostic criteria and the outcome of spontaneous pregnancy.

Braude: Presumably that information would also be important in assessing the relevance of the WHO Manual. We would be talking about objective data as opposed to subjective data.

Cooke: It is an expensive procedure to do automated analyses and the basic semen analysis could be regarded as a preliminary screening on which to concentrate, and so it is important not to neglect that. We have to be careful looking at Professor Aitken's data on the comparison of the computerised analysis with his own laboratory's manual analysis. He has an extremely sophisticated, well modulated analysis and we cannot compare that with the laboratories around the country where there is no quality control and there has been no self-educative process that often is generated by using the computerised analysis. It is important not to generalise.

Asch: Professor Nieschlag mentioned a number of treatments that were still valid, some that were doubtful and some that were obsolete, and in the last-named category he included treatment with gonadotrophins. In the US, a group in Norfolk is strongly suggesting high doses of FSH to improve spermatogenesis in people with oligoasthenospermia.

Nieschlag: The statement is based on the classical hCG–hMG treatment which has been used since the early 1960s for improvement of male infertility of the idiopathic type. That was used until the mid-1980s without any proper control studies. We did a control study that showed that where a placebo group is used, improvement is seen in the placebo group as well as in the hCG–hMG group, so it has clearly been shown that hCG–hMG is no better than placebo [2]. So this treatment is out.

The study I have seen using highly purified FSH is an uncontrolled study. They used a group of men, injected them with FSH, and said that their IVF results were better. It is a mistake to do such a study and these days all studies should be controlled studies. What this study has achieved is that large numbers of men, before they get into the IVF laboratory, are being injected with FSH, and I think that is almost unethical. One has to do a properly controlled study, and to my knowledge that has not yet been done.

Asch: I am sure the speakers are aware of the many recent studies looking at the effect of progesterone or of something in the follicular fluid that enhances sperm function.

Nieschlag: Most of this work comes from Norfolk and I am in no position to comment.

Aitken: There are data showing that some steroids can have effects on the cell membrane and not necessarily through the normal transcription mechanisms; neurosteroids are the obvious example. It is known that adding progesterone to sperm cells will generate calcium signals. We have done that ourselves and as have other groups. In our own hands we get no biological responses to those calcium signals; as I indicated in chapter 5, there is more than just a calcium spike to activate in these cells. So really there is still debate on the therapeutic value of adding progesterone to sperm cells.

Franks: On the subject of sperm autoantibodies. Medical therapy, as we have heard from Dr Nieschlag, is universally ineffective, or at least there have been no controlled studies that have shown that it works. Professor Hull, I think quite rightly, has drawn attention to the fallibility of the controlled studies of high dose steroids and the problems involved in using that treatment. What do our seminologists feel is the significance of sperm autoantibodies and what do they feel should be the treatment of choice?

Nieschlag: We have 3%–4% of patients who have sperm antibodies to a significant level and they have an infertility problem. We have noticed from IVF that many of them can be treated by washing the sperm and getting rid of the antibodies, and that seems to be a good therapy. Professor Hull mentioned the three studies with corticosteroids. In our study we showed, on a controlled level, that it does not help [3]. Offering the treatment over half a year, the pregnancy rate was equally low in the placebo group and in the treated group. The studies have to be done and I do not know why it is so difficult to convince andrologists to do controlled studies. They seem to be the last group in medicine to think that these should not be done and they are very difficult to convince that this is the way to go.

Franks: Are there controlled studies to show that sperm washing is more effective than leaving sperm alone for in vitro fertilisation?

Aitken: There is a paper by Robinson where they used immuno beads from the surface of sperm cells and they claimed that this leads to higher fertilisation rates [4].

Hull: There are perhaps three reports of in vitro fertilisation in cases of seminal antisperm antibodies and they produced fertilisation rates which are moderately reduced, just under 50% per oocyte, but pregnancy rates that are around 20%, which are really pretty respectable, implying that there is a good implantation rate once fertilisation is achieved. But those are the only three reports that I know about that are optimistic – of course uncontrolled, but nonetheless optimistic. They are better than has been achieved, it seems, with steroid therapy.

Gosden: We seem to be hearing that all the observations and the progress that has been made is pushing back the questions to the cell biology of spermatogenesis, and yet we have heard very little about this. I presume this is because of the great difficulty of making progress in this area, but ultimately we shall have to understand the conditions in which meiosis and spermatogenesis take place. One of the ways forward would be if we could mature spermatozoa in vitro, and this might also have clinical applications.

Nieschlag: Dr Gosden is quite right. In basic science in animal models we have made a lot of progress on the regulation of spermatogenesis, in particular at the cell to cell level. The proceedings of the last American Testis Workshop and the current European Testis Workshop report a very high standard of what is known about intratesticular regulation.

Slowly there are entities emerging, maybe coming from genetic control of spermatogenesis, and we see more and more genetic defects which are reflected in defects of spermatogenesis. Here we find small dents in the big block of unknown reasons, and similarly from paracrinology we should learn more. But the problem is that one cannot deal with human tissue, or it is very difficult to obtain it, and only when there is some real solid course from animal studies can we switch to humans and use either pathological material or biopsy material. But it is very difficult.

Aitken: There is a large number of people who would like to get spermatogenesis in a test tube. The problem is that it is complex. It is not just a single germ cell one has to get into culture, it is also all the associated cells that regulate its function, and the paracrine factors that have to be right for those cells to differentiate.

Gosden: So it is a very distant prospect?

Aitken: A very distant prospect, I think. The way that we are approaching it is to start with the sperm, define at a biochemical level what is wrong with the sperm, and then use that as a lantern to go back into the dark recesses of the testis looking for the origin of the defect.

References

1. Dunphy BC, Kay R, Barratt CL, Cooke ID. Is routine examination of the male partner of any prognostic value in the routine assessment of couples who complain of involuntary infertility? Fertil Steril 1989; 52:454–6.

2. Knuth UA, Hönigl W, Bals-Pratsch M, Schleicher G, Nieschlag E. Treatment of severe oligospermia with human chorionic gonadotropin/human menopausal gonadotropin: a placebo-controlled, double blind trial. J Clin Endocrinol Metab 1987; 65:1081–7.
3. Bals-Pratsch M, Dören M, Karbowski B, Schneider HPG, Nieschlag E. Cyclic corticosteroid immunosuppression is unsuccessful in the treatment of sperm antibody-related male infertility: a controlled study. Hum Reprod 1992; 7:99–104.
4. Grundy CE, Robinson J, Guthrie KA, Gordon AG, Hay DM. Establishment of pregnancy after removal of sperm antibodies in vitro. BMJ 1992; 304:292–3.

Chapter 6

Sperm Preservation: Fundamental Cryobiology and Practical Implications

P. F. Watson, J. K. Critser and P. Mazur

Human spermatozoa were first frozen successfully almost 40 years ago [1]. Although developments in dilution media, packaging and storing have occurred in the intervening period, little is known specifically about the effects of freezing and thawing on spermatozoa. This is largely due to the fact that a sufficient number of cells survive the challenge to give a reasonable expectation of fertilisation; but the shift to the exclusive use of cryopreserved semen for donor insemination, occasioned by the awareness of the risks particularly of AIDS transmission [2–4], has resulted in a renewed critical interest in the process of cryopreservation. This review will cover the more significant contributions over the past few years, and develop an argument for a sustained fundamental approach to sperm cryobiology.

Fundamental Cryobiology

Most sperm cryopreservation studies have been empirical, comparing a range of different treatments, usually by in vitro tests of cell survival. Although progress has undoubtedly been made, about 50% of cells are rendered immotile by the cryopreservation process, and there has been little improvement here.

We believe that further progress can only come from an understanding of the causes of cryoinjury. We are engaged in studies of fundamental cryobiology of spermatozoa to see if they respond as other cells. It is known that cells must be cooled sufficiently slowly to allow water to be removed from the intracellular environment, preventing intracellular ice crystal formation. It has been possible to describe mathematically the process of dehydration during cryopreservation and predict at what cooling rate one might expect to see evidence of cell death due to intracellular ice formation [5]. Our efforts are engaged to measure hydraulic conductivity and its activation energy, cell surface area and intracellular water volume in order to apply this mathematical model to spermatozoa [6–9].

The mathematical equations have been used to describe the survival of several widely differing cell types and have provided good agreement with practice [5]. If spermatozoa can be shown to be like other cells, it can be assumed that cell death is due to intracellular ice formation. If, on the other hand, the equations cannot match the empirical results, then other factors peculiar to spermatozoa will need to be identified.

The Measurement of Hydraulic Conductivity

The approach to this problem has been used previously for other cell types (e.g. erythrocytes) [10] and is two-staged. First, cells are exposed to increasing hypo-osmotic environments, and the critical value for cell lysis is determined; the rupture of cells (spermolysis) exposed to osmolalities below the critical value is then timed. We have measured the critical osmolality for several species [8,9,11; E. E. Noiles et al., unpublished observations; M. R. Curry and P. F. Watson, unpublished observations], and found some differences among species (Table 6.1). The calculations for human spermatozoa of hydraulic conductivity, L_p (2.84 μm min^{-1} atm^{-1} at 22°C), and activation energy, E_a (4.73 kcal mol^{-1} between 0° and 30°C) suggest a high water permeability and a low temperature dependence relative to other cell-types. However, between 0° and −3°C, L_p decreases rapidly and abruptly from 2.42 to 0.50 μm min^{-1} atm^{-1} and this low permeability was retained at temperatures down to −7°C, the lowest temperature it was possible to study the phenomenon [9]. The implication of these studies is that membrane water permeability during the cooling of spermatozoa changes discontinuously and dramatically, perhaps

Table 6.1. Critical osmolality, hydraulic conductivity, L_p, and activation energy, E_a

Species	n	Crit osm (mOsm)	Hyd cond (μm min^{-1} atm^{-1})	Act energy (kcal mol^{-1})	Reference
Fowl	5	17	2.1	4.4	[8]
Bull	6	36	10.8	3.0	[8]
Stallion	3	47	26.0	–	E. E. Noiles et al., unpublished
Human	10	55	2.84	4.73	[9]
Boar		95	–	–	E. E. Noiles et al., unpublished
Ram[a]	n	109	0.22	7.43	[7]

[a]Differences may be partly due to variations in techniques

related to membrane lipid phase changes. These values appear to imply that intracellular ice would form only at cooling rates far in excess of the upper limit of the empirically determined optimal range.

Nucleation Temperature

When cells are cooled below zero, ice crystallises out of the external medium creating an osmotic imbalance between the intra- and extracellular environments, and water is removed from the cell at a rate determined by the hydraulic conductivity of the membranes. If water cannot leave rapidly enough to keep the intracellular environment in osmotic equilibrium with the exterior; the intracellular water is progressively undercooled and becomes increasingly likely to crystallise intracellularly. However, this extent is unlikely to occur until the intracellular water is 10°C or more below its freezing point. This temperature is known as the nucleation temperature, and in large cells like mouse embryos it can be determined directly under a microscope equipped with a special cooling stage [12]. We have conducted experiments to determine the nucleation temperature indirectly by rapid cooling of spermatozoa in glass capillaries over carefully controlled temperature ranges. Intracellular freezing was determined by loss of cell membrane integrity as revealed by flow cytometry of cells stained with fluorescent viability stains. Cell death did not occur above −20°C in the presence of 0.85M-glycerol. Before intracellular ice crystal formation will occur the cells must contain undercooled water below −20°C. The likelihood of this event is then stochastically determined with a greater probability, the greater the degree of undercooling.

Surface Area and Water Volume

These two features are difficult to measure because of the peculiar shape of the human spermatozoon. Human sperm water volume has been estimated by electron spin reasonance studies as $21.5 \pm 2.1\ \mu m^3$ (mean ± SD) (F. W. Kleinhans et al., unpublished observations), a very similar estimate to that resulting from a double-isotope method [13], and representing total cell volumes of around 35 μm^3. These methods are appropriate because they are independent of cell shape, but do depend on estimates of cell concentration. However, particle size analysers give much lower values [14,15]. Surface area is even more difficult to determine and, combined with the pleomorphism of the human ejaculate, represents a continuing challenge.

Diluents and Cryoprotectants

A systematic study of eight diluent formulations used for the cryopreservation of spermatozoa [16] indicated a clear advantage to an egg yolk–citrate–tris diluent. Some have chosen to avoid egg yolk in human semen diluents for fear of potential disease and/or immune response risks and proposed more closely defined solutions (e.g. human sperm preservation medium, HSPM) [17]. Glycerol is universally regarded as the most suitable cryoprotectant [18,19].

However, glycerol has been demonstrated to have detrimental effects on the spermatozoa of several species [20] including the human [21,22]; and a few studies have proposed eliminating it from cryopreservation diluents [23]. The detrimental effects of glycerol on human spermatozoa have been variously described as "toxic" or "osmotic"; however, distinguishing between these two mechanisms of damage has not been critically examined [22]. Our recent studies of the permeability of human spermatozoa to glycerol gave a value of P_g at 22°C of 2.21×10^{-3} cm min^{-1} and an Arrhenius temperature dependence (E_a) of 11.76 kcal mole^{-1} based on measurements of the time to spermolysis in hypotonic media (D. Y. Gao et al., unpublished observations); similar values have been obtained using electron spin resonance techniques (J. Du et al, unpublished observations). These values are comparable to those previously found for human red blood cells (P_g: 2.41×10^{-4} cm min^{-1} and an E_a of P_g: 7.2 kcal mole^{-1}) [24] and suggest that glycerol penetrates human spermatozoa readily at ambient temperature but that permeability declines rapidly with temperature. As glycerol is added to the sperm suspension, the cells undergo a transient reduction in volume, as water initially leaves the spermatozoa in response to the aniosmotic conditions. Preliminary data (D. Y. Gao, et al., unpublished observations) suggest that this transient reduction in sperm cell volume may be irreversibly damaging if the extent of the shrinkage is too great. The implication for sperm cryopreservation protocols is that glycerol should be added in such a manner as to minimise this sperm cell shrinkage, by introducing it "slowly enough" (as determined by the P_g value). Moreover, since P_g decreases so markedly with temperature, the temperature of addition becomes a relevant consideration.

After warming, the issue of cryoprotectant removal becomes important and is directly related to the sperm membrane permeabilities to water and glycerol. However, cryoprotectant removal is of most concern in essentially spherical cell types (e.g. embryos). When the cells are exposed to media with cryoprotectants and with a physiological osmolarity (i.e. ~300 mosmol), they will swell to a volume proportional to the ratio of internal:external osmolality (assuming the cell responds as an ideal osmometer). For example, if a cell is initially equilibrated with 1 osmol glycerol and then placed into isosmotic medium (say 300 mosmol) it will attempt to swell to approximately 4.3 times its original volume (1300/300). The maximal tolerated volume expansion in many cell types is between 1.8 and 2.5 times the isosmotic volume, and abrupt placement of cells equilibrated with cryoprotectants into isosmotic conditions will cause cells to swell beyond this "critical volume" and lyse. Spermatozoa in general, and human spermatozoa specifically, however, are capable of expanding between four and nine times their isosmotic volume depending on species [8] (F. W. Kleinhans et al., unpublished observations). This is because the sperm cell has an highly specialised morphology which is more elliptical than spherical. As the spermatozoon swells, it changes shape from an elongated ellipse to a sphere, and in this geometric configuration, it is capable of encompassing a much larger volume. Ultimately, therefore, concentrations of glycerol commonly used for the cryopreservation of spermatozoa (< 1 osmol) appear not to make the cells particularly susceptible to dilution-related lysis.

Contained in that argument are several assumptions, some of which are as yet untested. First, the extent of the shrinkage and swelling is dependent on the difference in the relative permeabilities of water and glycerol. In practice, glycerol is not impermeable and neither is water freely permeable, thus the calculated volume changes represent theoretical maxima. Second, it is assumed that as long as shrinkage or swelling do not exceed certain limits no permanent cell damage is caused. The limits have not been ascertained and may be less than those determined by the accommodation of the plasma membrane; the spermatozoon has several membrane-bound subcompartments (e.g. mitochondria, acrosome) which are essential to cell function and which might be more susceptible to distortion. Third, the foregoing discussion has focused on osmotic injury caused by glycerol. Alternatively, glycerol may be chemically toxic to spermatozoa which would favour improvements in cryosurvival when glycerol is added at lower temperatures [22].

Cooling Rates

Most cells exposed to the multiple factors associated with cryopreservation demonstrate "inverted U-shaped" survival curves as a function of cooling rate below the freezing point of the extracellular medium, i.e. low cryosurvival resulting both from sub- and supraoptimal cooling rates [25]. A similar relationship with cooling rate has recently been shown to be applicable to spermatozoa (Fig. 6.1) [26–28]. In general, the survival of spermatozoa is considerably less sensitive to cooling rate than that of most other cell types; optimal cooling rates for spermatozoa have been reported between 10°C and 170°C min^{-1} [20]. On the other hand, the overall survival at optimum cooling rates differs between cell types. This may be because multiway interactions exist among cooling rate, warming rate and cryoprotectant concentrations [20].

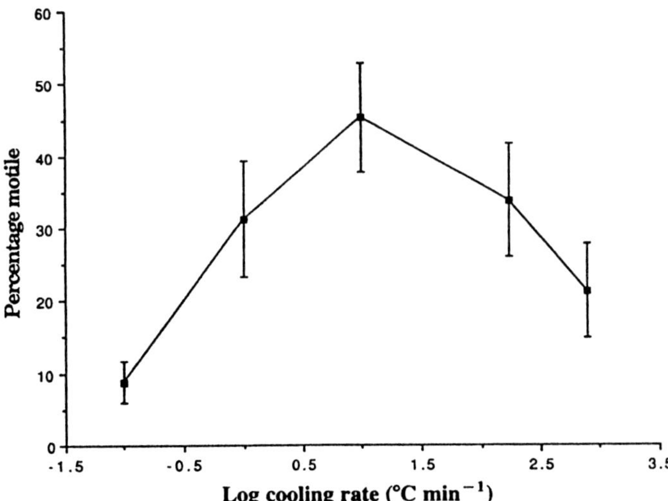

Fig. 6.1. The motility of human spermatozoa frozen in plastic straws at various cooling rates, and thawed rapidly. The spermatozoa were diluted in a 7.5% glycerol HSPM diluent and frozen at 10°C min^{-1}.

For human spermatozoa, optimal freezing rates have been reported to range between 1°C min^{-1} [29] to 16–25°C min^{-1} [21,30,31]. It is important to note that many of the cooling rates recorded in earlier literature were performed with techniques which do not produce linear cooling. The values cited therefore are often average cooling rates and should be carefully interpreted [20]. Compared with other mammalian spermatozoa, human spermatozoa are relatively resistant to cryoinjury, and tolerate a wide range of cooling rates (Fig. 6.1). Many laboratories freeze spermatozoa in plastic straws in the vapour phase over liquid nitrogen (J. K. Critser and N. A. Ruffing, unpublished observations). This method gives no direct control over the cooling rate and in small vessels is not very repeatable. The advent of controlled-rate freezing devices, introduced into IVF laboratories for embryo freezing, has resulted in protocols being adopted for spermatozoa [32,33]. Often, although the cooling rates may not be optimal, the control provides a superior environment to that achieved in the vapour phase in small liquid nitrogen dewars [34].

Insufficient attention has been given to the importance of the sample dimensions in determining both the actual cooling rate affecting the sperm cells and the removal of heat from the specimen (i.e. the thermal properties of the sample itself and not just the dewar or cooling chamber environment). Also, the criterion of success has invariably been the percentage of motile cells. In view of the limitations of motility as an indicator of cryosurvival (see later) further work is necessary to determine optimal cooling rates and the potential interactions with thawing rate and cryoprotectant concentration.

Critser et al. [33] considered the importance of eliminating the temperature fluctuation resulting from the evolution of the latent heat of crystallisation (seeding), and found no effect on post-thaw motility, but seeded samples demonstrated a higher frequency of zona-free hamster ova penetration, a higher frequency of membrane integrity, but also a higher frequency of acrosomal damage compared with non-seeded samples. Similar results have been reported for boar semen [35].

Warming and Thawing

The effects of warming rate depend on the prior cooling rate and on the cell type. For human spermatozoa there are relatively few data regarding the optimal warming rates. Previous reports have indicated a wide range of rates usually determined by pragmatic considerations, including thawing in ambient air, in water baths with temperatures ranging between 5°C and 37°C, in coat pockets and by thawing the samples in the hand [36]. Mahadevan [36] conducted a more comprehensive study in which warming rates between 9.2°C min^{-1} and 2140°C min^{-1} utilising coat pocket, hands and air (5°, 20° or 37°C) or water baths (5°, 20°, 37°, 55° or 75°C) were examined in combination with a standard freezing rate of 10.0°C min^{-1} from +5°C to −80°C. The results of that study indicated that slow thawing in 20°C or 35°C air resulted in more nearly optimal cryosurvival of human spermatozoa in terms of both motility and supravital staining; although there were significant differences in post-thaw motilities only among the fastest two rates (1837° and 2140°C min^{-1}) and all the other (twelve) rates. Additionally, Mahadevan [36] examined the interaction between freezing and thawing rates. Slow thawing was reported optimal for samples frozen slowly; however, there was little effect of thawing rate when a fast freezing rate was employed.

Although it is common among many practitioners to use relatively slow methods for warming frozen human spermatozoa, e.g. ambient air (J. K. Critser and N. A. Ruffing, unpublished observations), it is clear that further investigation of the multiple interaction among cooling rate, cryoprotectant concentration and warming rate will be required before optimal conditions can be determined for human spermatozoa. This is supported by the several animal studies demonstrating the advantage of more rapid warming [37–40], and the evidence presented above regarding glycerol permeability at low temperatures.

Fertility of Frozen-Thawed Semen

The minimum number of competent spermatozoa to achieve maximum chances of conception in humans is unknown, and commonly 10^7–10^8 spermatozoa are inseminated with an average cryosurvival of 50–60%; this would yield approximately 5×10^6–5×10^7 motile spermatozoa. Species differences in the numbers of spermatozoa deposited by natural coitus and the site of deposition make comparisons of actual numbers of spermatozoa required for artificial insemination invalid. However in cattle and pigs the reduction in fertility with decreasing sperm numbers occurs over a relatively narrow range [41,42]. More than double the numbers of cryopreserved spermatozoa compared with fresh spermatozoa appear to be required to achieve comparable fertility and this cannot be explained solely in terms of numbers of motile cells [18]; other factors, such as membrane damage to a proportion of the surviving motile population [43], are also important.

That the fertility of frozen–thawed semen is poorer than fresh semen seems hardly in doubt, although obtaining data which satisfy the point statistically is more difficult. Barwin [44] found that overall success rates (percentage of patients achieving pregnancy) were 76% (n = 56) for fresh semen and 63% (n = 64) for cryopreserved semen; a difference which is not significant. Similar figures, 76% (n = 112) with fresh semen and 64% (n = 69) for cryopreserved semen, were given by Steinberger et al. [45] and these were also not significant. A more sensitive parameter is mean numbers of inseminations per pregnancy (5.7 for fresh semen and 10.3 for cryopreserved semen) [46]; these figures were assembled from different published studies and were not directly comparable. However, when the Steinberger data quoted above were expressed as numbers of cycles to achieve pregnancy (4.2 ± 0.3 vs 9.3 ± 1.5; mean ± SEM), the difference was significant ($P < 0.05$). Although this parameter may permit a more sensitive measurement of relative fertility the problem of uncontrolled population variables (discontinuity with treatment, age, female fertility, etc.) still exists.

Cramer et al. [47] introduced the concept of fecundability (monthly probability of conception), a life-table analysis method which takes these factors into account. The fecundability rate for frozen semen (5.0–17.0%) was always lower than that for fresh semen (12.0–27.4%) in studies where both fresh and frozen semen were compared [36,48–52] although some studies reported no significant difference between the groups [36,50,51]. These figures may be broadly compared with a 20–30% probability of conception per ovulatory cycle in the normal population for unprotected intercourse 2–3 times per week [53].

The wide variability in the results of these studies is probably due to differences in insemination practice (e.g. numbers of insemination per cycle, number of motile spermatozoa per insemination, methods for timing insemination in respect of ovulation), and in patient populations (e.g. age range, indications for donor insemination).

Selection of "Viable" Cells

Both in the course of preservation and in the assessment of the survival there is an element of selection which may have practical implications. It is inherent in the understanding of cell loss during freezing and thawing that a random untoward effect, such as intracellular ice formation [54], and/or membrane rupture by osmotic gradient [55] affects a proportion of the cell population as they undergo a single freeze–thaw event. However, the likelihood of these events being lethal is dependent on the extent of cell dehydration, and this may be determined, not simply by random events but by characteristics of cell membranes of individual cells. The question thus raised is whether there are subpopulations of peculiarly susceptible cells. For spermatozoa this has never been resolved, and in view of the recognised heterogeneity and, especially in the human ejaculate, the pleomorphic range of spermatozoa, perhaps it would be surprising if some element of selection did not exist. Such a situation might explain the substantial cell loss irrespective of the optimisation process of empirical research, and it might also explain the variation in cryosurvival between spermatozoa of different individuals.

Kincade et al. [28] studied this question by looking at the survival after each of a number of freeze–thaw cycles. If cryoinjury is random, a similar reduction is expected after each cycle, but if it is non-random, a resistant subpopulation should be selected. The result (Fig. 6.2) suggested that a random event affected the majority of the cells, but some 10% of the population represented a more resistant subgroup. However, the criterion of survival in this case was solely motility, assessed visually; it would be helpful to know if a similar pattern is reflected in other parameters of cell viability. The implications of this study for the expectation of success in cryopreservation, depend both on the minimisation of the cell loss per freeze–thaw cycle and whether the resistant subpopulation is in fact also fertile.

The Effects of Cryopreservation

Spermatozoa are recognised to be involved in a maturational journey commencing in the testis and ending at the egg surface with the acrosome reaction and fertilisation [56]. The assumption that this process is merely interrupted for an

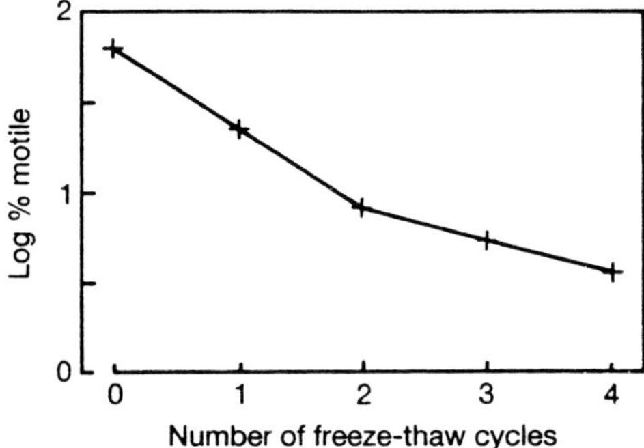

Fig. 6.2. The survival of human spermatozoa in plastic straws after one or more freeze–thaw cycles. The spermatozoa were diluted in a 7.5% HSPM diluent, frozen at 10°C min^{-1} and thawed rapidly.

extended interval during the cryopreservation process (Fig. 6.3) now seems to be in doubt. A growing body of evidence suggests that frozen–thawed spermatozoa may be able to undergo an acrosome reaction and fertilisation sooner than uncapacitated fresh spermatozoa.

Readiness to undergo maturational development may relate to glycerol exposure since glycerol may have more actions on the cell than simply its colligative effect on water [57]. Slavik [58] showed that exposure of ram spermatozoa to glycerol resulted in their penetrating eggs more rapidly.

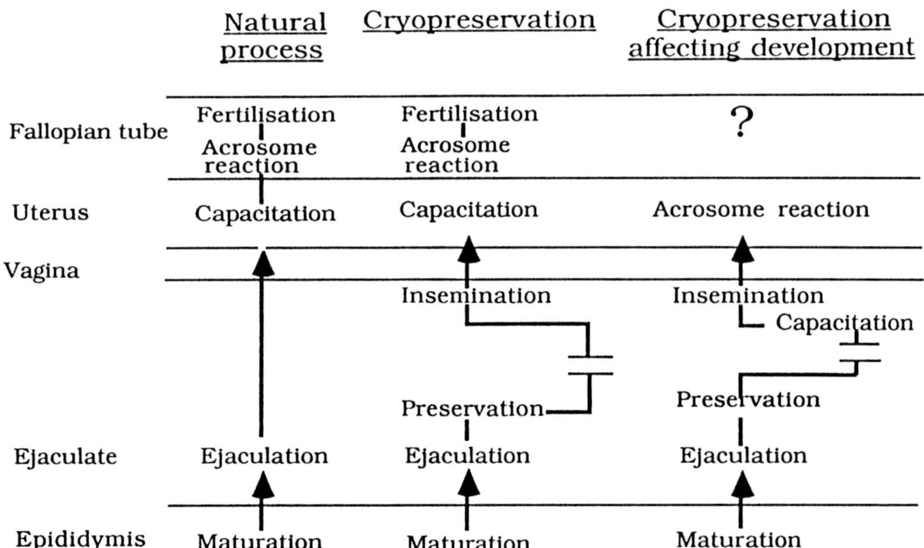

Fig. 6.3. Diagrammatic comparison of the life history of human spermatozoa during the natural insemination and fertilisation process (left), and the cryopreservation, artificial insemination and fertilisation process (centre and right). The representation (right) presupposes that cryopreservation either induces accelerated maturation or bypasses the need for maturation (see text).

Both bull [59] and ram spermatozoa [60,61] accumulate calcium ions when cooled, and egg yolk, a known protectant during cooling, reduced the accumulation of calcium [62]. It is well recognised that an increase in intracellular calcium ion concentration occurs at the end of capacitation and is a prelude to the acrosome reaction. Frozen–thawed spermatozoa have been cooled and have suffered altered permeability to calcium; they may, therefore, be more permeated by calcium at the time they are thawed. A subsequent readiness to undergo the acrosome reaction may simply reflect their already elevated intracellular calcium pools.

Dilution, cooling and holding rabbit [63] and human serum [64] overnight resulted in spermatozoa more readily penetrating zona-free hamster eggs. Also, Critser et al. [65] found that frozen spermatozoa were capable of maximal penetration of zona-free hamster eggs immediately after thawing, whereas fresh spermatozoa required some 24 h incubation in capacitating conditions to achieve maximal penetration (Fig. 6.4); the significant difference in chronology was not attributable to a difference in motility although the overall penetration rate was lower for frozen semen [65].

These studies point towards the conclusion that cryopreservation advances spermatozoa closer to their functional end-point, or so affects the membranes that the maturational stages may be, to some degree, by-passed (Fig. 6.3). If this suggestion is borne out by further study it has important implications for

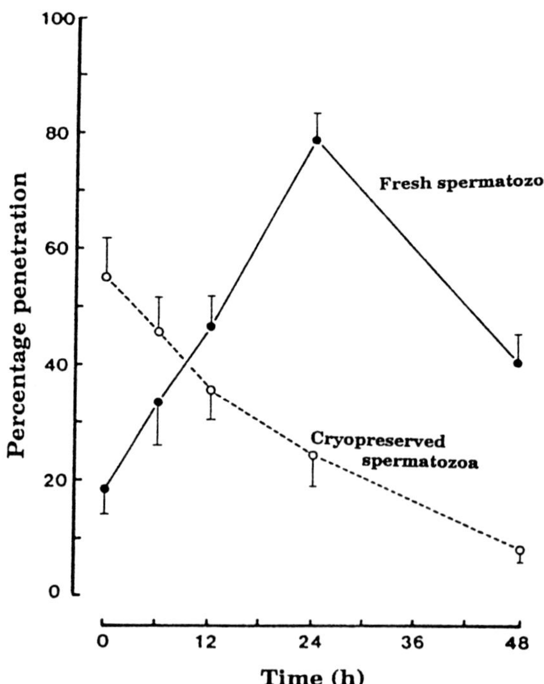

Fig. 6.4. The penetration of zona-free hamster eggs by fresh or cryopreserved human spermatozoa after incubation in media for up to 48 h. (Modified from Critser et al. [65], by permission).

the fertility of frozen–thawed spermatozoa. It also has serious implications for the criteria we choose to evaluate freeze–thaw success. If methods are chosen which favour the greatest proportion of cells appearing like unfrozen cells, there is a risk of selecting for unresponsive cells rather than fertile cells.

The site and timing of insemination, and number of spermatozoa inseminated, are also influenced by these considerations. It is acknowledged that cryopreserved spermatozoa have a shorter survival time in the female tract than fresh spermatozoa [66,67]. However, studies in sheep have shown that high fertility can be achieved with very modest numbers of cryopreserved spermatozoa if they are inseminated into the uterine horn laparoscopically, carefully timed to coincide with ovulation [68,69]. These results might all be interpreted in the light of a shorter time to acrosome reaction, survival after which is reckoned to be limited [56].

As a result of the studies reviewed here suggesting the means of selection of subpopulations of hardy cells and the maturational advancement of the survivors, the future for human sperm cryopreservation lies in the need to adapt our techniques to accommodate these observations. Perhaps with the development of in vitro fertilisation with cryopreserved spermatozoa it may be possible to develop further selection of the surviving population to obtain cells which have a very high likelihood of achieving fertilisation under those conditions. For cervical insemination, greater attention is needed to predict the time of ovulation, while possibly intrauterine insemination, would make more fertile cells available for fertilisation.

Acknowledgements. Much of the work referred to here was funded by various research grants: AFRC (AG 48/67) and MRC (G8912737SB) to PFW, NIH (1 RO 1 HD25949-01) and USDA (89-37240-4681) to JKC and PM, and their support is gratefully acknowledged.

References

1. Bunge RG, Keettel W, Sherman J. Clinical use of frozen semen, report of 4 cases. Fert Steril 1954; 5:520–9.
2. Stewart G, Tyler JPP, Cunningham AL, Barr JA, Driscoll GL, Gold J, Lamont BJ. Transmission of human T-cell lymphotropic virus type III (HTLV-III) by artificial insemination by donor. Lancet 1985; ii:581–4.
3. DHSS. Acquired immune deficiency syndrome (AIDS) and artificial insemination – guidance for doctors and AI clinics. London: Department of Health and Social Security, 1986.
4. American Fertility Society. New guidelines for the use of semen donor insemination: 1990. Fert Steril 1990; 53 (Suppl 1):1S–13S.
5. Mazur P. Freezing of living cells: mechanisms and implications. Am J Physiol 1984; 247:C125–C142.
6. Critser JK, Kleinhans FW, Mazur P. Cryopreservation of human sperm. Cryobiology 1991; 28:525–6.
7. Duncan AE, Watson PF. Predictive water loss curves for ram spermatozoa during cryopreservation: comparison with experimental observations. Cryobiology 1992; 29:95–105.
8. Watson PF, Kunze E, Cramer P, Hammerstedt RH. A comparison of critical osmolality, hydraulic conductivity and its activation energy in fowl and bull spermatozoa. J Androl 1992; (in press).
9. Noiles EE, Mazur P, Boldt HD, Kleinhans FW, Critser JK. Hydraulic conductivity (L_p) and its activation energy (E_a) in human sperm. J Androl 1992 (Suppl): P-30.
10. Dick DAT. Cell water. Washington DC: Butterworths, 1966.
11. Noiles EE, Ruffing NA, Kleinhans FW et al. Critical tonicity determination of sperm using

dual fluorescent staining and flow cytometry. In: Johnson LA, Rath D, eds. Boar semen preservation II. Berlin, Hamburg: Paul Parey Scientific Publishers, 1991; 359–64.

12. Rall WF, Mazur P, McGrath JJ. Depression of the ice-nucleation temperature of rapidly cooled mouse embryos by glycerol and dimethyl sulfoxide. Biophys J 1983; 41:1–12.
13. Ford WCL, Harrison A. D-[1-^{14}C]mannitol and [U-^{14}C]sucrose as extracellular space markers for human spermatozoa and the uptake of 2-deoxyglucose. J Reprod Fertil 1983; 69:479–87.
14. Laufer N, Segal S, Yaffe H, Svartz H, Grover NB. Volume and shape of normal human spermatozoa. Fertil Steril 1977; 28:456–8.
15. Jeyendran RS, Karuhn RF, Van der Ven HH, Perez-Pelaez M. Volumetric analysis of human spermatozoa. Andrologia 1987; 19:54–7.
16. Weidel L, Prins GS. Cryosurvival of human spermatozoa frozen in eight different buffer systems. J Androl 1987; 8:41–7.
17. Mahadevan M, Trounson AO. Effects of cryoprotective media and dilution methods on the preservation of human spermatozoa. Andrologia 1983; 15:355–66.
18. Watson PF. Artificial insemination and the preservation of semen. In: Lamming GE, ed. Marshall's physiology of reproduction. Edinburgh: Churchill Livingstone, 1990; 747–869.
19. Sherman JK. Current status of clinical cryobanking of human semen. In: Paulson JD, Negro-Vilar A, Lucena E, Martini L, eds. Andrology: male fertility and sterility. Orlando: Academic Press, 1986; 517–47.
20. Watson PF. The preservation of semen in mammals. In: Finn CA, ed. Oxford reviews of reproductive biology. Oxford: Oxford University Press, 1979; 283–350.
21. Sherman JK. Questionable protection by intracellular glycerol during freezing and thawing. J Cell Comp Physiol 1963; 61:67–83.
22. Critser JK, Huse-Benda AR, Aaker DV, Arneson BW, Ball GD. Cryopreservation of human spermatozoa. III. The effect of cryoprotectants on motility. Fertil Steril 1988; 50:314–20.
23. Jeyendran RS, Van der Ven HH, Kennedy W, Perez-Pelaez M, Zaneveld LJD. Comparison of glycerol and a zwitter ion buffer system as cryoprotective media for human spermatozoa. J Androl 1984; 5:1–7.
24. Mazur P, Miller RH. Permeability of the human erythrocyte to glycerol in 1 and 2M solutions at 0 and 20°C. Cryobiology 1976; 13:507–22.
25. Mazur P. Cryobiology: the freezing of biological systems. Science 1970; 168:939–49.
26. Fiser PS, Fairfull RW. The effect of glycerol concentration and cooling velocity on cryosurvival of ram spermatozoa frozen in straws. Cryobiology 1984; 21:542–51.
27. Watson PF, Duncan AE. Effect of salt concentration and unfrozen water fraction on the viability of slowly frozen ram spermatozoa. Cryobiology 1988; 25:131–42.
28. Kincade RS, Colvin KE, Kleinhans FW, Critser ES, Mazur P, Critser JK. The effects of cooling rate and repeated freezing on human sperm cryosurvival. J Androl 1989 (Suppl):48P.
29. Freund M, Wiederman J. Factors affecting the dilution, freezing and storage of human semen. J Reprod Fertil 1966; 11:1–17.
30. Trelford JD, Mueller F. Observations and studies on the storage of human sperm. Can Med Assoc J 1969; 100:62–5.
31. Watson PF. Recent advances in sperm freezing. In: Thompson W, Joyce DN, Newton JR, eds. In vitro fertilization and donor insemination. London: Royal College of Obstetricians and Gynaecologists, 1985; 261–7.
32. Serafini P, Marrs RP. Computerised staged-freezing technique improves sperm survival and preserves penetration of zona-free hamster ova. Fertil Steril 1986; 45:854–8.
33. Critser JK, Huse-Benda AR, Aaker DV, Arneson BW, Ball GD. Cryopreservation of human spermatozoa. I. Effects of holding procedure and seeding on motility, fertilizability, and acrosome reaction. Fertil Steril 1987; 47:656–63.
34. McLaughlin EA, Ford WCL, Hull MGR. A comparison of the freezing of human semen in the uncirculated vapour above liquid nitrogen and in a commercial semiprogrammable freezer. Hum Reprod 1990; 5:724–8.
35. Fiser PS, Hansen C, Underhill KL, Shrestha JNB. The effect of induced ice nucleation (seeding) on the post-thaw motility and acrosomal integrity of boar spermatozoa. Anim Reprod Sci 1991; 24:293–304.
36. Mahadevan MM. Cryobiological and biochemical studies of human serum. PhD Thesis, Monash University, Melbourne, 1981.
37. Rodriguez OL, Berndtson WE, Ennen BD, Pickett BW. Effect of rates of freezing, thawing and level of glycerol on the survival of bovine spermatozoa in straws. J Dairy Sci 1975; 41:129–36.
38. Senger PL, Becker WC, Hilliers JK. Effect of thawing rate and post-thaw temperature on

motility and acrosomal maintenance in bovine semen frozen in plastic straws. J Anim Sci 1976; 42:932–6.
39. Fiser PS, Ainsworth L, Lanford G. Effect of osmolality of skim-milk diluents and thawing rate on cryosurvival of ram spermatozoa. Cryobiology 1981; 18:399–403.
40. Cochran JD, Amann RP, Froman DP, Pickett BW. Effects of centrifugation, glycerol level, cooling to 5°C, freezing rate and thawing rate on the post-thaw motility of equine sperm. Theriogenology 1984; 22:25–38.
41. Pickett BW, Berndtson WE. Preservation of bovine spermatozoa by freezing in straws: a review. J Dairy Sci 1974; 57:1287–301.
42. Polge C. Fertilization in the pig and horse. J Reprod Fertil 1978; 54:561–70.
43. Wilmut I, Polge C. The low temperature preservation of boar semen. III. The fertilizing capacity of frozen and thawed boar semen. Cryobiology 1977; 14:483–91.
44. Barwin BN. Artificial insemination. In: Paulson JD, Negro-Vilar A, Lucena E, Martini L, eds. Andrology: male fertility and sterility. Orlando: Academic Press, 1986; 461–74.
45. Steinberger E, Rodriguez-Rigau LJ, Smith KD. Comparison of results of AID with fresh and frozen semen. In: David G, Price WS, eds. Human artificial insemination and semen preservation. New York: Plenum Press, 1980; 283–94.
46. Richardson DW. Factors influencing the fertility of frozen semen. In: Richardson DW, Joyce D, Symonds EM, eds. Frozen human semen. London: Royal College of Obstetricians and Gynaecologists, 1979; 33–58.
47. Cramer DW, Walker AM, Schiff I. Statistical methods in evaluating the outcome of infertility therapy. Fertil Steril 1979; 32:80–6.
48. Smith KD, Rodriguez-Rigau LJ, Steinberger E. The influence of ovulatory dysfunction and timing of insemination on the success of artificial insemination donor (AID) with fresh or frozen semen. Fertil Steril 1981; 36:496–502.
49. Richter MA, Haning RV, Shapiro SS. Artificial donor insemination: fresh versus frozen semen; the patient as her own control. Fertil Steril 1984; 41:277–80.
50. Hammond MG, Jordan S, Sloan CS. Factors affecting pregnancy rates in a donor insemination program using frozen semen. Am J Obstet Gynecol 1986; 155:480–5.
51. Bordson BL, Ricci E, Dickey RP, Dunaway H, Taylor SN, Curole DN. Comparison of fecundability with fresh and frozen semen in therapeutic donor insemination. Fertil Steril 1986; 46:466–9.
52. Brown CA, Boone WR, Shapiro SS. Improved cryopreserved semen fecundability in an alternating fresh-frozen artificial insemination program. Fertil Steril 1988; 50:825–7.
53. Leridon H. The efficacy of natural insemination: a comparative standard for AID. In: David G, Price WS, eds. Human artificial insemination and semen preservation. New York: Plenum Press, 1980; 191–6.
54. Mazur P. The role of intracellular freezing in the death of cells cooled at supraoptimal rates. Cryobiology 1977; 14:251–72.
55. Muldrew K, McGann LE. Mechanisms of intracellular ice formation. Biophys J 1990; 57:525–32.
56. Yanagimachi R. Mammalian fertilization. In Knobil E, Neill JD, eds. The physiology of reproduction. New York: Raven Press, 1988; 135–85.
57. Hammerstedt RH, Graham JK, Nolan JP. Cryopreservation of mammalian sperm: what we ask them to survive. J Androl 1990; 11:73–88.
58. Slavik T. Effect of glycerol on the penetrating ability of fresh ram spermatozoa with zona-free hamster eggs. J Reprod Fertil 1987; 79:99–103.
59. Karagiannidis A. The distribution of calcium in bovine spermatozoa and seminal plasma in relation to cold shock. J Reprod Fertil 1976; 46:83–90.
60. Robertson L, Watson PF. Calcium transport in diluted or cooled ram semen. J Reprod Fertil 1986; 77:177–85.
61. Simpson AM, White IG. Effect of cold shock and cooling rate on calcium uptake of ram spermatozoa. Anim Reprod Sci 1986; 12:131–43.
62. Robertson L, Watson PF. The effect of egg yolk on the control of intracellular calcium in ram spermatozoa cooled and stored at 5°C. Anim Reprod Sci 1987; 15:177–87.
63. Parrish JJ, Foote RH. Fertility of cooled and frozen rabbit sperm measured by competitive fertilization. Biol Reprod 1986; 35:253–7.
64. Chan SYW, Li SQ, Wang C. TEST-egg yolk buffer storage increases the capacity of human sperm to penetrate hamster eggs *in vitro*. Int J Androl 1987; 10:517–24.
65. Critser JK, Arneson BW, Aaker DV, Huse-Benda AR, Ball GD. Cryopreservation of human spermatozoa. II. Post-thaw chronology of motility and of zona-free hamster ova penetration. Fertil Steril 1987; 47:980–4.

66. Mattner P, Entwistle KW, Martin ICA. Passage, survival and fertility of deep-frozen ram semen in the genital tract of the ewe. Aust J Biol Sci 1969; 22:181–7.
67. Keel BA, Black JB. Reduced motility longevity in thawed human spermatozoa. Arch Androl 1980; 4:213–15.
68. Maxwell WMC. Artificial insemination of ewes with frozen–thawed semen at a synchronized oestrus. I. Effect of time of onset of oestrus, ovulation and insemination on fertility. Anim Reprod Sci 1986; 10:301–8.
69. Maxwell WMC. Artificial insemination of ewes with frozen–thawed semen at a synchronized oestrus. II. Effect of dose of spermatozoa and site of intrauterine insemination on fertility. Anim Reprod Sci 1986; 10:309–16.

Chapter 7

Sperm–Egg Manipulation for the Treatment of Male Infertility

S. B. Fishel and J. Timson

Introduction

Male infertility, or defective sperm function, has been described as the single, largest defined cause of human infertility [1]. However, a precise definition of the infertile male is elusive. Ironically, it is more difficult to define defective sperm function with the advent of assisted conception, and high technology micromanipulation procedures, as the dual nature of the infertile couple is more thoroughly appreciated.

Infertility in the male can be divided into three broad categories:

1. Discernible problems: abnormal semen parameters associated with somatic dysfunction (endocrine, structural, chromosomal or genetic).
2. Abnormal semen parameters unassociated with a somatic abnormality.
3. Apparently normal semen characteristics (count, motility, morphology and biochemistry), but an inability to procure conception.

Problems have arisen in predicting the fertility of a male simply on sperm density, motility and morphology because it is difficult to assess these in isolation from female factors. Many oligozoospermic and oligoasthenozoospermic males ($<10 \times 10^6$ ml^{-1} and less than 40% motility) father children and it is not infrequent that males with a high density of spermatozoa and motility with greater than 70% normal forms, are unable to achieve fertilisation in vitro.

Clearly, the ability of a spermatozoan to fertilise an oocyte resides more in its physiological and biochemical competence, than the sheer weight of numbers, its progressive activity or, perhaps, its overall shape. However, given that any of these problems exist in an ejaculate, and that this might reduce the chances of conception in vivo, the spermatozoa and oocytes can be manipulated in vitro in an attempt to procure fertilisation.

Various methods may be used to concentrate the spermatozoa. The gametes may be placed in immediate contact and either transferred directly to the fallopian tube (gamete intrafallopian transfer, GIFT), or inseminated in vitro (in vitro fertilisation, IVF). Should the numbers of spermatozoa be too few, the motility too weak, or, for other submicroscopic reasons, unable to bind to or penetrate the zona pellucida, the gametes may be micromanipulated to bring the spermatozoan and the oolemma into immediate proximity.

Micromanipulative Techniques

The concept of micromanipulation, or microsurgical techniques is not new. Since the invention of the microscope all sorts of microdissections carried out in a microscopic field have been devised. One of the first researchers to report the use of microscopic dissection of tissues was H. D. Schmidt in 1895 [2]. Microsurgical or micromanipulation techniques were first used to study the living eggs of mammals, including rat [3] and human [4], during the 1940s and 1950s. Some of the most important studies on the technical and experimental nature of egg micrurgy was performed by T. P. Lin [5,6] during the mid 1960s.

The early events of fertilisation (videlicet; membrane fusion between homologous and heterologous gametes, activation of the ooplasm and formation of the pronuclei) were investigated by the injection of a spermatozoan directly into the oocyte [7,8]. The specificity of fertilisation was further studied using interspecies experiments between, for example, rat and mouse, rabbit and fish, and frog and human [9].

Apart from the enormous contribution to our understanding of the biological mechanisms of fertilisation, this work led to the appliance of techniques for animal production. For the first time it was perceived that production and efficient use could be made of genetically valuable, but biological defective male gametes from domestic and wild animal species. During the 1980s advances were being made in increasing the incidence of fertilisation in cleavage using micromanipulative techniques, with the eventual birth of live offspring in the rabbit and cow from direct injection of spermatozoa, intact and epididymal, into the oocyte [10–12].

It was against this backdrop of nearly three decades of research on the technique of micromanipulation of whole, part or single cells of embryos and almost four decades on the manipulation of animal gametes, that embryologists began to utilise these techniques for the alleviation of infertility.

The clinical application of the microinsemination techniques for procuring fertilisation in vitro was first reported about 7 years ago [13]. Since then a number of techniques have been developed (Fig. 7.1). In 1987 Laws-King et al. [14] reported the insertion of single sperm into the perivitelline space with subsequent fertilisation. This report used capacitated spermatozoa from donors with a normal semen profile, and sperm from donors with semen profiles of low quality, including those with coiled and twin tails. Lanzendorf et al. [15]

Direct injection into ooplasm

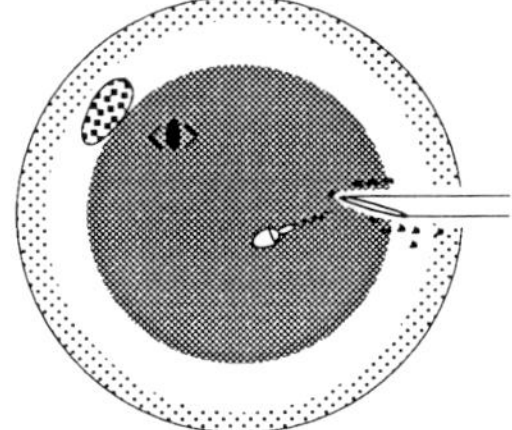

Injection into perivilline space

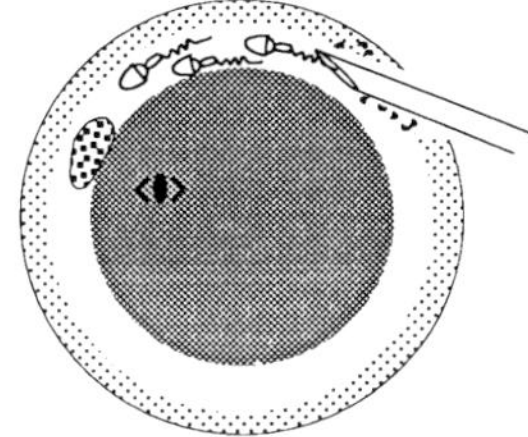

Opening the Zona pellucida

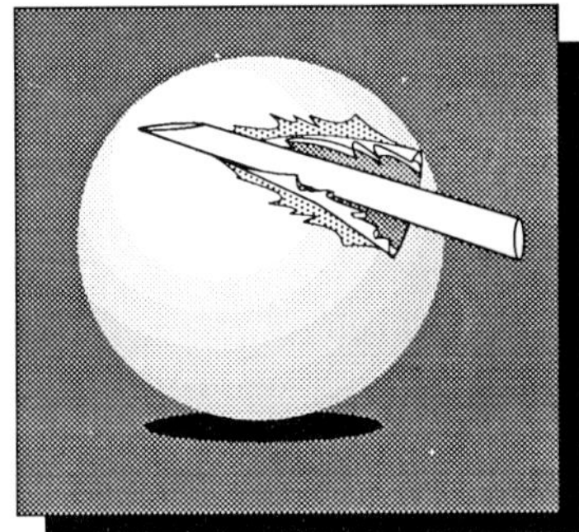

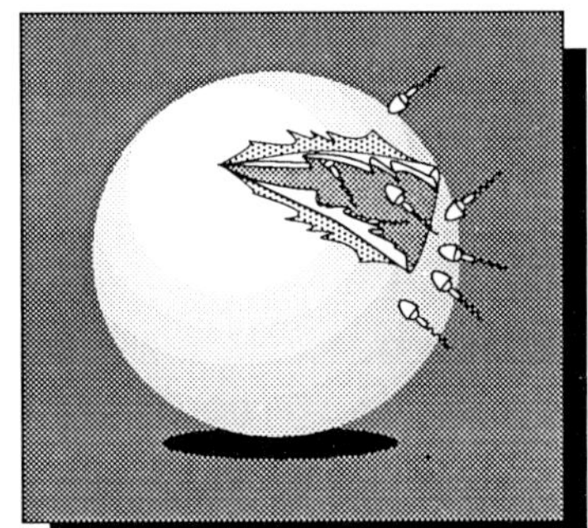

Zona drilling

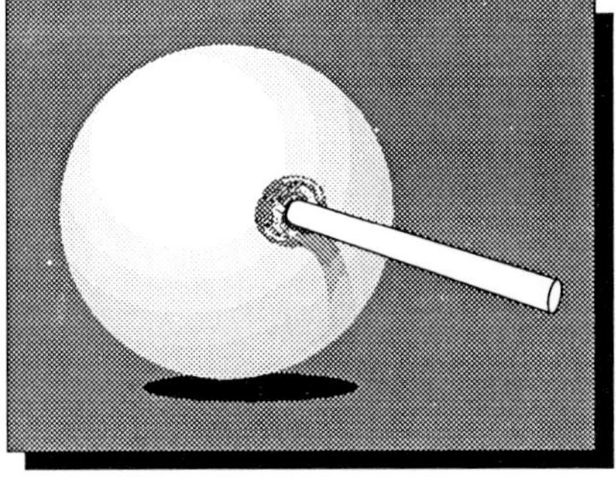

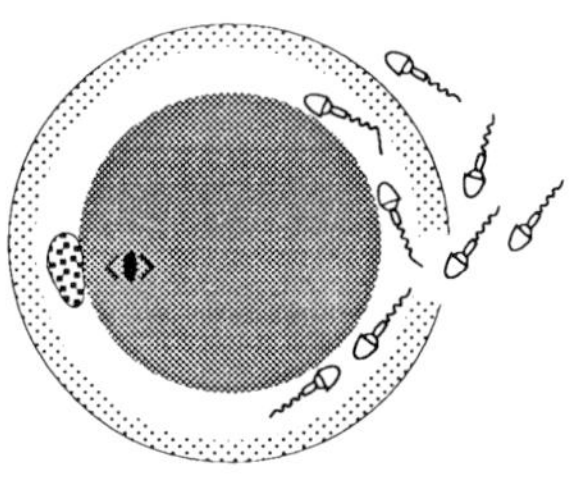

Fig. 7.1. Description of microinsemination techniques.

injected sperm from infertile males directly into the ooplasm with 58% of patients having an oocyte developing with a male pronucleus. Gordon et al. [16] used acid Tyrodes solution for "drilling" a hole in the zona pellucida of the human oocyte before transferring them to a suspension of sperm. Insemination of these oocytes was with semen from patients who had failed to achieve fertilisation by conventional IVF. Of the 63 eggs from 10 couples, 32% of those "zona drilled" fertilised, compared with 25% of the non-drilled. Five diploid eggs were transferred, but no pregnancies were achieved. In the same year, Cohen et al. [17] reported their technique of partial zona dissection (PZD). This mechanical approach, which causes a breach in the zona pellucida by the microneedle piercing opposing ends of the zona pellucida, followed by rubbing the latter up against the hole in the pipette until it tears (Fig. 7.2), resulted in 2 PZD and a single control (fertilised by conventional IVF) embryos being transferred into two patients. Both patients became pregnant, each with a set of twins. Further work resulted in the establishment of a number of pregnancies from only micromanipulated eggs, and one patient was reported to have delivered healthy dizygotic twins [18].

A few years earlier a group in Singapore [19] reported studies in the human which followed as a result of an observation by Aitkin et al. which was published in 1983 [20]. These workers demonstrated that sperm from men with immotile cilia (Kartagener) syndrome can fuse, penetrate and undergo pronuclear formation in the egg cytoplasm. It was reported by Ng et al. [19] that fertilisation of oocytes and the formation of embryos occurred from a patient with immotile cilia syndrome after the spermatozoa had been deposited in the perivitelline space. Continuing efforts resulted in a report of the first human pregnancy after subzonal insemination [21], and this was followed by a report on the birth of a set of normal, healthy twins and a singleton after subzonal insemination (SUZI) for men with severe oligoasthenozoospermia [22].

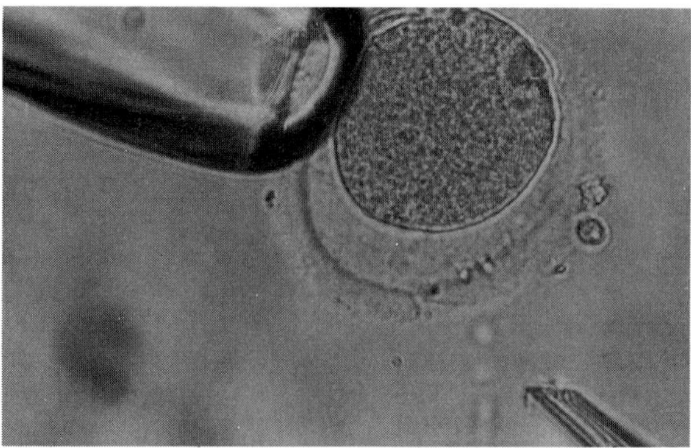

Fig. 7.2. Creating a break in the human zona pellucida. Original magnification × 200.

Preparation of Spermatozoa for Micromanipulation

For IVF, GIFT and micro-assisted fertilisation (MAF) procedures it is essential to separate the spermatozoa from seminal plasma to permit capacitation and the expression of the biochemical events on which fertilising ability depends. For MAF it is also necessary to use a technique which can concentrate spermatozoa into a small volume. Various methods have been used for the preparation of spermatozoa, particularly for assisted conception procedures not involving micromanipulation. The two main techniques are Ficoll [23] and Percoll [24]. The former, which has also been called "sperm entrapment" [23], involves the centrifugation of spermatozoa prior to the layering of a Ficoll solution on to the concentrated suspension of spermatozoa. Motile sperm swim into the Ficoll layer, and remain trapped there. The Ficoll is removed by centrifugation and fresh sperm suspension introduced into the microinjection microneedle.

The other technique, which is the method of choice of this author, is centrifugation through a discontinuous gradient of 90% and 45% Percoll. The seminal plasma is layered on top of the 45% Percoll and, after centrifugation, sperm cells, depleted of the cellular and debris contamination from seminal plasma, are recovered from the 90% Percoll at the base of the centrifuge tube. This sperm suspension is mixed with culture medium, re-centrifuged, the supernatant removed, and the pellet resuspended in a tiny volume of fresh culture medium. The volume of culture medium added to the sperm pellet will depend on the concentration of spermatozoa, but can be as little as 20 μl.

Concerns about the generation of reactive oxygen species (ROS, superoxide and hydroxyl radicals) [25] from "debris" in the pellet of centrifuged seminal plasma, including immotile sperm cells, and especially from men with severe seminal defects, has necessitated the preparation of spermatozoa by methods which separate out the latter from the seminal components. Any method which involves the separation of spermatozoa from the seminal plasma such as discontinuous Percoll or Nycodenz [26] is acceptable, as are migration methods using, for example, Sperm Select [27] or Tea-Jondet tubes [28]. However, the latter methods depend on the density and progressive activity of the available spermatozoa.

Micro-Assisted Fertilisation (MAF)

The indications for using MAF can be divided into four major categories.

1. "Normal" semen/spermatozoal characteristics where fertilisation in vitro has been repeatedly unsuccessful.
2. Immunologically derived infertility, such as antisperm autoantibodies, with failure to achieve fertilisation in vitro.
3. Defects in the quality of spermatozoa, but where the density was sufficient to try IVF but this had repeatedly failed.

4. Insufficient spermatozoa or severely defective spermatozoa, such as 100% immotility or spermatozoa extracted from the epididymis because of obstructive azoospermia.

Techniques which rely on creating a breach in the zona pellucida, which have been termed partial zona dissection (PZD), zona puncture (ZP), zona cutting (ZC), zona tearing (ZT), or zona drilling (ZD), rely on the inherent motility of the spermatozoa. Given that there is a failure in the spermatozoa either to bind or to penetrate the zona pellucida, but they still retain fertilisation potential, clinical results have shown that the size of the breach is extremely important. A large breach may cause a high incidence of polyspermy and care must be taken to reduce the concentration of spermatozoa during insemination. Therefore, the assessment of the particular technique for a specific indication is crucial if a high incidence of fertilisation and a low incidence of polyspermy is to be achieved. Cohen et al. [29] and Malter et al. [30] demonstrated that fertilisation can occur after PZD in patients who have repeatedly failed to achieve fertilisation in vitro.

After a series of animal studies, the first zona breaching technique to be applied to clinical studies was ZD [31]. Fertilisation was frequently achieved, but in many instances embryonic cleavage was abnormal. Many centres have attempted ZD, but no pregnancies have been reported to date. In addition to the acidic methods used for breaching the zona pellucida, enzymatic methods, using chymotrypsin or trypsin have been used. Both these methods have proved to be less effective with human oocytes than with the mouse. This approach has ceased as a viable option in humans.

SUZI, rather than creating a major breach in the zona pellucida, by-passes the latter depositing the spermatozoa in the vicinity of the oolemma, in the perivitelline space. Therefore, the number of spermatozoa injected could be controlled and this approach might reduce the incidence of polyspermy in high risk cases.

Results

Overall Data for SUZI: Patients and Oocytes

A total of 307 patients underwent oocyte recovery, 47.9% achieved fertilisation with SUZI, and 43.6% had embryo transfer. The 31 pregnancies established represented an overall incidence of pregnancy of 23.1% per transfer and 10.1% of all oocyte recoveries. Approximately 54% of patients had one embryo transferred compared with 28% and 17% having two and three embryos transferred respectively.

The number of oocytes available for SUZI was 1384, 21% of these fertilised, 92.4% cleaved and 77.3% were replaced; this represented 16.3% of the total number of oocytes undergoing SUZI. Of the fertilised oocytes, 9% were multipronucleate. The highest order of multipronucleate eggs observed was five per nucleus which arose in 1% of oocytes. The incidence of parthogenetic activation by SUZI was observed at 0.58%.

The mean number of spermatozoa injected for all oocytes was 5.13 ± 1.06 (SEM). The incidence of fertilisation increased as the number of sperm injected into the perivitelline space increased (Fig. 7.3). The majority of oocytes were injected with either three (23.2%) or four (26.4%) spermatozoa.

SUZI Versus IVF with Sibling Oocytes

To establish whether the SUZI technique per se is more effective than IVF it is necessary to use sibling oocytes with the spermatozoa from the same ejaculate. This was possible in 191 patients, and the data demonstrated that significantly more patients, 44%, achieved fertilisation with SUZI compared to those who achieved fertilisation with both SUZI and IVF (8.4%). Less than 2% of patients achieved fertilisation with IVF only.

In this group of 191 patients the incidence of fertilisation per egg with SUZI was significantly higher than with IVF, 26.8% compared with 4.4% respectively.

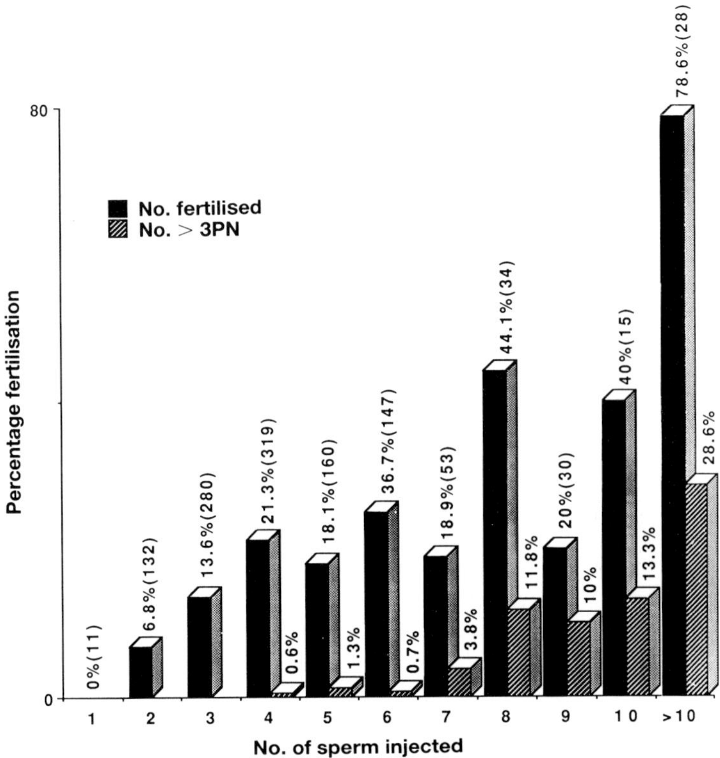

Fig. 7.3. Assessment of fertilisation versus number of sperm injected. The values in parentheses are the numbers of oocytes.

Assessment and Classification of Semen Parameters

Six parameters were used to assess the semen: volume (ml), total count (TC $\times 10^6$), total motility (TM, %), total count $\times 10^6$ ml^{-1} (TC ml^{-1}), total motile count $\times 10^6$ (TMC) and progression (P, on a scale 0–4).

In early studies the semen parameters according to four groups of patients: those patients in whom fertilisation did not occur (non-fertilised); where fertilisation occurred but no cleavage; the occurrence of fertilisation; and those patients in whom pregnancy occurred, it was found that only the total motile count and progression showed a significance difference [32]. There was a higher total motile count (TMC) in the group of patients that achieved fertilisation compared with those which failed to achieve fertilisation, and similarly for a rating of progression of spermatozoa for the same two groups. Although progression is highly subjective and observer dependent, this might suggest an advantage for considering computerised tracking of spermatozoa as a prognosticator of fertilisation after MAF procedures.

The previous data established that SUZI was beneficial in achieving fertilisation where IVF failed. However, one of the main aims of this work is to try to establish which patients will benefit from these procedures. In an attempt to do this the incidence of fertilisation was assessed according to two classifications of seminal parameters. The first, using the standard World Health Organisation (WHO) characteristics was difficult as these parameters do not consider the various degrees of severity of asthenozoospermia and oligozoospermia, nor the various combinations. For example, a classification based on asthenozoospermia and oligozoospermia only, resulted in 16 groups ranging from "normal" to a combination of very severe oligozoospermia/very severe asthenozoospermia. Such a large division presents difficulties for a meaningful analysis.

The second classification utilised total motile count only, which considers volume, density and percentage motility of spermatozoa; providing information on a total number of spermatozoa available. This classification utilised five groups according to TMC. Groups 1 to 5 had a TMC of <1, 1–5, 6–10, 11–20 and >20 $\times 10^6$, respectively. The incidence of fertilisation with SUZI, according to this classification, is shown in Figs. 7.4 and 7.5. The majority, 68%, of patients were in groups 1 and 2. The incidence of fertilisation per patient was approximately 40% (Fig. 7.4), and per oocyte approximately 21% (Fig. 7.5). There was a trend to decreasing fertilisation per patient, with increasing total motile count. In groups 1 and 2, significantly more patients achieved fertilisation with SUZI than with IVF. In group 1, none of the 72 patients had fertilisation with IVF. Evaluating the data on a per egg basis, all groups, except in group 4, had a significant increase in the incidence of fertilisation with SUZI compared with IVF.

This is in slight contrast to earlier published reports [32] that patients presenting with a TMC in group 1 had a significantly reduced chance of fertilisation, both on a per egg and per patient basis. However, since those early data, an increasing number of spermatozoa have been injected into the perivitelline space in this group of patients. This has resulted in an increase in the incidence of fertilisation without a significant increase in the incidence of polyspermy.

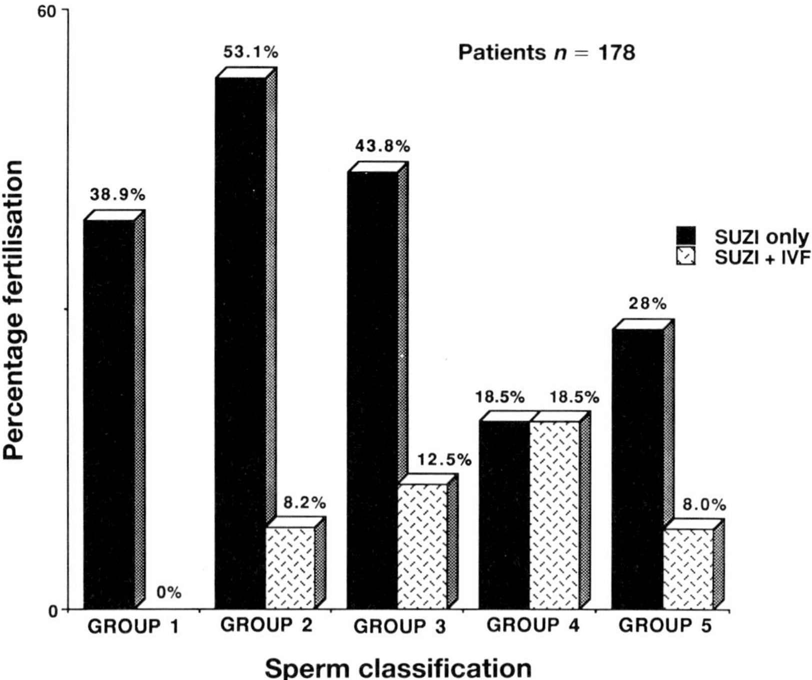

Fig. 7.4. SUZI versus IVF and sperm classification 2 (patients).

Sperm Morphology and SUZI

The strict morphological criteria for assessing spermatozoa is shown in Fig. 7.6. Assessing the data overall, there was a significant reduction in the incidence of fertilisation, multipronucleate oocytes and cleavage, and a significant increase in cytoplasmic fragments in patients with >95% abnormal forms (Fig. 7.7). Assessing the data compared with Classification 2 confirmed that patients with the severest morphology (>95% abnormal forms) tended to result in a lower incidence of fertilisation between groups and within each group. The incidence of pronuclear stage arrest and, possibly, the level of cytoplasmic fragments, appeared to be related more to the degree of abnormal forms than the actual TMC.

SUZI Versus PZD and IVF

Tables 7.1 and 7.2 present the published results comparing SUZI with PZD. There is great difficulty in assessing these data on the basis of matched controls and, therefore, in comparing the efficacy of the various procedures described. In all but one study it is clear that the number of patients described is the same as the number of cycles (to get a measure of success it is important to

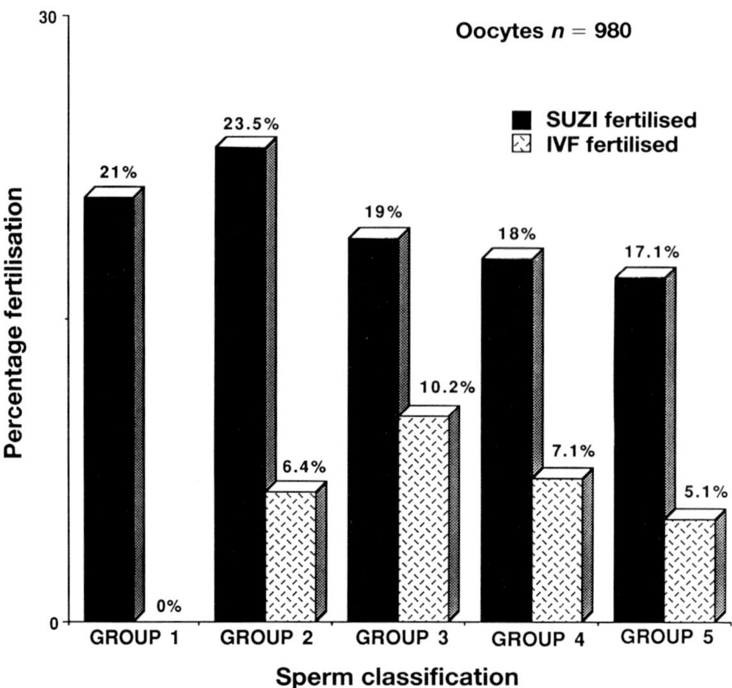

Fig. 7.5. SUZI versus IVF and sperm classification 2 (oocytes).

Table 7.1. Comparative fertilisation with SUZI and PZD: Patients

Reference	SUZI			PZD		
	Number of patients (? cycles)	Number with fertilisation E.T. (%)	Percentage pregnancy	Number of patients	Number with fertilisation E.T. (%)	Percentage pregnancy
Ng et al. [42]	131 (?)	58 (44.3)	6.9[a] 3.1[b]			
Palermo et al. [45][c]	44	34 (77.3)	20.6[a] 15.9[b]			
Payne et al. [46][c]				23	15 (65.2)	0
Cohen et al. [38]	47	33 (70.2)	15.2[a] 10.6[b]	57	38 (66.7)	10.5[a] 7.0[b]
Fishel et al. [32]	307	134 (43.6)	231. 10.1			
Fishel et al. [32][d]	131	56 (42.7)	?[e]	131	40 (30.5)	?[e]

[a]Pregnancy per transfer.
[b]Pregnancy per cycle (patient).
[c]Not clear cases of severe male infertility.
[d]Sibling oocytes used for each case study.
[e]Pregnancy resulted from mixed embryos so cannot be assessed.

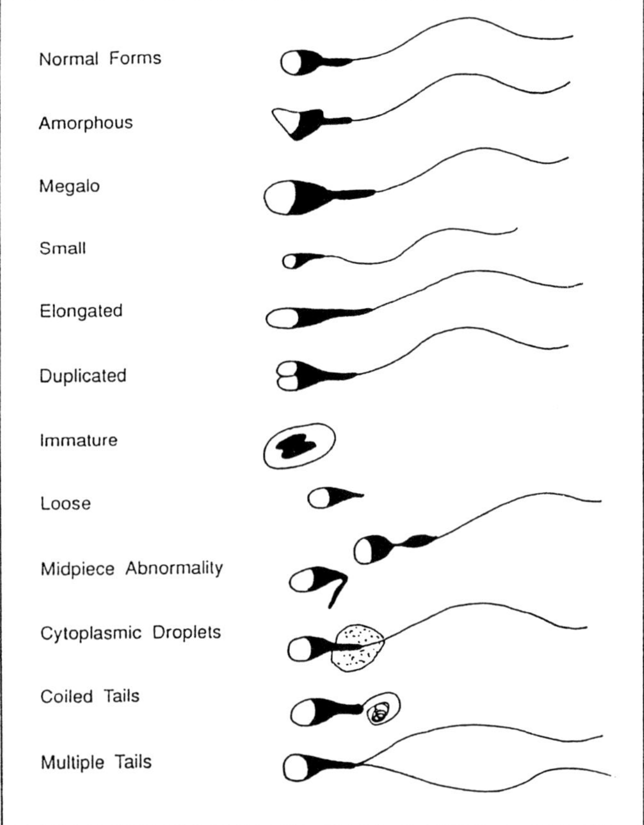

Fig. 7.6. Sperm morphology evaluation.

Table 7.2. Comparative fertilisation with SUZI and PZD: Oocytes

Reference	SUZI			PZD		
	Number of oocytes	Number fertilised (%)	>2PN	Number of oocytes	Number fertilised (%)	>2PN
Ng et al. [42]	771	128 (16.6)	1.9%			
Palermo et al. [45][a]	433	134 (30.9)	3%			
Payne et al. [46][a]				147	47 (32)	6.4%
Cohen et al. [38]	254	65 (25.6)	20%	267	97 (36.3)	24%
Fishel et al. [32]	1384	291 (21)	8.9%			
Fishel et al. [32][b]	319	101 (31.7)	10.9%	300	93 (31)	21.5%

[a]Not clear cases of severe male infertility.
[b]Sibling oocytes used for each case study.

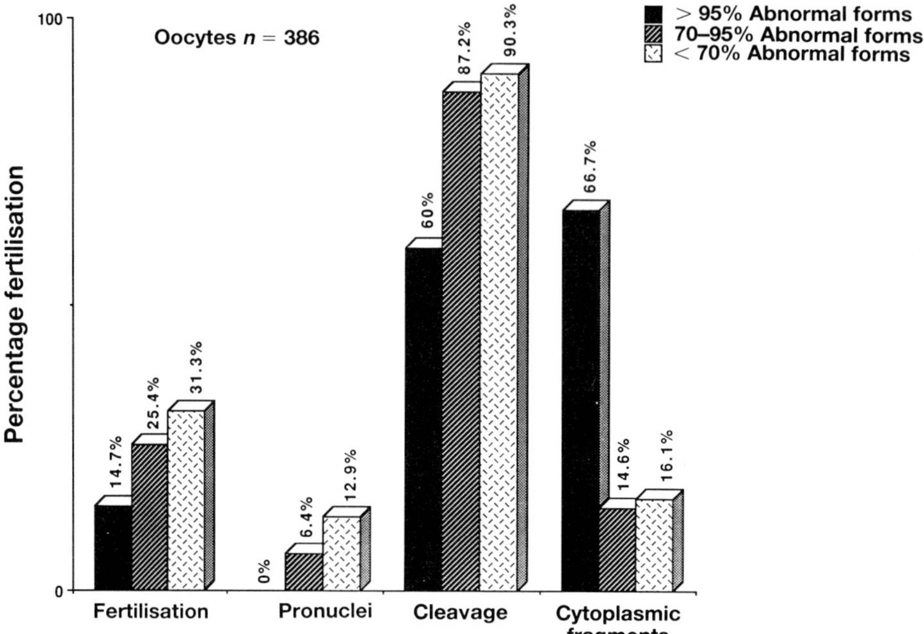

Fig. 7.7. SUZI and sperm morphology versus fertilisation, pronuclei, cleavage and cytoplasmic fragments.

know whether data are presented on a per patient basis or per cycle basis). In many of the control studies presented a majority of patients had mixed embryos transferred, i.e. embryos from PZD and/or SUZI and/or IVF were transferred to a single patient. In assessing fertilisation, it is not always clear that sibling oocytes have been used, as percentages of overall incidence of fertilisation are often presented. Furthermore, there is no clearly defined assessment of the indication for micro-assisted fertilisation, for example, the number of times the patients have failed IVF. Specific criteria on the seminology is often not presented, only broad ranges. For example, the comparative data presented by Cohen et al. [38] purports to compare SUZI and PZD, but with 47 patients having SUZI and 57 with PZD! It was not entirely clear whether all oocytes were directly compared on a sibling oocyte/same ejaculate basis.

In the data presented in Tables 7.1 and 7.2 the range of fertilisation and embryo transfer per patient was 43%–77% with SUZI compared to 31%–67% with PZD. The pregnancy rate with PZD varied from 0 (and the author has had this confirmed from a number of unpublished studies from various groups representing a total of more than 194 cases), to 11% per transfer. The incidence of fertilisation per egg for SUZI was 17%–32% compared to 31%–36% for PZD (Table 7.2). However, the incidence of polyspermy was 2%–20% with SUZI compared to 6%–24% with PZD. Overall, the incidence of polyspermy was higher with PZD than with SUZI.

A direct comparative study, by this author, used sibling oocytes from 131 patients to compare various MAF procedures. Each cohort of oocytes had a proportion subjected to either SUZI, PZD or IVF. These data were further compared according to the TMC. In the group 1 patients, 45.2% achieved fertilisation with SUZI only, compared with 4.8% having PZD. Four (9.5%) of patients achieved fertilisation with both SUZI and PZD. No patient achieved fertilisation with PZD. As the concentration of TMC increased, the incidence of fertilisation per patient with PZD increased, but overall significantly more patients achieved fertilisation with SUZI only, compared to PZD only. In the group of patients with antisperm autoantibodies, five of the eight achieved fertilisation, two with SUZI only, one with PZD only, two with both PZD and SUZI, and none with IVF.

Assessing fertilisation on a per egg basis (Fig. 7.8), significantly more oocytes from patients in TMC groups 1 and 2 fertilised with SUZI compared with PZD. The overall incidence of fertilisation was generally higher with SUZI than PZD in TMC groups 1 to 5, and, including a group of patients with antisperm autoantibodies, significantly more oocytes were fertilised with SUZI than with PZD. The incidence of fertilisation with the group expressing antisperm autoantibodies was, however, higher with PZD (68%) than with SUZI (44%). There was a general trend indicating a lower incidence of fertilisation with SUZI but a higher incidence with PZD as the TMC increased.

As the TMC increased, the incidence of polysperm increased with SUZI and PZD, although this was more marked in the latter group, 10.9% versus 24.6%, respectively. Conversely, there seemed to be a higher incidence of moderate to severe cytoplasmic fragments in embryos from SUZI when sperm from TMC groups 1 and 2 were used. Overall there was a higher incidence of cytoplasmic

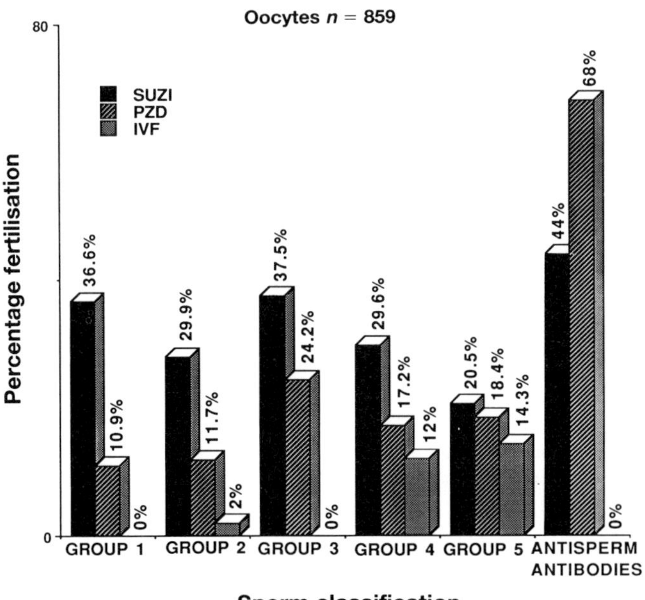

Fig. 7.8. Fertilisation and cleavage in sibling oocytes with SUZI, PZD and IVF.

fragments in embryos resulting from SUZI (31%) compared to PZD (19.6%), although this was not statistically significant. However, this may reflect the high incidence of abnormal sperm morphology often seen in TMC groups 1 and 2. The incidence of cleavage in zygotes resulting from SUZI or PZD was similar, 86.1% and 89.2% overall, respectively; but this is generally lower than that observed after routine IVF.

When these comparative data between MAF procedures was reassessed according to whether the patients had a previous single or multiple failure of IVF a different pattern emerged. Combining TMC groups 2 to 5 (group 1 patients have too few sperm for IVF) resulted in 123 patients, 34.1% of whom had failed IVF on only one occasion and 65.8% of whom had multiple failures of IVF. In all cases the incidence of fertilisation was less with both MAF techniques in the latter group of patients. This was more evident with oocytes undergoing PZD. Fertilisation after PZD was greatly reduced in patients having more than one failure after IVF.

Discussion and Conclusions

The data presented in this series, and that published by other workers, have demonstrated that by micromanipulating sperm and oocyte it is possible to achieve fertilisation in patients who have previously failed to attain conception in vitro by IVF, or who have been repeatedly rejected for IVF based on severe defects in their seminal parameters. The procedures per se have not resulted in a significant increase in abnormal embryos [33], nor in the birth of abnormal babies [33]. However, the efficiency of the technique and the type of patient who may benefit has yet to be established. The efficacy of any micro-assisted fertilisation procedure can only be assessed where (a) the use of in vitro insemination with sibling oocytes is used for comparison; and (b) a clear classification of the sperm is used.

The relevance of sperm morphology to micro-assisted fertilisation procedures, especially to SUZI, needs to be clarified. A mechanistic relationship, that is, sperm function, to particular morphological abnormalities remains unclear. Sperm morphology has been shown to be of value as a prognosticator of sperm function, with the zona free hamster egg assay [34,35] and human oocyte fertilising ability [36,37]. However, the particular relevance of sperm morphology in relation to micro-assisted fertilisation is unclear. Recently Cohen et al. [38] demonstrated that in cases of extreme teratozoospermia embryos derived from SUZI implant at a significantly higher rate than those resulting from PZD. Another study has demonstrated that one-third of spermatozoa located in the perivitelline space after SUZI from men with severe oligospermia, were abnormal [39]; with defects of the nucleus, acrosome, midpiece and axoneme having been observed in the spermatozoa penetrating the ooplasm after SUZI [40].

From the current study, in which the incidence of fertilisation was clearly affected by the severity of the TMC and morphology of spermatozoa, it would appear unnecessary to determine the optimum numbers of spermatozoa used for SUZI in each particular group. For example, with TMC Group 1 and >95%

abnormal forms, the group with the lowest incidence of fertilisation, would it be advantageous to use more spermatozoa for SUZI? The incidence of polyspermy was lowest in this group, but the incidence of cleavage arrest was highest. It may be that in this group the incidence of viable embryos produced by SUZI is lower. Whether this is a result of achieving fertilisation with abnormal sperms (and whether there is an association with abnormal phenotype and genotype is unclear), or whether it is related to the supernumerary sperm, active, senescent or dead, within the perivitelline space, remains to be established. Obtaining a balance between the incidence of polyspermy, which was also a function of TMC, and the effective maximum number of spermatozoa that should be deposited in the perivitelline space, above which there may be a negative correlation with fertilisation, must also be determined.

Although it is assumed that the only block to polyspermy, in the human, existed at the level of zona pellucida, it is clear from studies with MAF procedures that there is some degree of block to polyspermy at the level of the oolemma. However, the main block to polyspermy still resides at the level of the zona pellucida. The existence of slow and fast blocks to polyspermy and the respective roles of the oolemma and zona pellucida in human oocytes has been discussed extensively elsewhere [32,39,41].

Summary

Techniques of sperm preparation have improved as an understanding of the pathophysiology of sperm function and fertilisation has increased. Careful and precise preparation of the spermatozoa is essential with severe cases of male infertility, and for the use of spermatozoa in the micro-assisted fertilisation procedures.

Zona breaching and subzonal insemination procedures are effective means of achieving fertilisation in certain cases of sterility. The method using enzymes or acid to "drill" a hole in the zona pellucida is insuccessful to date in humans.

If patients had failed IVF on more than one occasion PZD would probably be less successful than SUZI. It is evident that sperm morphology is relevant and in conjunction with an acceptable sperm classification such as TMC, may be predictive of fertilisation by either SUZI or PZD.

Current data suggest that PZD should the first approach with antisperm autoantibodies, but care should be given to the insemination concentration to prevent a high incidence of polyspermy. However, for patients with very low TMC SUZI would be the preferred approach.

Even during this early stage in the use of MAF procedures, normal pregnancies and births have been achieved, but there is an underlying concern because of the incidence of abnormal embryos. Whether this increases after SUZI, or PZD, due to by-passing selection criteria for spermatozoa at the level of the zona pellucida is unclear. There is some suggestion, especially in cases with very severely reduced TMC and after using SUZI, that there may be an increase in the demise of resulting embryos. Despite nature having its own peri-implantation screening, this concern cannot be dismissed. However, the numbers of normal, ongoing pregnancies and births to date, have alleviated many of the initial concerns.

Little mention has been made of the direct injection of spermatozoa into the cytoplasm for clinical use. There has been reported fertilisation [42] and the occasional mention of a pregnancy with this procedure [43]. However, no data reflecting the possible efficacy or success of this procedure are available for analysis. Clearly, there is a higher risk of damage to the oocyte than with PZD or SUZI, and, as it would only be used in the severest of cases, this work should be reserved until more research data are available. A recent study [44] reported a high incidence of chromosomal abnormalities in human sperm haploid complements after direct ooplasmic injection into hamster oocytes. The frequency of structural chromosomal abnormalities was up to 10 times higher than that from the same donors used for fertilisation of zona-free hamster oocytes.

Evaluation of the fertility of a given sperm sample, and optimising the efficacy of MAF procedures is essential. An important step forward would be the development of acceptable and reproducible sperm function test(s). The current assessment of sperm function, especially in relation to micro-assisted fertilisation, which is based on TMC and morphology, is a crude assessment. More sophisticated sperm function tests, which might include computerised sperm tracking, an assessment of the incidence of acrosome-reacted sperm in a given population, and perhaps sperm binding and fusion assays, must be obtained.

Acknowledgement. The author is very grateful to Louise Gore for the preparation of this manuscript, and in such a remarkably short time. The author also wishes to acknowledge his co-workers from the previously published work which, combined, is presented here. My gratitude to: Judy Timson, and Drs Severino Antinori, Franco Lisi, Leonardo Rinaldi and Caterina Versaci.

References

1. Hull MGR. Infertility: nature and extent of the problem. In: Bock G, O'Connor M, eds. Human embryo research: yes or no. London: Ciba Foundation/Tavistock Publications, 1986; 24–38.
2. Schmidt HD. Minute structure of the hepatic lobules particulary with reference to the relationship between capillary blood vessels and the hepatic cells. Am J Med Sci 1895; 37:13–40.
3. Nicholas JS, Hall BV. Experiment on developing rats. II. The development of isolated blastomeres and fused eggs. J Exp Zool 1942; 90:441–58.
4. Duryee WR. Micro-dissection studies on human ovarian eggs. Trans NY Acad Sci 1954; 17:103–8.
5. Lin TP. Micro-injection of mouse eggs. Science 1966; 151:33–7.
6. Lin TP. Micro-pipetting cytoplasm from the mouse eggs. Nature 1967; 216:162–3.
7. Hiramoto Y. Micro-injection of live spermatozoa into sea urchin eggs. Exp Cell Res 1962; 27:416–26.
8. Uehara T, Yanagimachi R. Micro-surgical injection of spermatozoa into hamster eggs with subsequent transformation of sperm nuclei into male pronuclei. Biol Reprod 1976; 15:467–70.
9. Ohsumi K, Katagiri C, Yanagimachi R. Development of pronuclei from human spermatozoa injected microsurgically into (*Xenopus*) eggs. J Exp Zool 1986; 237:319–25.
10. Iritani A, Utsumi K, Miyake M, Hosoi Y, Saeki K. In-vitro fertilization by a routine method and micro-manipulation. Ann NY Acad Sci 1988; 541:583–90.
11. Keefer CL. Fertilization by sperm injection in the rabbit. Gamete Res 1989; 22:59–69.

12. Younis AI, Keefer CL, Brackett BG. Fertilization of bovine oocytes by sperm injection. Theriogenology 1989; 31:276 (Abstract).
13. Metka M, Haromy T, Huber J, Schurz B. Artificial insemination using a micro-manipulator. Fertilitat 1985; 1:41–7.
14. Laws-King A, Trounson A, Sathananthan H, Kola I. Fertilization of human oocytes by micro-injection of a single spermatozoan under the zona pellucida. Fertil Steril 1987; 48:637–42.
15. Lanzendorf E, Maloney MK, Veeck LL, Slusser J, Hodgen GD, Rosenwaks Z. A preclinical evaluation of human spermatozoa into human oocytes. Fertil Steril 1988; 49:435–42.
16. Gordon JW, Grunfeld J, Garrisi GJ, Talansky BE, Richards C, Laufer N. Fertilization of human oocytes by sperm from infertile males after zona pellucida drilling. Fertil Steril 1988; 50:68–73.
17. Cohen J, Malter M, Fehilly C, Wright G, Elsner C, Kort H, Massey J. Implantation of embryos after partial opening of oocyte zona pellucida to facilitate sperm penetration. Lancet 1988; ii:162.
18. Malter H, Talansky B, Gordon J, Cohen J. Monospermy and polyspermy after partial zona dissection of reinseminated human oocytes. Gamete Res 1989; 23:377–86.
19. Ng S-C, Sathananthan AH, Edirisinghe WR, Kum Chue JH, Wong PC, Ratnam SS, Sarla G. Fertilization of a human egg with sperm from a patient with immotile cilia syndrome: case report. Adv Fertil Steril 1987; 4:71–6.
20. Aitken RJ, Ross A, Lees MM. Analysis of sperm function in Kartageners syndrome. Fertil Steril 1983; 40:696–8.
21. Ng S-C, Bongso TA, Ratnam SS et al. Pregnancy after transfer of multiple sperm under the zona. Lancet 1988; ii:790.
22. Fishel SB, Antinori S, Jackson P, Johnson J, Lisi F, Chiariello F, Versaci C. Twin birth after subzonal insemination. Lancet 1990; ii:722.
23. Bongso A, Ng S-C, Mok H, Lim MN, Teo HL, Wong PC, Ratnam SS. Improved sperm concentration, motility and fertilization rates following Ficoll treatment of sperm in an human in-vitro fertilization programme. Fertil Steril 1989; 51:850–4.
24. Aitken RJ, Clarkson JS. Significance of reactive oxygen species and antioxidants in defining the efficacy of sperm preparation techniques. J Androl 1988; 9:367–76.
25. Aitken RJ, Clarkson JS, Fishel SB. Generation of reactive oxygen species, lipid peroxidation, and human sperm function. Biol Reprod 1989; 40:183–97.
26. Gellert-Mortimer ST, Clarke GN, Baker HWG, Hyne RV, Johnston WIH. Evaluation of Nycodenz and Percoll density gradiants for the selection of motile human spermatozoa. Fertil Steril 1988; 49:335–41.
27. Wikland M, Wik O, Steen Y, Qvist K, Soderlund B, Janson PO. A self-migration method for preparation of sperm for in-vitro fertilization. Human Reprod 1987; 2:191–5.
28. Lucena E, Lucena C, Gomez M et al. Recovery of motile sperm using the migration-sedimentation technique in an in-vitro fertilization-embryo transfer programme. Hum Reprod 1989; 4:163–5.
29. Cohen J, Malter H, Wright G, Kort H, Massey J, Mitchell D. Partial zona dissection of human oocytes when failure of zona pellucida penetration is anticipated. Hum Reprod 1989; 4:435–42.
30. Malter HE, Talansky BE, Gordon J, Cohen J. Monospermy and polyspermy after partial zona dissection of re-inseminated human oocytes. Gamete Res 1989; 23:377–86.
31. Gordon JW, Grunfeld L, Garrisi GJ, Talansky BE, Richards C, Laufer N. Fertilization of human oocytes by sperm from infertile males after zona pellucida drilling. Fertil Steril 1988; 50:68–73.
32. Fishel SB, Timson J, Lisi F, Rinaldi L. Evaluation of 225 patients undergoing subzonal insemination for the procurement of fertilisation in-vitro. Fertil Steril 1992; 57:840–9.
33. Kola I, Lacham O, Jansen RPS, Turner M, Trounson A. Chromosomal analysis of human oocytes fertilised by micro-injection of spermatozoa into the perivitelline space. Hum Reprod 1990; 5:575–7.
34. Shalgi R, Dor J, Rudak E, Lusky A, Goldman B, Mashiach S, Nabel L. Penetration of sperm from teratospermic men into zona-free hamster eggs. Int J Androl 1985; 8:285–94.
35. Marsh SK, Bolton VN, Braude PR. The effect of morphology on the ability of human spermatozoa to penetrate zona free hamster oocytes. Hum Reprod 1987; 2:499–503.
36. Kruger TF, Acosta AA, Simmons KF, Swanson RJ, Matta JF, Oehninger S. Predicted value of abnormal sperm morphology in IVF. Fertil Steril 1988; 49:112–17.
37. Menkveld R, Stander FSH, Kotze TJvW, Kruger TF, Van Zyl JA. The evaluation of morphological characteristics and human spermatozoa according to stricter criteria. Hum Reprod 1990; 5:586–92.

38. Cohen J, Alikani M, Malter HE, Adler A, Talansky BE, Rosenwaks Z. Partial zona dissection or subzonal sperm insertion: microsurgical fertilisation alternatives based on evaluation of sperm and embryo morphology. Fertil Steril 1991; 56:696–706.
39. Sathananthan AH, Ng S-C, Trounson A, Bongso A, Laws-King A, Ratnam SS. Human micro-insemination by injection of single or multiple sperm: ultrastructure. Hum Reprod 1989; 4:574–83.
40. Sathananthan AH, Trounson A, Wood C. Atlas of fine structure of human sperm penetration, eggs and embryos cultured in-vitro. Philadelphia: Praeger Scientific, 1986.
41. Fishel SB, Jackson P, Antinori S, Johnson J, Grossi S, Versaci C. Subzonal insemination for the alleviation of infertility. Fertil Steril 1990; 54:828–35.
42. Ng S-C, Bongso A, Ratnam SS. Micro-injection of human oocytes: a technique for severe oligoasthenoteratozoospermia. Fertil Steril 1991; 56:1117–23.
43. Palermo G, Joris H, Devroey P, van Steirteghem AC. Pregnancies after intracytoplasmic injection of single spermatozoon into an oocyte. Lancet 1992; 340:17–18.
44. Martin RH, Ko E, Rademaker A. Human sperm chromosome compliments after micro-injection of hamster eggs. J Reprod Fertil 1988; 84:179–83.
45. Palermo G, Joris H, Devroey P, van Steirteghem AC. Induction of acrosome reaction in human spermatozoa used for subzonal insemination. Hum Reprod 1992; 7:248–54.
46. Payne D, McLaughlin KJ, Depypers HT, Kirby CA, Warnes GM, Matthews CD. Experience with zona drilling and zona cutting to improve fertilisation rates of human oocytes in vitro. Hum Reprod 1991; 6:423–31.

Chapter 8

Epididymal Surgery

S. J. Silber and R. H. Asch

Male factor infertility has undergone revolutionary changes in the last decade. The human male, and the gorilla, have the poorest sperm production of any animal. Whereas most animals produce 20 to 25 million sperm per gram of testicular tissue per day, the human produces only 4 million [1,2]. Even the very fertile human male has terrible sperm when compared to most other animals. The large number of abnormal forms, debris, and non-motile sperm found in human semen is not seen in other animals. This chapter will be divided into three sections: (a) the treatment of male factor infertility with IVF or GIFT; (b) microsurgery for obstructive azoospermia; and (c) IVF with epididymal sperm for irreversible obstructive azoospermia and congenital absence of the vas.

IVF and GIFT for "Male Factor" Infertility

Great excitement has been generated about the use of IVF in couples with very low or poor quality sperm counts. Fertilisation rates in most IVF centres are clearly poorer in couples with oligospermia, but fertilisation and pregnancy have been regularly obtained in severely oligospermic couples using IVF, GIFT and ZIFT [3]. Theoretically IVF or GIFT allows the fewer number of available sperm a greater opportunity for direct contact with the ovum [4,5]. In fact, if the partner's eggs fertilise in vitro with the husband's few sperm, the pregnancy rate is no different from what it is in couples with normal semen analysis.

Dr Silber unfortunately had to cancel his attendance at the Study Group for personal reasons. He was therefore unable to present his chapter for discussion by the group.

In cases of congenital absence of the vas, a collaborative study has been made to aspirate sperm from the blocked epididymis, fertilise the partner's oocytes in vitro, and place the resultant embryos into the fallopian tubes (Zygote Intrafallopian Transfer, ZIFT) [6–8]. Resultant pregnancies led to the conclusion that even sperm with severely reduced motility can often fertilise the oocyte if other obstacles are eliminated.

Rodriguez-Rigau et al. [9] assessed the semen of a large number of couples who did or did not achieve a pregnancy with GIFT (Table 8.1). They found a correlation of the pregnancy rate with standard semen parameters such as count and motility. A similar finding was also reported by Kruger et al [10]. Morphology of spermatozoa was examined very carefully in partners of patients undergoing IVF. Any spermatozoa with a slight defect such as neck droplets, bent necks, abnormal heads, etc., were considered to be abnormal. Remarkably, provided that at least 4% of spermatozoa demonstrated perfectly normal morphology, fertilisation in vitro was achieved. Once again, this suggests that a critical visual exmination of the husband's spermatozoa is a predictor of the likelihood of fertilisation.

The great success in treating many severely oligospermic couples with IVF, GIFT or ZIFT, has required a whole new definition of "male factor" so as to distinguish those who readily fertilise their partner's oocytes despite oligospermia from those who do not. With IVF in "male factor" cases, the pre-wash motile sperm count in the semen is not a heavily significant determinant of fertilisation, or pregnancy. The pregnancy rate with IVF or GIFT is, of course, lower in men with low sperm counts than in men with high sperm counts, but what really determines the pregnancy rate is the total motile sperm count after washing. In our experience, when the total motile sperm count recovered from a Percoll or mini-Percoll preparation is greater than 1.5×10^6 million motile sperm, the fertilisation rate is not significantly different from patients with higher numbers of recoverable sperm. When the motile sperm count after washing is less than this, the fertilisation rate and pregnancy rate are greatly reduced.

Table 8.1. Relationship of motile sperm count and linearity of sperm movement to GIFT pregnancy rate

Total motile count per ejaculate ($\times 10^6$)	Motility index	Clinical pregnancy (%)
25		7.7
25–100		15.2
100		24.1
	0.8	12.2
	0.8	27.6

From Rodriguez-Rigau et al. [9].

Processing "Male Factor" Sperm for IVF, GIFT and ZIFT

When the total number of motile spermatozoa in the ejaculate is less than 5×10^6 (classified as severe oligoasthenozoospermia) [11], the conventional methods of sperm preparation such as swim-up [12], wash and resuspension [13], sedimentation [14] and Percoll [15] are not very effective in allowing the recovery of a sample that is clean with a sufficient number of normal motile spermatozoa. This, in turn, leads to very poor IVF results [16–18]. We use a modified mini-Percoll technique [19,20], which consists of a reduced volume, discontinuous, Percoll gradient.

Semen samples are diluted 1:2 with culture medium and centrifuged at 200 ***g*** for 10 min. After centrifugation, pellets are suspended in 0.3 ml of medium and layered on a discontinuous Percoll gradient consisting of 0.3 ml each of 50%, 70% and 95% isotonic Percoll (mini-Percoll). An isotonic solution of Percoll is obtained by mixing nine parts of Percoll (Pharmacia, Sweden) with one part Ham's F-10 (10 ×) (Gibco, RI), and adding 2.1 g/l of sodium bicarbonate. To obtain the 95%, 70% and 50% layers, this isotonic solution is diluted with human tubal fluid (HTF) and HEPES. The discontinuous gradient is established by carefully pipetting 0.3 ml of 95%, 70% and finally 50% isotonic Percoll into a 15-ml centrifuge tube, centrifuging at 300 ***g*** for 30–45 min. Following centrifugation of the gradient, the 95% Percoll layer is removed, washed twice and resuspended in 1 ml of HTF and 10% human cord serum (HCS) and incubated until the time of insemination.

Several advantages can be associated with the use of a mini-Percoll gradient. First, the reduced volume of each Percoll layer allows better migration of spermatozoa; second, the volume of 0.3 ml per layer still retains the "cleaning" function, as in other reported techniques using Percoll, filtering out all the cells and debris that are usually present in severe oligoasthenozoospermic samples and represent one of the limiting factors for successful fertilisation; and third, the use of mini-Percoll allows recovery of a high proportion of the normal and motile spermatozoa present in the sample.

Other aspects of IVF are beyond the scope of this chapter except for epididymal sperm aspiration, which will be discussed at the end.

Microsurgery for Obstructive Azoospermia

An understanding of how to obtain high success rates with vasectomy reversal will eventually lead to more successful vasoepididymostomy results in postinflammatory obstruction, and finally to success with sperm aspiration and in vitro fertilisation for congenital absence of the vas.

Vasectomy Reversal

Vasectomy is the most popular method of birth control in the world today [21]. For many years the pregnancy rate after surgical reanastomosis of the vas had been very low, and a variety of explanations has been offered for this [22–24]. With the advent of microsurgical techniques pregnancy rates improved considerably, suggesting that purely micromechanical factors were associated with the low success rates [25–27]. Yet there were still many cases of technically perfect vasovasostomies followed by complete azoospermia or severe oligoasthenospermia. It was then found that the pressure increase after vasectomy had led to secondary epididymal obstruction which was the cause of failure of otherwise successful vasovasostomy. The greater the duration of time since vasectomy the greater the chance of either "blowouts" or inspissation in the epididymis with failure to achieve fertility. Thus, vasoepididymostomy was required in many cases of vasectomy reversal in order to obtain a high success rate.

Theories for the consistently poor results with vasectomy reversal had included development of sperm antibodies, damage to the deferential nerve, and testicular damage [28–35]. However, any major correlation between sperm antibodies and subsequent fertility after vasovasostomy has been questioned [36,37]. It has been established that the deleterious effect of pressure increase subsequent to vasectomy was not in the testis, but on epididymal dilatation, perforation and sperm inspissation and blowouts in the epididymis, causing secondary epididymal obstruction, which is the major problem in readily returning fertility to vasectomised men [25,27]. Despite the finding of no sperm in the vas fluid at the time of vasovasostomy the testicular biopsy of such patients had always appeared normal [38,39]. This deleterious effect of pressure increase is always on the epididymis, not on the testis, in humans. In fact, the secondary epididymal obstruction caused by vasectomy leads us to recommend that the testicular end of the vas not be sealed at the time of vasectomy, so as to lessen the pressure build-up, and possibly increase the ease of reversibility later (notwithstanding the potentially damaging immunological consequences) [40–42].

What is the fertility rate in the favourable group of patients undergoing vasovasostomy who have suffered no secondary epididymal damage (as shown by sperm being present in the vas fluid at the time of vasovasostomy)? Ten years ago a group of such patients was studied [37]. A total of 326 men who had been previously vasectomised underwent vasovasostomy and received extensive long-term follow-up. In 44 of those men, no sperm was found in the vas fluid. All such patients have been found to be azoospermic after vasovasostomy and required vasoepididymostomy later.

The vasovasostomy involved a meticulous, two-layer microsurgical technique performed by the same surgeon with accurate mucosa-to-mucosa approximation [25]. Almost all the patients had proven prior fertility as evidenced by previous fatherhood. All patients were followed for nine or ten years.

The overall, long-term pregnancy rate is summarised in Table 8.2. None of the azoospermic patients got their partners pregnant. If azoospermic patients are excluded, 88.4% of patients with sperm patency postoperatively eventually impregnated their partners. This compares to Vessey's expected pregnancy rate of 96% for previously fertile couples discontinuing contraception (1978).

The frequency distribution of semen parameters postoperatively in men who did and did not get their partners pregnant is summarised in Tables 8.3 and 8.4. There was remarkably little difference in pregnancy rate among men with low or high sperm counts. The pregnancy rate was somewhat lower with

Table 8.2. Overall long-term pregnancy rates in patients undergoing vasovasostomy: 10-year follow-up (sperm seen in vas fluid)

	Combined 1975 and 1976–77 series	Original 1975 series
Total patients	282 (100%)	42 (100%)
Total pregnant	228 (81%)	32 (76%)
Azoospermic	24 (9%)	5 (12%)

Table 8.3. Pregnancy rate according to distribution of motile sperm count in men with sperm patency following vasovasostomy (10-year follow-up)

Total motile sperm count (per ejaculate) (10^{-6})	Total patients	Pregnant	Not pregnant
>0–10	32 (12%)	25 (78%)	7
10–20	31 (12%)	27 (87%)	4
20–40	32 (12%)	30 (93%)	2
40–80	79 (31%)	68 (86%)	11
80	84 (33%)	78 (92%)	6
Totals	258 (100%)	228 (88%)	30

Table 8.4. Pregnancy rate according to percentage sperm motility in men with sperm patency following vasovasostomy (10-year follow-up)

Motility	Total patients	Pregnant	Not pregnant
>0–20	24	18 (75%)	6
20–40	70	66 (94%)	4
40–60	82	71 (86%)	11
60–80	62	55 (88%)	7
80	20	18 (90%)	2
Total	258 (100%)	228 (88%)	30 (100%

motility of less than 20%. Above 20% motility, the pregnancy rate was not seriously affected by low semen parameters. These postoperative semen parameters in patent cases were not very different from previously reported prevasectomy semen parameters [43].

As shown in Table 8.5, a left-sided varicocele was clinically apparent in 42 of the 282 patients (14.8%). Varicoceles were not operated on, and yet the pregnancy rate was not significantly different in patients with varicocele as opposed to patients without varicocele. Table 8.6 summarises the relationship

Table 8.5. Lack of effect of varicocele (not operated on) on pregnancy rate following vasovasostomy

	No. patients	Patients with varicocele	Patients without varicocele
Pregnant	228 (80.9%)	33 (78.5%)	195 (81.2%)
Not pregnant	54 (19.1%)	9 (21.4%)	45 (18.8%)
Totals	282 (100.0%)	42 (14.8%)	240 (85.2%)

Table 8.6. Relationship of serum sperm antibody titres to pregnancy rate after vasovasostomy

	Total studied	Immobilising titre (Isojima)		Agglutinating titre (Kibrick)	
		>2	>10	>0	>20
Husband not azoospermic:					
Wife pregnant	75	29 (39%)	18 (24%)	42 (56%)	30 (40%)
Wife not pregnant	11	4 (36%)	2 (16%)	6 (54%)	6 (54%)
Husband azoospermic	12	5 (42%)	3 (25%)	7 (58%)	5 (42%)
Entire group studied	98	38 (39%)	23 (24%)	56 (57%)	41 (42%)

of preoperative serum antisperm antibody titres to the pregnancy rate after vasovasostomy. As with varicocele, the presence of high immobilising or agglutinating titres in serum had no influence on the pregnancy rate.

Reason for High Pregnancy Rate in Patients with No Secondary Epididymal Blockage

The high pregnancy rate in this group of patients requires some explanation. Many reasons have been suggested for the failure to achieve fertility after reversal of vasectomy, including autoimmune changes and damage to the testis. Our study suggested that the eventual pregnancy rate in patients who have patency accurately re-established without epididymal damage is not significantly less than a normal population of couples. Vessey demonstrated that among couples with proven prior fertility, 96.5% conceive within four years of discontinuing contraception (1978). In our couples with patent results after vasovasostomy who had no evidence of epididymal pressure damage, 88% conceived with long-term follow-up. Patients with secondary epididymal blockage require a completely different approach.

It has been shown that the success rate of vasovasostomy decreases with the duration of time since vasectomy [25]. This decrease is directly related to the absence of sperm in the vas fluid at the time of vasovasostomy, and this is caused by the interruption of epididymal patency by pressure-induced sperm extravasation, and inspissation [38]. The incidence of this pressure mediated interruption of epididymal patency is reduced dramatically by the presence of a sperm granuloma at the vasectomy site which serves as a release valve to prevent the pressure increase that would otherwise occur proximal to the vasectomy site [27,40,41]. When there are no spermatozoa in the vas fluid, vasoepididymostomy proximal to the site of epididymal blockage is required [44,45].

It thus appears that the fertility rate and pregnancy rate are quite high in patients with no epididymal blockage who undergo technically "successful" vasovasostomy.

Vasoepididymostomy

When vasectomy has produced secondary epididymal blockage, or in cases of postinflammatory obstructive azoospermia, very precise microsurgical tubule-to-tubule vasoepididymal anastomosis is required. But as important as precise microsurgical technique is a practical understanding of epididymal physiology [26,27,44–46].

In every animal that has been studied, spermatozoa from the caput epididymis are capable of only weak circular motion at most, and are not able to fertilise [47]. In previous studies, spermatozoa from the corpus epididymis could occasionally fertilise, but the pregnancy rate was still low. Spermatozoa were simply aspirated from specific regions of the epididymis, and then promptly inseminated [48–50]. In some studies where the epididymis was ligated to determine if time alone could allow spermatozoal maturation, the obstructed environment was so pathological that no firm conclusions could be reached. Thus the outlook seemed theoretically poor for vasoepididymostomy.

In 1969, Orgebin-Crist [47] pointed out that it was not known with certainty from any of these animal studies whether the factors governing the maturation process of spermatozoa are intrinsic to the spermatozoa themselves and just require time, or whether spermatozoa must transit through most of the epididymis in order to mature. It was entirely possible that ageing alone might mature the spermatozoa, and that spermatozoa might not need to pass through all of the epididymis in order to develop the capacity to fertilise. Yet because of the animal studies, and poor results in humans using non-microsurgical techniques, it has always been assumed that epididymal blockage carries a poor prognosis [51–54].

However, as far back as 1931, Young's experiments in guinea pigs with ligation at various levels of the epididymis indicated "that the time consumed by spermatozoa in passing through the epididymis is necessary for a completion of their development; that the changes undergone during this period represent a continuation of changes which start while the spermatozoa are still attached to the germinal epithelium, and are not conditioned by some specific epididymal secretion" [55]. In fact, he observed the same inversion of regions of sperm motility and non-motility in the obstructed epididymis that we have noted in clinical obstructive azoospermia. The more distal regions have the poorest motility and the more proximal regions have the best motility. Young concluded that in an obstructed epididymis the more distal sperm are senescent, whereas the more proximal sperm have had time to mature despite having not traversed the epididymis. Our clinical experience with specific tubule vasoepididymostomy supports Young's original thesis [56].

All vasoepididymostomies are performed with the "specific tubule" technique described, which involves either an end-to-end or an end-to-side anastomosis of the inner lumen of the vas to the epididymal tubule, mucosa-to-mucosa in a leakproof fashion [25,44,46]. Because of the high rate of technical failure with older surgical methods, reliable data on the fertility of spermatozoa from the epididymis in the past has been difficult to obtain.

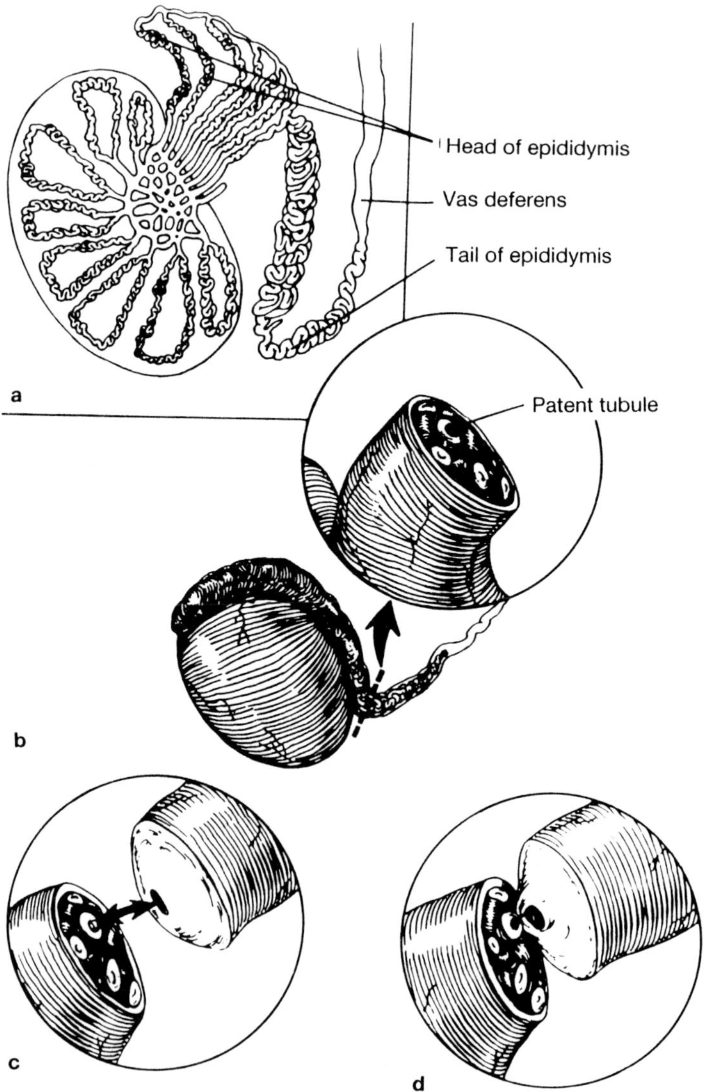

Fig. 8.1. Specific tubule-to-end anastomosis of the vas lumen to the epididymis proximal to site of obstruction.

The anastomosis of the vas to the epididymis is performed at the transition point from no spermatozoa to the point where there is an abundant amount of spermatozoa in the fluid coming from the epididymis tubule (Figs. 8.1 and 8.2a and b). Usually five or six 10-0 nylon interrupted sutures complete the leakproof end-to-end anastomosis, and then the outer muscular layer of the vas is separately sutured to the outer epididymis tunic with 9-0 nylon interrupted sutures.

Of the cases of epididymal anastomosis 72% have resulted in eventual pregnancy [56]. The younger the partner, the higher was the pregnancy rate. The pregnancy rate was not related to the numerical sperm count but was related to the motility. The fact that pregnancy occurred in cases patent to the caput indicates that transit beyond the head of the epididymis is not an absolute requirement for spermatozoa to attain fertilising capacity.

Recent clinical cases have demonstrated that it is possible in some circumstances for spermatozoa which have never transited any length of epididymis to fertilise the human egg. In two cases of anastomosis between the vasa efferentia and the vas deferens, the postoperative ejaculate contained normally motile sperm, and the partners became pregnant [57]. In addition, pregnancy from aspiration of epididymal sperm combined with in vitro fertilisation and ZIFT in cases of unrepairable obstruction gives further evidence that transit through the epididymis is not a mandatory requirement for fertilisation [6,7].

Newer studies of epididymal sperm transport in the human indicate that the human epididymis is not a storage area, and spermatozoa transit the entire human epididymis very quickly, in two days, not eleven days as was previously thought [58]. Thus, it is possible that in the human, the epididymis may not be as essential to spermatozoal development and fertility as it appears to be in most animals.

Congenital Absence of the Vas Deferens and Sperm Aspiration with IVF

Congenital absence of the vas deferens accounts for 11%–50% of cases of obstructive azoospermia, and has until now been considered untreatable [59]. This is a large and frustrating group of patients who have been shown on testicular biopsies to have normal spermatogenesis, and are theoretically making sperm quite capable of fertilising an egg. Yet treatment up until now has been very poor [60].

Dr Ricardo Asch and I have collaborated to develop a treatment protocol involving microsurgical aspiration of sperm from the proximal region of the epididymis, combined with IVF and ZIFT, which now offers very good results in this previously frustrating group of couples [7,8,25,57].

Induction of Follicular Development and Oocyte Retrieval

The female partners of men with azoospermia caused by congenital absence of the vas undergo induction of multiple follicular development with the following

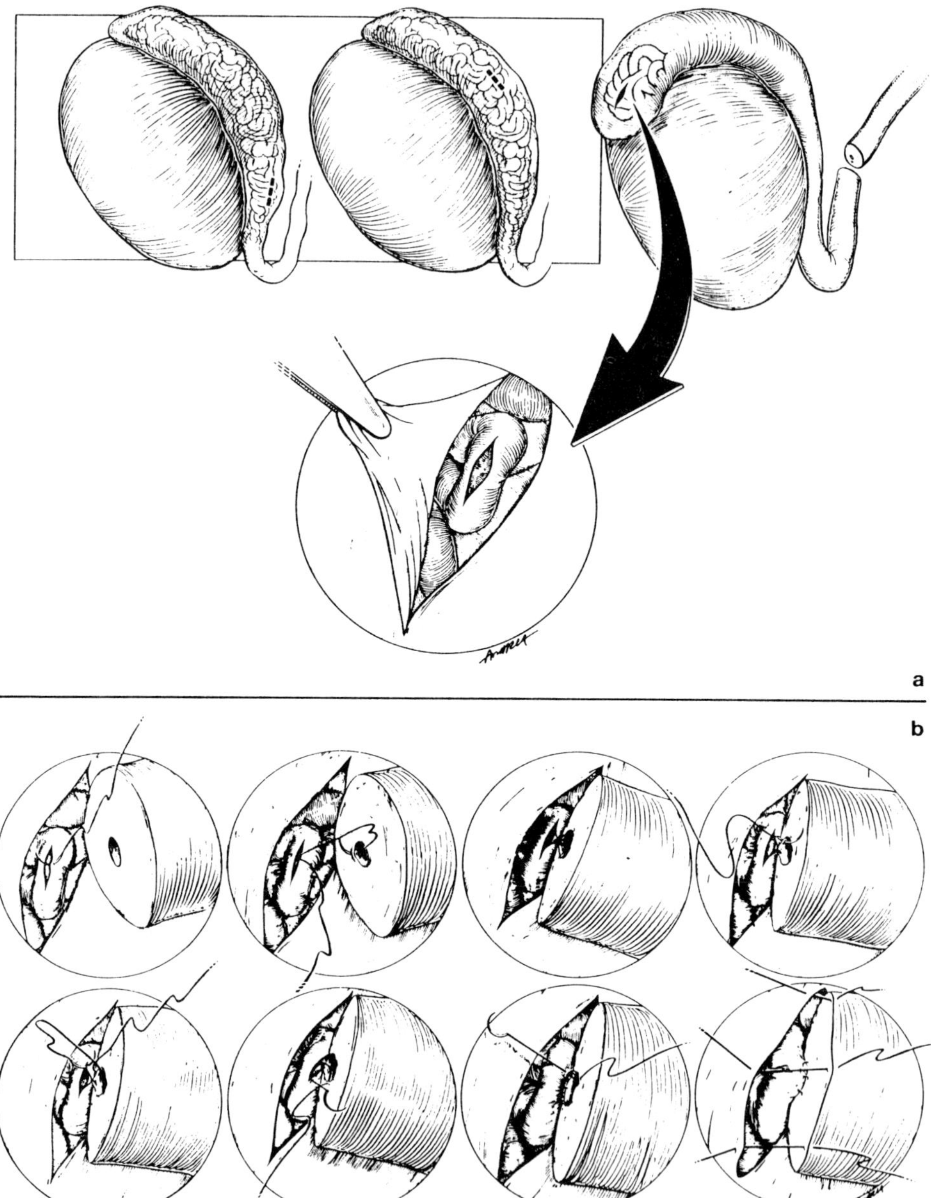

Fig. 8.2. **a** Small openings are made in the epididymal tunic beginning distally and moving proximally. After a longitudinal slit is made in the epididymal tubule with the microscissors, the distal-most level at which motile sperm are found is used for the anastomosis. **b** The end-to-side specific tubule anastomosis of the vas lumen to the epididymal tubule requires first a posterior row of three 10-0 nylon interrupted sutures followed by an anterior row of three 10-0 nylon interrupted sutures. The muscularis of the vas is then sutured to the outer epididymal tunic with 9-0 nylon interrupted sutures.

protocol: leuprolide acetate (Lupron, TAP Pharmaceuticals, North Chicago, IL) 1 mg subcutaneously daily until the day of follicular aspiration (Fig. 8.3). Patients then receive human follicle stimulating hormone (FSH) (Metrodin, Serono Laboratories, Inc., Randolph, MA) and human menopausal gonadotropins (hMG) (Pergonal, Serono) 150 IU intramuscularly (IM) daily from day 2 of the menstrual cycle until many follicles of 2.0 cm were noted on ultrasound. Then human chorionic gonadotropin (hCG) (Profasi, Serono, Randolph, MA) 10 000 IU is administered IM.

At 36 h after hCG administration, the patients undergo follicular aspiration using a transvaginal probe (GE H4222 TV) adapted to an ultrasound system (GE RT 3000 General Electric Company, Milwaukee, WI) with a needle set for ovum aspiration and connected to a Craft Suction Unit (Rocket USA, Branford, CT) (no. 33-100) at a maximum vacuum pressure of 120 mmHg. The follicular fluid is given immediately to the embryology laboratory adjacent to the operating room.

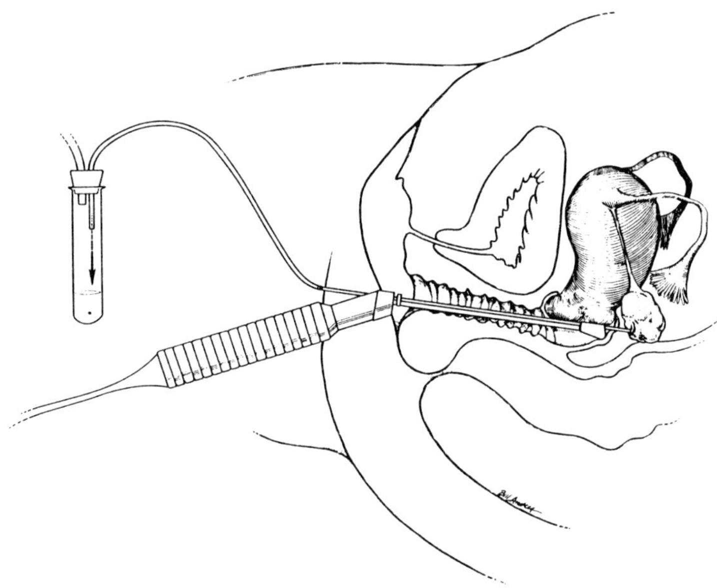

Fig. 8.3. Illustration of placement of the ultrasound probe for transvaginal needle aspiration of eggs.

Epididymal Sperm Aspiration, Washing Methodology, and IVF

At the same time the husband undergoes scrotal exploration in order to aspirate sufficient numbers of motile spermatozoa for use for IVF of the aspirated eggs, with subsequent transfer into the partner's fallopian tube.

The surgical technique (Fig. 8.4) in the male is as follows: scrotal contents are extruded through a small incision, the tunica vaginalis is opened, and the epididymis is exposed. Under 10-40× magnification with an operating microscope, a tiny incision is made with microscissors into the epididymal tunic

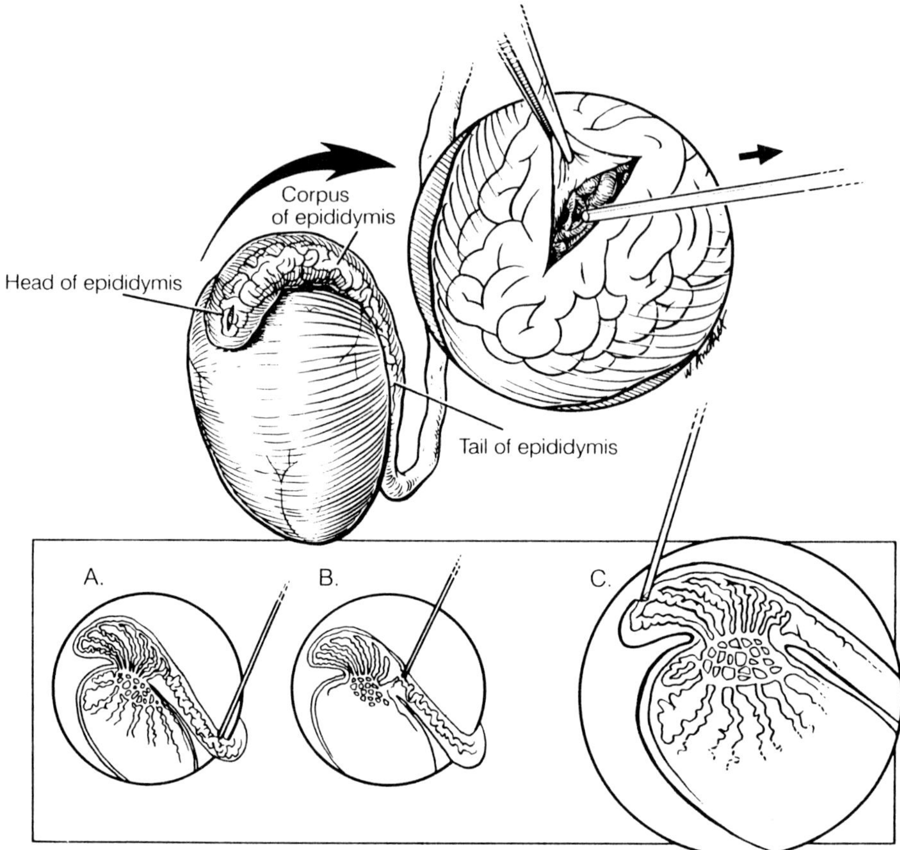

Fig. 8.4. Technique for epididymal sperm aspiration which begins in the distal corpus region of the epididymis, and moves proximally until motile sperm are recovered. In most cases, motility is observed only in the most proximal region of the epididymis.

to expose the tubules in the distal-most portion of the congenitally blind-ending epididymis. Sperm are aspirated with a no. 22 Medicut on a tuberculin syringe directly from the opening in the epididymal tubule. Great care is taken not to contaminate the specimen with blood, and careful haemostasis is achieved with microbipolar forceps. The epididymal fluid is immediately diluted in Hepes-buffered medium, and a tiny portion examined for motility and quality of progression. If there is no motility or poor motility, another aspiration is made 0.5 cm more proximally. We thus obtain sperm from successively more and more proximal regions until progressive motility is found. In all cases, motile sperm were not obtained until the proximal-most portion of the caput epididymis or even the vasa efferentia was reached. This is the inverse of what might have been anticipated (Fig. 8.5).

In the laboratory the epididymal sperm are concentrated into a volume of 0.3 ml, layered on a discontinuous mini-Percoll gradient, and centrifuged for 30 min. The entire 95% fraction is then washed twice and inseminated with all

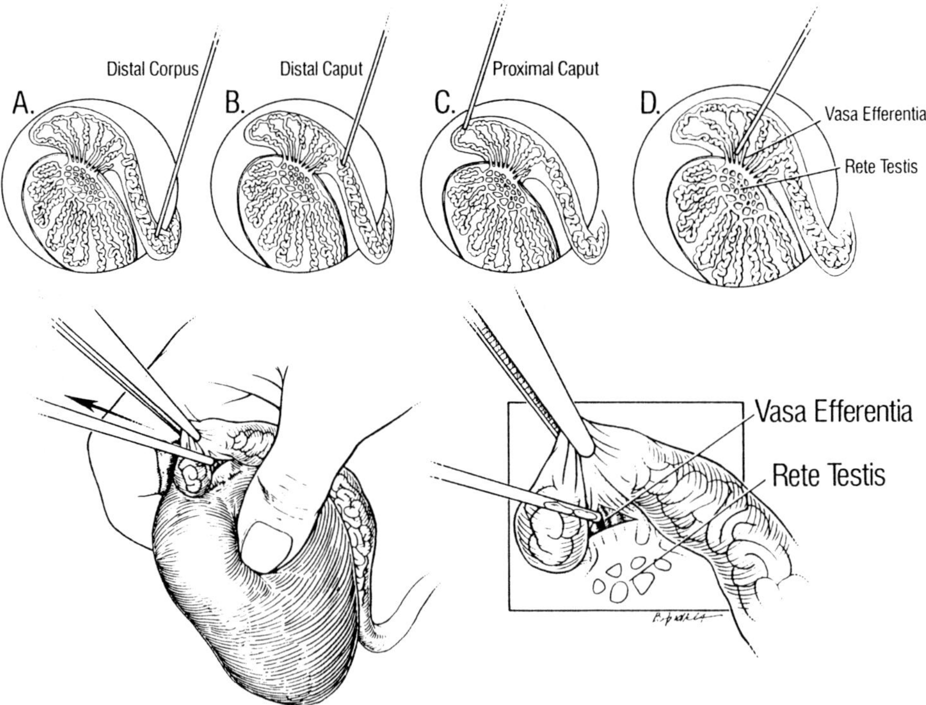

Fig. 8.5. The most motile sperm are found very proximally, usually in the vasa efferentia or rete testis.

of the eggs in a Falcon mini-test tube with 1 ml of HTF culture medium and incubated at 37°C with 5% CO_2 in air [47] (Fig. 8.6).

Two days after insemination, embryos are transferred to the fallopian tubes, via minilaparotomy using a technique similar to that used for gamete intrafallopian transfer (GIFT), via a Tomcat catheter (Monoject, St Louis, MO) 2.5 cm inside the fimbrial ostium (Fig. 8.7). The patients are discharged the next day and undergo fairly painless postoperative recovery. The wives receive progesterone in oil, 50 mg IM/day beginning with the day of embryo transfer.

Results

At present, of 115 cases, there have been 24 pregnancies, with six miscarriages, i.e. a pregnancy rate of 21% and a live-baby rate of 16% (Table 8.7).

Pregnancies which have occurred readily after vasoepididymostomy to the caput epididymis (and even in some cases to the vasa efferentia) suggest that immature sperm which have not had a chance to transit the epididymis might mature on their own during storage in the vas deferens [6,8]. If this theory were true, it might explain why we have been able to achieve success by aspirating more proximally, not being limited (because of theoretical

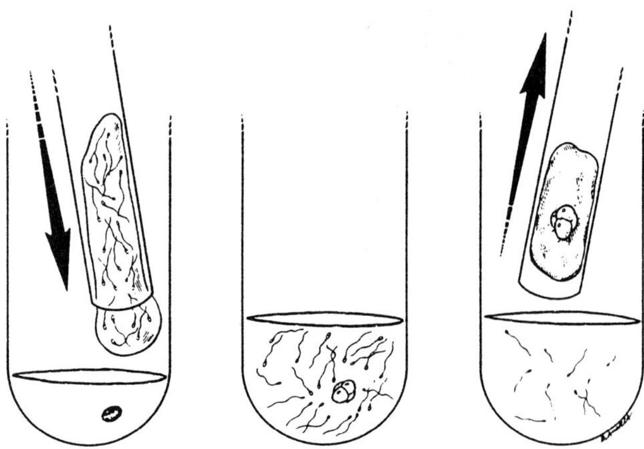

Fig. 8.6. Sperm being placed in test tube with egg, resulting in 4-cell embryo two days later, and retrieval of 4-cell embryo for transfer back to patient.

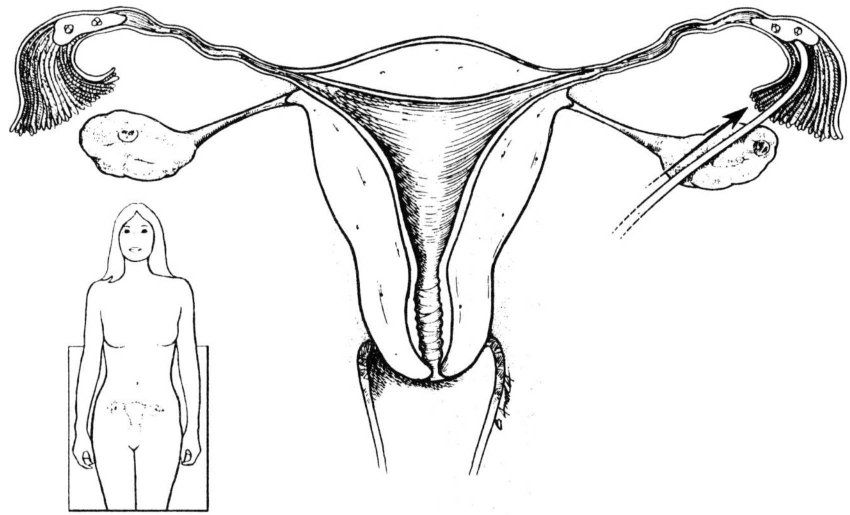

Fig. 8.7. In the ZIFT procedure eggs would have already been fertilised and growing for the previous 2 days in the laboratory and placed in the fallopian tube in a similar fashion to the GIFT procedure.

considerations) to distal regions of the epididymis where the sperm are generally senescent and non-motile in the chronically obstructed state.

Other factors in the success of this technique which may be equally important are: (1) obtaining large numbers of oocytes in order to increase the odds of fertilisation; (2) obtaining sperm which are clean and free of erythrocytes; (3) incubation of sperm outside the milieu of the obstructed epididymis; and (4) transfer of the embryos into the fallopian tube (ZIFT) rather than into the uterus.

Table 8.7. First 100 cases of in vitro fertilisation for congenital absence of vas: pregnancy rates

Series no.	No. of sperm aspiration cycles	No. pregnant (term pregnancy)	Pregnancy rate per cycle
1	32	10 (7)	31% (22%)
2	16	2 (1)	12% (6%)
3	21	5 (4)	24% (19%)
4	13	0 (0)	0 (0)
5	18	5 (4[a])	28% (22%)
Totals	100	22 (16)	22% (16%)

[a] Ongoing pregnancies not yet delivered.

Problems that remain in the way of achieving a higher pregnancy rate are that even the most proximal epididymal sperm with the best motility, have poor fertilisation rates in 70% of cases, and no fertilisation in 40% of cases. It is our suspicion that these obstructed sperm are senescent, but slightly less senescent than the dead, distal epididymal sperm. If that hypothesis is true, better fertilisation rates might be achieved by first aspirating the epididymis of all old sperm, and then at a later interval, re-aspirating to get sperm for IVF.

Although improvements are needed, for the moment it is safe to conclude that:

1. Sperm from the proximal-most caput epididymis are capable of fertilising the human egg in vitro.
2. Passage of time after emergence from the testicle may be adequate for sperm maturation in some cases without the absolute need for transit through the rest of the epididymis.
3. There is now a non-experimental, valuable approach for achieving pregnancy in couples with a previously dismal condition, congenital absence of the vas deferens.

References

1. Chowdhury AK, Marshall G. Irregular pattern of spermatogenesis in the baboon (*Papicanubis*) and its possible mechanism. In: Steinberger A, Steinberger E, eds. Testicular development, structure, and function. New York: Raven Press, 1980; 129–138.
2. Austin CR, Short RV. The evolution of reproduction. London: Cambridge University Press, 1976.
3. Cohen J, Edwards R, Fehilly C et al. In vitro fertilization: a treatment for male infertility. Fertil Steril 1985; 43:422–32.
4. Matson PL, Blackledge DG, Richardson PA, Turner SR, Yovich JM, Yovich JL. The role of gamete intrafallopian transfer (GIFT) in the treatment of oligospermic infertility. Fertil Steril 1987; 48:608–12.
5. McDowell JS, Veeck LL, Jones HW Jr. Analysis of human spermatozoa before and after processing for in vitro fertilization. J In Vitro Fertil Embryo Transfer 1985; 2:23–6.

6. Silber SJ, Ord T, Borrero C, Balmaceda J, Asch R. New treatment for infertility due to congenital absence of the vas deferens. Lancet 1987; ii:850–1.
7. Silber SJ, Asch R, Balmaceda J, Borrero C, Ord T. Pregnancy with sperm aspiration from the proximal head of the epididymis: a new treatment for congenital absence of the vas deferens. Fertil Steril 1988; 50:525–8.
8. Silber SJ, Ord T, Balmaceda J, Patrizio P, Asch RH. Congenital absence of the vas deferens: the fertilizing capacity of human epididymal sperm. N Engl J Med 1990; 323:1788–92.
9. Rodriguez-Rigau LJ, Ayala C, Grunert GM et al. Relationship between the results of sperm analysis and GIFT. J Androl 1989; 10:139–44.
10. Kruger TF, Acosta AA, Simmons KF et al. A new method of evaluating sperm morphology with predictive value for human in vitro fertilization. Urology 1987; 30:248–51.
11. Yovich JL, Blackledge DG, Richardson PA, Matson PL, Turner SR, Draper R. Pregnancies following pronuclear stage tubal transfer. Fertil Steril 1987; 48:851–7.
12. Wong PC, Balmaceda JP, Blanco JD, Gibbs S, Asch RH. Sperm washing and swim-up technique using antibiotics remove microbes from human semen. Fertil Steril 1986; 45:97–100.
13. Schlaff WD. New ways to prepare semen for IUI. Contemp Obstet Gynacol 1987; 4:79–86.
14. Purdy JM. Methods for fertilization and embryo culture in vitro. In: Edwards RG, Purdy JM, Human conception in vitro. London: Academic Press, 1982; 135–48.
15. Berger T, Marrs RP, Moyer DL. Comparison of techniques for selection of motile spermatozoa. Fertil Steril 1985; 43:268–73.
16. Yates CA, de Kretser DM. Male factor infertility and in vitro fertilization. J In Vitro Fertil Embryo Transfer 1987; 4:141–7.
17. Gellert-Mortimer ST, Clarke GN, Baker HW, Hyne RV, Johnson WI. Evaluation of Nycodenz and Percoll density gradients for the selection of motile human spermatozoa. Fertil Steril 1988; 49:335–41.
18. Guerin JF, Mathieu C, Lornage J, Pinatel MC, Boulieu D. Improvement of survival and fertilizing capacity of human spermatozoa in an IVF programme by selection on discontinuous Percoll gradients. Hum Reprod 1989; 4:798–804.
19. Ord T. A different approach to sperm preparation in severe male factor for assisted conception techniques. In: Capitanio GL, Asch RH, Croce S, DeCecco L, eds. GIFT: from basics to clinics. New York: Serono Symposia Publications, Raven Press, 1989; Vol. 63, p. 229.
20. Ord T, Patrizio P, Marello E, Balmaceda JP, Asch RH. Mini-Percoll: a new method of semen preparation for IVF in severe male factor infertility. Hum Reprod 1990; 5:987–9.
21. Liskin L, Pile JM, Quillin WF. Vasectomy – safe and simple. Popul Rep 1983; 4:63–100.
22. O'Connor VJ. Anastomosis of the vas deferens after purposeful division for sterility. JAMA 1948; 136:162.
23. Phadke GM, Phadke AG. Experiences in the reanastomosis of the vas deferens. J Urol 1967; 98:888–90.
24. Middleton RG, Henderson D. Vas deferens reanastomosis without splints and without magnification. J Urol 1978; 119:763–4.
25. Silber SJ. Microscopic vasectomy reversal. Fertil Steril 1977; 28:1191–202.
26. Silber SJ. Vasectomy and its microsurgical reversal. Urol Clin North Am 1978; 5:573–84.
27. Silber SJ. Vasectomy and vasectomy reversal. Fertil Steril 1978; 29:125–40.
28. Sullivan NJ, Howe GE. Correlation of circulating antisperm antibodies to functional success of vasovasostomy. J Urol 1977; 117:189–91.
29. Fowler JE Jr, Mariano M. Immunoglobin in seminal fluid of fertile, infertile, vasectomy and vasectomy reversal patients. J Urol 1983; 129:869–72.
30. Middleton RG, Urry RL. Vasovasostomy and semen quality. J Urol 1980; 123:518.
31. Ansbacher R. Sperm agglutinating and sperm immobilizing antibody in vasectomized men. Fertil Steril 1971; 22:629–32.
32. Linnet L, Hgort T. Sperm agglutinins in seminal plasma and serum after vasectomy: correlation between immunological and clinical findings. Clin Exp Immunol 1977; 30:413–20.
33. Pabst R, Martin O, Lippert H. Is the low fertility rate after vasovasostomy caused by nerve resection during vasectomy? Fertil Steril 1979; 31:316–20.
34. Brickel D, Bolduan J, Farah R. The effect of vasectomy–vasovasostomy on normal physiologic function of the vas deferens. Fertil Steril 1982; 37:807–10.
35. Jarow JP, Budin RE, Dym M et al. Quantitative pathologic changes in the human testis after vasectomy: a controlled study. N Engl J Med 1985; 313:1252–6.
36. Thomas AJ et al. Microsurgical vasovasostomy: immunologic consequences in subsequent fertility. Fertil Steril 1981; 35:447–50.
37. Silber SJ. Pregnancy after vasovasostomy for vasectomy reversal: a study of factors affecting

long-term return of fertility in 282 patients followed for 10 years. Hum Reprod 1989; 4:318–22.
38. Silber SJ. Epididymal extravasation following vasectomy as a cause for failure of vasectomy reversal. Fertil Steril 1979; 31:309–15.
39. Silber SJ, Rodriguez-Rigau LJ. Quantitative analysis of testicle biopsy: determination of partial obstruction and prediction of sperm count after surgery for obstruction. Fertil Steril 1981; 36:480–5.
40. Silber SJ. Sperm granuloma and reversibility of vasectomy. Lancet 1977, ii:588–9.
41. Shapiro EI, Silber SJ. Open-ended vasectomy, sperm granuloma, and post vasectomy orchialgia. Fertil Steril 1979; 32:546–50.
42. Alexander NJ, Schmidt SS. Incidence of antisperm antibody levels in granulomas in men. Fertil Steril 1977; 28:655–7.
43. Zuckerman Z, Rodriguez-Rigau LJ, Smith KD et al. Frequency distribution of sperm counts in fertile and infertile males. Fertil Steril 1977; 28:1310–13.
44. Silber SJ. Microscopic vasoepididymostomy: specific microanastomosis to the epididymal tube. Fertil Steril 1978; 30:565–71.
45. Silber SJ. Microsurgery for vasectomy reversal and vasoepididymostomy. Urology 1984; 23:504–24.
46. Silber SJ. Diagnosis and treatment of obstructive azoospermia. In: Santen RJ, Swerdloff RS, eds, Male reproductive dysfunction. New York: Marcel Dekker, 1986; 479–517.
47. Orgebin-Crist MC. Studies of the function of the epididymis. Biol Reprod 1969; 1:155–75.
48. Glover TD. Some aspects of function in the epididymis. Experimental occlusion of the epididymis in the rabbit. Int J Fertil 1969; 14:216–21.
49. Gaddum P, Glover TD. Some reactions of rabbit spermatozoa to ligation of the epididymis. J Reprod Fertil 1965; 9:119–30.
50. Paufler SK, Foote RH. Morphology, motility and fertility of spermatozoa recovered from different areas of ligated rabbit epididymides. J Reprod Fertil 1968; 17:125–37.
51. Hanley HC. The surgery of male subfertility. Ann R Coll Surg 1955; 17:159.
52. Hotchkiss RS. Surgical treatment of infertility in the male. In: Campbell NF, Harrison HS, eds, Urology, 3rd edn, Philadelphia: W. B. Saunders, 1970; 671.
53. Amelar RD, Dubin L. Commentary on epididymal vasostomy, vasovasostomy and testicular biopsy. In: Current operative urology, New York: Harper and Row, 1975; 1181–5.
54. Schoysman R, Drouart JM. Progres recents dans la chirurgie de a sterilite masculine et feminine. Acta Chir Belg 1972; 71:261–80.
55. Young MC. A study of the function of the epididymis, III. Functional changes undergone by spermatozoa during their passage through the epididymis and vas deferens in the guinea pig. J Exp Biol 1931; 8:151–62.
56. Silber SJ. Results of microsurgical vasoepididymostomy: role of epididymis in sperm maturation. Hum Reprod 1989; 4:298–303.
57. Silber SJ. Pregnancy caused by sperm from vasa efferentia. Fertil Steril 1988; 49:373–5.
58. Johnson L, Varner DD. Effect of daily spermatozoan production but not age on transit time of spermatozoa through the human epididymis. Biol Reprod 1988; 39:812–17.
59. El-Itreby AA, Girgis SM. Congenital absence of the vas deferens in male sterility. Int J Fertil 1961; 6:409–16.
60. Temple-Smith PD, Southwick GJ, Yates CA et al. Human pregnancy by in vitro fertilization (IVF) using sperm aspirated from the epididymis. J In Vitro Fertil Embryo Transfer 1985; 2:119–22.

Discussion

Aitken: I would like to know the mechanism by which cooling has an effect on sperm–oocyte fusion. If we freeze and store cells, they are much more fusogenic than the fresh cell. What I have always imagined to be the mechanism is that when a cell is cooled and it gets down below its phase transition temperature, big areas of lipid crystallise out and push the intercalating proteins to one side, and when the cell is thawed out or warmed up, those intercalating proteins do not flow back into the lipid-rich areas of membrane, they stay as just areas of lipid.

Watson: Not all data confirm that. Some studies suggest there is a phased separation, which is reversed when the cells have been warmed. So that may be part of the story. Our thawing rates are much more rapid than our cooling rates, and it may well be that there is not the time during thawing for that redistribution to occur, so that we may well get cells at body temperature which do have these large areas. It may also be that the process of fusion takes a very short time.

Aitken: I am sure that is the case. How is it possible to differentiate in those cooled/warm cells between increased permeability to calcium and a decreased ability to pump out calcium? I had always imagined that the sperm cell was relatively permeable to calcium. What the intact cell does is to continually bail out the cation.

Watson: I think that is right. As the cell cools below 15°C or so, the pumps shut off and the cell then accumulates calcium simply because it is stopping the process of bailing out.

Abdalla: In our clinic, patients who have artificial insemination with frozen donor sperm have a low pregnancy rate, compared to those where GIFT is carried out using frozen sperm.

Watson: The GIFT procedure provides the optimal position for fertilisation. If the sperm are closer to that fertilisation event in terms of the membrane maturational changes, they may be able to achieve that fertilisation before they run out of steam. An acrosome reacted sperm has a relatively short lifespan. We have also referred to the importance of the acrosome reaction taking place close to or attached to the zone surface.

Kerin: We are told that the frozen–thawed sperm have a short time in which to fertilise the egg. Is there evidence from clinical studies that intrauterine insemination for these reasons using frozen sperm might be better than pericervical insemination?

Watson: I can only answer in terms of my reading, the literature is conflicting.

Asch: What is Dr Watson's impression, based on the physiology of how cryoinsemination works, of freezing severely oligoasthenospermic specimens? Is there any technique that would allow concentration of many specimens from an individual.

Watson: I do not have any specific techniques for dealing with that situation, but I do believe if we can come to grips with the causes of cryoinjury in normal sperm to get a higher percentage of cells that survive the freeze–thaw process, then that will have its repercussions for a situation where only a few cells are present. But the fertility, as I understand it, of oligospermic individuals, is lower in terms of function and therefore the chances of achieving reasonable fertility by that procedure is reduced.

Fishel: It strikes me from the data just presented on behalf of Dr Silber that the pregnancy rate is probably influenced by the number of embyros replaced. But the incidence of implantation or pregnancy, given the low fertilisation rate, may also be a function of the type of sperm – as we also see in microinsemination. Are there data to suggest a higher miscarriage rate or an earlier miscarriage rate among these patients?

Asch: We have very few miscarriages in these patients. I do not know why. Perhaps we take particularly good care of them. The take home lesson for me is that if we take sperm from the same place, in patients with the same condition, we can get four totally different populations of sperm, from sperm that are severely asthenospermic all the way to sperm that are almost like ejaculated sperm, those that fertilise 66% of the time.

Winston: Surely the problem could be solved by letting us know the implantation rate per single embryo transferred. That should be a useful statistic for us to have and we can then evaluate this. It is a statistic in general by which we should be trying to evaluate IVF more and more, because that would certainly answer that problem.

Dr Fishel reported on 307 patients. How many cycles? Of the 31 clinical pregnancies, how many of those have progressed to something looking like a pregnancy which will establish successfully.

Fishel: I did not have time to show all the data. The 307 is all cycles and it also represents patients. Most of the microinsemination data that are being produced at the moment are cycles and patients. However, the miscarriage rate is definitely high.

Aitken: Professor Asch mentioned that they stimulated the sperm; he mentioned pentoxyfilline and I think they also use 2-deoxyadenosine. Do they do that globally or do they have some sort of selection criteria by which to apply these stimulants?

Asch: We do it in all cases, but we do not do it in all sperm.

Aitken: Has Dr Fishel tried any stimulatory drugs with these sperm?

Fishel: Yes, I have. It is a little preliminary and I do not have too much to show at the moment. Actually I am disappointed with pentoxyfilline. My finding at the moment is that there is some increase in the incidence of fertilisation with IVF, but once we start to move to microinsemination techniques, they will either fertilise or they will not, and pentoxyfilline at the moment, from the data that I have, does not seem to make that much difference.

We are discussing the use of drugs for stimulating fertilisation. It is my understanding that these drugs still have to be used in the UK on a named patient basis, but that the HFEA may give clarification on their use.

Whittall: The line that the authority takes on drugs such as pentoxyfilline and their clinical use in IVF is that if a centre intending to use it can demonstrate effectiveness, either through literature or through its own work, then the authority would consider it. Generally it will take the line on the basis of an individual centre applying for use, but there will come a point eventually with a particular aspect of treatment that it will become generally accepted. I do not think that we want to put down a set of rules, but just to recognise that there is a process that each particular drug will go through in becoming accepted.

Hull: My reading of the information available seems to show a rapid change in the last few years in which fertilisation rates by standard IVF – this is in controlled studies, sister oocytes– have been of the order of 20%–30% and higher. Jacques Cohen has published fertilisation rates of 20% or 30% with other methods of micromanipulation [1]. What is becoming apparent is that by better methods of sperm preparation, we are now doing a lot better. Certainly we have controlled studies, as yet unpublished, showing marked improvement with Percoll.

Fishel: Only a few people are doing microdrop IVF, but to do this with Percoll-type preparations increases the chances of IVF fertilisation. One of the big problems with comparative data is that there are enormous differences in the populations of males being assessed. Certainly the high incidence of fertilisation in some of the controlled studies says a lot about the types of males that are undergoing micro manipulation.

Braude: Dr Fishel was rather disparaging about direct intracytoplasmic injection, yet four pregnancies in patients with two failed SUZI attempts have been reported.

Fishel: I do not view it disparagingly. I think it is a valuable technique which will come. But at this stage, apart from Palermo's data there are no other studies to suggest an advantage.

Braude: If one was summarising the state of the art in terms of micro-assisted fertilisation, then surely the way we should be progressing is towards direct intracytoplasmic injection, which eliminates all the other problems if it can be made to succeed. I cannot see the disadvantage of direct injection over any of the other methods, but I can see advantages.

Fishel: I do not think I am arguing against it at this stage but there will have to be a far greater evaluation of that technique. I do not deny it may be of use and eventually better than SUZI.

Templeton: Dr Fishel, I wanted clarification of Group 1, which included patients with extremely low sperm motility. Was it always necessary that there was at least some residual motility for fertilisation to take place, or were there patients with no sperm motility where fertilisation occurred?

Fishel: It would only be fair to assess patients that have zero motility as a separate group, which I have done. Again I have been singularly unsuccessful in achieving fertilisation with 100% immotile sperm.

Reference

1. Cohen J, Alikani M, Malter HE, Adler A, Talansky BE, Rosenwaks Z. Partial zona dissection or subzonal sperm insertion: microsurgical fertilization alternatives based on evaluation of sperm and embryo morphology. Fertil Steril 1991; 56:696–706.

Section III

Investigation and Treatment of the Female

Chapter 9

Tubal Physiology and Function

H. J. Leese and C. J. Dickens

The emphasis in this review will be on the human Fallopian tube.

Research on the tube has tended to be overshadowed by that on the uterus. As Fig. 9.1 shows, the number of papers published between 1980 and 1990 on the uterus (basic and clinical aspects) far exceeded those on the rest of the female tract combined; mainly the Fallopian tube, cervix and vagina.

As techniques for assisted conception have become more widespread, publications in the category; "Egg, fertilisation and early embryonic development" have increased dramatically (Fig. 9.1). However, IVF and related procedures such as GIFT remain relatively unsuccessful, and this has led to renewed interest in the tube, much of it very recent and not reflected in Fig. 9.1. A particularly lively area of research is that of "co-culture", in which preimplantation embryos are grown together with oviduct cells in vitro. This is proving especially popular in domestic animals, such as the cow, sheep and pig, where there are commercial advantages to be had from the improvement of embryo culture techniques.

Tubal Anatomy

The Fallopian tube may be divided anatomically into an outer, myosalpinx, responsible for tubal contraction and an inner, endosalpinx, responsible for creating the environment in the lumen. Research on tubal physiology and biochemistry has tended to reflect this anatomical distinction, with papers on tubal contractility rarely considering the endosalpinx and those on tubal secretions ignoring the myosalpinx. Such a rigid allocation of function in terms of anatomy is unwise. Contractions of the myosalpinx may influence events in

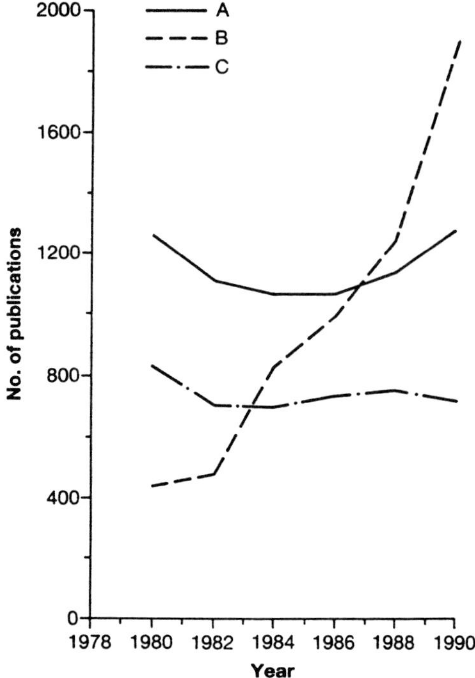

Fig. 9.1. Number of publications per annum listed in the "Bibliography of Reproduction" under the headings: A, egg, fertilisation and early embryonic development; B, uterus: basic and clinical; C, reproductive tract (excluding uterus): basic and clinical.

the lumen by minimising the build-up of unstirred layers and there may be subtle co-ordination of the response of the endo- and myosalpinx via the sympathetic nervous system. For example, in the rabbit, β-adrenergic stimulation increases ovarian arterial pressure, stimulates oviduct fluid secretion (unpublished observations) and causes relaxation of the myosalpinx [1]. Analogous effects may be shown in the anatomically related small intestine [2].

A second way in which the tube may be subdivided anatomically is into its regions: infundibulum, ampulla, ampullary-isthmic junction, isthmus and uterotubal junction. The classification might be seen as arbitrary on anatomical grounds, since its most obvious basis, the proportion of endosalpinx to myosalpinx, does not change abruptly, but continuously along the length of the tube. There is a parallel with the anatomy of the small intestine, in the gradual manner by which the jejunum becomes the ileum. However, the existence of different regions along the length of the tube may be justified on physiological grounds, since it is generally thought that the isthmus acts as a sperm reservoir, fertilisation occurs in the lower ampulla above the ampullary-isthmic junction and that the resulting preimplantation embryos are retained at this point, before moving rapidly through the isthmus into the uterus.

Cell Biology of the Endosalpinx

The tubal lumen is lined by two main cell types: ciliated and non-ciliated, also called secretory. Together, they comprise the tubal epithelium. As with the proportion of myo- to endosalpinx, so the proportion of ciliated to non-ciliated cells also changes continuously rather than abruptly along the length of the tube, with the ciliated predominating in the infundibulum and non-ciliated in the isthmus [3].

The turnover of the epithelium is very low, and there appears to be no population of stem cells. Cell division is seldom observed, either as mitotic indices in histological sections [4] or as the incorporation of bromodeoxyuracil (unpublished observations). This has clinical implications since the tube regenerates only slowly after being damaged.

In contrast to the low overall cell turnover, there are well-established, hormonally controlled changes in the morphology of individual epithelial cells, well-summarised by Jansen [5], and, for the primate oviduct, by Brenner and Maslar [6]. Epithelial height and degree of ciliation reach a maximum during the late follicular phase in women, with some atrophy and deciliation at the end of the luteal phase [7]. About 10%–20% of the cells of the fimbriae and ampulla then form new cilia during the early follicular phase.

The cell biology of the hormonal response of the primate oviduct is intriguing. Using indirect immunocytochemistry to localise oestrogen and progestin receptors in the macaque oviduct, Brenner et al. [8] have shown convincingly that some of the steroid hormonal effects exhibited by the epithelial cells are mediated via receptors in the underlying stroma. The data, summarised in Fig. 9.2, show that ciliated cells lack detectable oestrogen and progestin receptors at all times of the cycle, whereas these are both present in the secretory cells during the follicular phase but not the luteal phase. The data require the existence of soluble factors to transmit hormonal signals from the stroma to the epithelium. These paracrine effects are not unique to the oviduct, but are also found in the primate endometrium, prostate and seminal vesicles.

Tubal Fluid

Tubal fluid has a number of origins, as outlined in Fig. 9.3 [9]. It is mainly formed by the selective passage of molecules from the plasma, with specific contributions from the epithelial cells. At ovulation, there may be important additions from follicular fluid and the cumulus oophorus. Up to 90% of oviduct fluid in rabbits, sheep and cows is thought to be lost into the peritoneal cavity [10], following the path of least resistance. Bellve and McDonald [11] showed in the sheep, that when embryos are traversing the uterotubal junction, much of the flow, reduced at this stage (as in the rabbit, [10]), is into the uterus. Such a phenomenon would tend to prevent ectopic pregnancies. Perhaps such a reversal of oviduct fluid movement is less well developed in women, since ectopics are rare in other species.

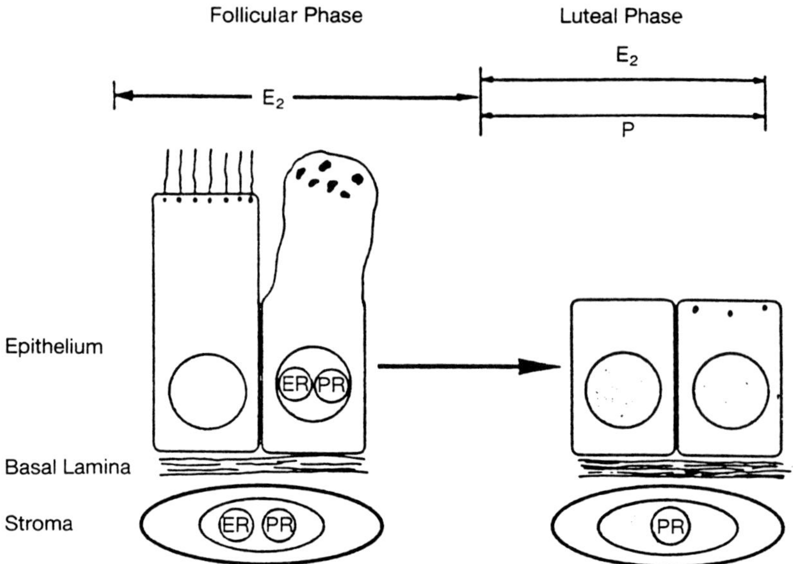

Fig. 9.2. Hormonal regulation in the oviduct. The figure summarises the relationships between oestrogen receptor (ER), progestin receptor (PR) and the growth and differentiation of ciliated and secretory cells in the oviduct. At all times during the cycle, ciliated cells lack detectable ER and PR. Secretory cells develop both ER and PR under oestrogen but lose both receptors when progestins dominate. Stromal cells develop ER and PR under oestrogen influence and lose ER, but not PR during progestin action. (Reproduced with permission, from Brenner et al. [8].)

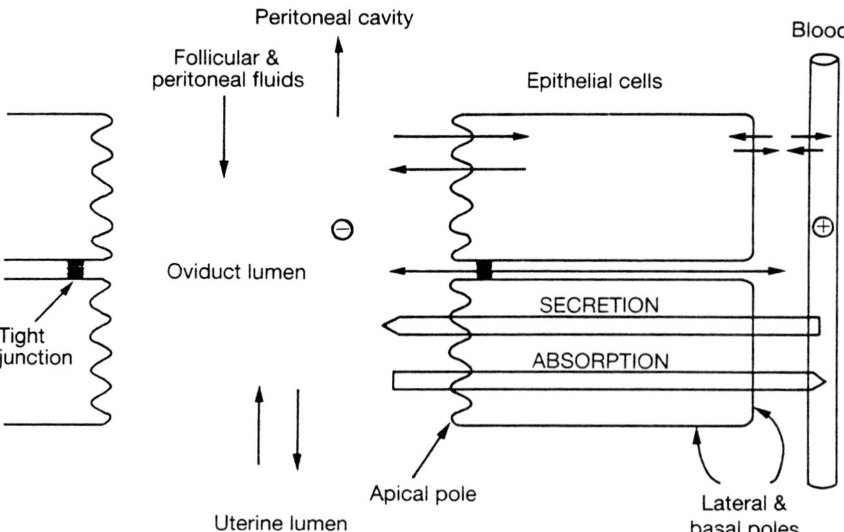

Fig. 9.3. Generalized diagram of the potential routes by which molecules may move into or out of the oviduct lumen. (Reproduced with permission from Leese [9].)

Movement of Large Molecules into Oviduct Fluid

The rate of passage of plasma proteins into oviduct fluid in rabbits is inversely proportional to their molecular weight [12]. The data shown in Fig. 9.4 have great significance; they imply that all plasma proteins will be present, in varying degrees in oviduct fluid. Thus, although serum albumin is a very large molecule, it is the most abundant protein in the plasma and also in oviduct fluid. When relatively small proteins such as insulin have been sought in oviduct fluid, they have been found [13]. In other words, gametes and embryos are likely to be exposed, at all times, to the proteins present in plasma. The same argument applies to molecules such as steroid hormones, which, since the ovarian vein and artery lie close to one another, may also pass from the ovary to the tube via a countercurrent mechanism [14]. These considerations all argue for the inclusion of patient's serum in the media used for in vitro fertilisation.

Very few studies have addressed the question of the route by which plasma proteins traverse the endosalpinx. The most likely possibility is by endocytosis at the basal membrane of the epithelium, with the passage of membrane vesicles across the cells and their release into the lumen by exocytosis. Evidence comes from studies on the local mucosal immune system in the oviduct [15]. For example, Parr and Parr [16] detected IgA and IgG in epithelial cell vesicles of the mouse preampulla; Oliphant et al. [17] have reported the presence of immunoglobulins in rabbit oviduct fluid at about one-tenth their concentration in plasma, and Yang et al. [18] found IgM in rhesus monkey fluid. This is a potentially important area of research. Does the secretion of epithelial cell-derived proteins, such as those involved in the mucosal defence system of the tube, follow the same pathway of exocytosis as that used by plasma proteins?

It seems that the oviduct of all species studied, secrete a small number of glycoproteins [19]. In some species, such as the rabbit, the secretion is very pronounced such that the proteins form a conspicuous mucus coat which

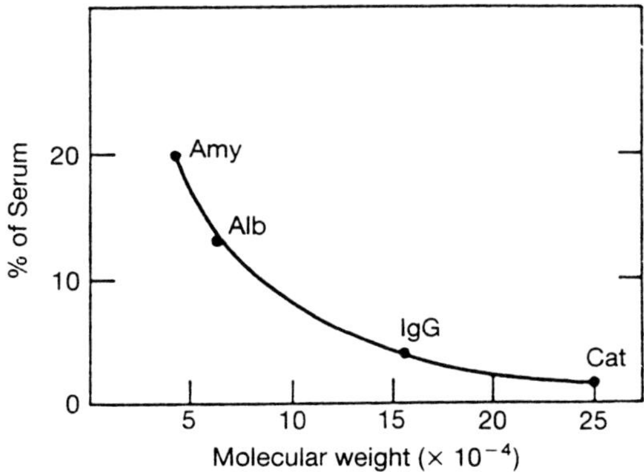

Fig. 9.4. The passage of serum proteins into rabbit oviduct fluid. The amount (%) of serum proteins present in oviduct fluid plotted against the molecular weight of these specific proteins; amylase (Amy), albumin (Alb), immunoglobulin G (IgG) and catalase (Cat). (Reproduced, with permission, from Oliphant et al. [12].)

envelops the embryos. In most cases, however, the secretion is not so apparent. The glycoproteins seem to be derived from the non-ciliated cells and are produced by the oviducts of the mouse [20], hamster [21], rabbit [22–24], cow [25–27], pig [28], sheep [29], non-human primate [30] and human [31].

At the time of writing, these are proteins looking for a function(s). They have been shown to associate with the zona pellucida and/or perivitelline space of the eggs and embryos in a number of species [32–38]. Do they act to mediate embryo–maternal dialogue? Are they important at implantation? Two cautionary notes are required. The first is to guard against these molecules being seen as magic ingredients which will solve the problems of embryo culture, and the failure of implantation. The phenomenon of implantation is daunting in its complexity with at least 20 factors implicated [39]. The second is that it has been known for many years that oocytes can take up macromolecules such as albumin and horseradish peroxidase in a non-specific manner [40,41].

The Movement of Small Molecules into Tubal Fluid

For the purposes of this review, two types of small molecule will be considered: ions and nutrients.

Ions in Tubal Fluid

Borland et al. [42] used the technique of electron probe microanalysis applied to picolitre-sized samples to determine the ionic composition of human Fallopian tubal fluid. The most significant result was that K^+ ions were present in 4–5-fold higher concentration (21.2 mM) than in human serum (3.6–5.0 mM). It may also be significant that the Cl^- concentration was slightly higher (132 mM) than in serum (102–113 mM) and that of Ca^{2+} (1.13 mM) was about half that in serum (2.27–2.72 mM). The physiological significance of these observations is unknown. The ionic composition of tubal fluid may obviously influence gamete and embryo survival but could also be a consequence of the mechanism by which tubal fluid is secreted.

This aspect of tubal physiology has been explored in our laboratory. Using a preparation for the combined vascular and luminal perfusion of the rabbit oviduct [43], evidence was produced which strongly suggested that the tube was lined by a chlorine ion-secreting epithelium [44] and that the rate of secretion could be controlled by second messenger molecules such as cyclic AMP. We have followed up these observations using a pure preparation of rabbit oviduct epithelial cells grown on collagen-impregnated filters in primary culture [45]. The advantage of this technique is that the cells re-form tight junctions with one another and establish a polarised cell monolayer with their basal surface attached to the collagen substrate and their apical surface with microvilli and cilia projecting into the culture medium. This enables the transepithelial flux of ions to be measured readily. The results with this preparation with respect to chloride ions are summarised in Fig. 9.5 which shows a net secretion of Cl^- from the basal to apical side of the cells, which is inhibited by cyclic AMP.

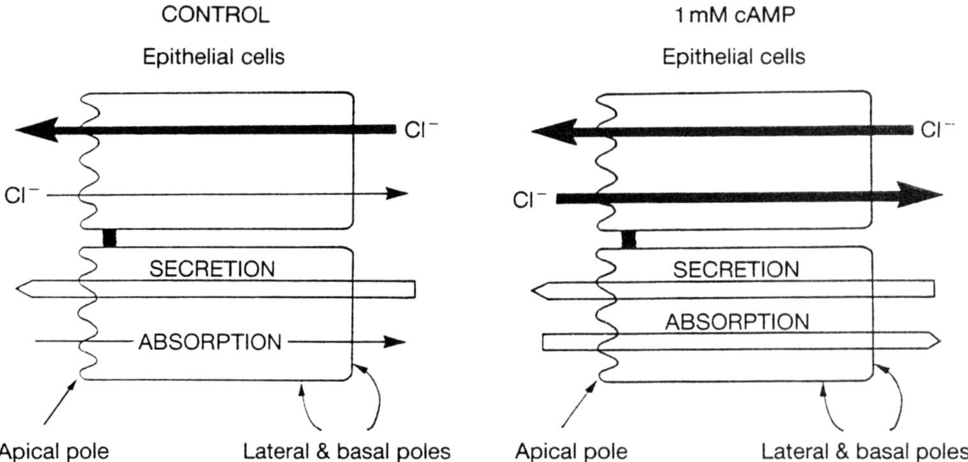

Fig. 9.5. The role of chloride ions in mediating rabbit oviduct fluid secretion. Net Cl^- flux is normally in the secretory direction; 1 mM-dibutyl cyclic AMP (cAMP) abolishes net Cl^- flux and oviduct fluid formation.

Nutrients in Tubal Fluid

The nutrients of greatest interest in tubal fluid are those which are utilised by gametes and preimplantation embryos: pyruvate, lactate, glucose and amino acids, in particular, glutamine. They are ultimately derived from the plasma [43] and may be metabolised as they cross the oviduct; indeed glucose is a major fuel for the human and rabbit endosalpinx [46,47]. Nutrients may also be supplied to unfertilised and fertilised oocytes by cumulus cells [48].

The nutrient composition of tubal fluid is known for a variety of species, the most reliable data being for those animals which secrete copious amounts, such as the rabbit [49]. In the human, tubal fluid is present in very small quantities. Samples of the order of 1 μl were obtained with the minimum of disturbance, and when analysed, gave values for pyruvate, lactate and glucose concentrations of 0.24, 2.16, and 1.0 mM respectively (Leese and Winston, unpublished observations). These values were almost identical to concentrations reported for human hydrosalpinx fluid [50]. The glutamine concentration in hydrosalpinx fluid is 0.30 mM. Such information is useful in devising human embryo culture media; spare human embryos developed successfully in a medium whose nutrient composition was similar to that of tubal fluid [51].

Functional Aspects

These fall into three categories:

1. Sperm transport, egg/cumulus and preimplantation embryo transport

2. Sperm, egg and preimplantation embryo nutrition
3. Defence against infection

Gamete and Preimplantation Embryo Transport

The detailed mechanisms for moving sperm in one direction along the Fallopian tube and eggs and embryos in the opposite direction are poorly understood. The most noteworthy of recent studies are those which have sought to examine the complex manner by which sperm reach the ampullary-isthmic junction, the site of fertilisation, and the manner by which ovulation and sperm transport are synchronised [52,53]. Research on the pig, cow and rabbit has established that sperm are sequestered in a number of "reservoirs" along the female reproductive tract, the one closest to the site of fertilisation being the tubal isthmus. Sperm in this location undergo cycles of attachment and detachment from the epithelial cells lining the isthmic lumen; the ciliated cells seem more important that the non-ciliated in this respect [54–56]. Such sperm exhibit further characteristics; hyperactivated motility, the completion of capacitation and onset of the acrosome reaction. As regards the human, very little is known of sperm physiology in the upper reaches of the female reproductive tract and the extrapolation of these observations on animals has, at present, to be tentative, although there is evidence for a sperm reservoir in the isthmus in the non-human primate [52]. Comparable studies on possible interactions between embryos and the tubal epithelium have not been carried out.

Knowledge on the location of preimplantation embryos in the human female genital tract has been summarised by Ortiz and Croxatto [57]. The number of eggs and early embryos that have been recovered from women is limited but it seems that development occurs up to the 8-cell stage in the Fallopian tube with morulae and blastocysts being recovered from the uterus. Tubal embryos were recovered from the ampulla or ampullary-isthmic junction, implying that transit through the isthmus is rapid.

Sperm, Egg and Preimplantation Embryo Nutrition

This topic has been referred to earlier. In summary, sperm can use a wide variety of substrates; in most mammals, unfertilised and fertilised oocytes have a requirement for pyruvate; The later preimplantation stages can utilise glucose; glutamine may also be an important substrate; lactate is a prominent end-product. It is important to note that all this information is derived from the study of gametes and embryos in vitro; it remains an act of faith that these nutrients are consumed in vivo [58]. One topic yet to be pursued is how changes in the concentrations of tubal fluid constitutents, in response to the presence of gametes and embryos, are brought about, together with their physiological significance.

Defence Against Infection

The oviduct mucosal immune system, mentioned in the context of macromolecular secretion, is likely to be crucial to the maintenance of an appropriate tubal environment, but remains little studied. Kutteh et al. [59] reported the presence of IgA-containing plasma cells in the stroma and secretory component-producing cells in the epithelium, providing a first line of local defence against tubal infection. This is, however, one of few papers to have explored the operation or control of the tubal mucosal immune response.

Co-culture

Preimplantation embryo development is retarded in vitro; typically being delayed by one or two cell cycles over a 5-day period. One way of overcoming this retarded development is to culture embryos in the presence of oviduct cells maintained either as monolayers or in suspension. In some cases, medium "conditioned" by exposure to oviduct cells has been used. The number of reports of such "co-culture" has increased dramatically over the past two years [60,61], largely because the practice is widely used commercially to grow the embryos of farm animals. At the present time, research in this area is characterised by a bewildering number of claims and counter-claims of the success of co-culture. The consensus is that the procedure does confer some advantage to embryo development, but not to the extent that in vivo rates are restored. Most studies report only limited information on embryo phenotype such as the proportion which reach the blastocyst stage and their cell number and the basis of the contribution made by the supporting oviduct cells is unknown. This research area is likely to be very active over the next few years, and lead directly, or indirectly to answers to the types of questions which follow.

Basic Research Questions

1. By what mechanisms are large and small molecules and water transported into tubal fluid?
2. Are there different mechanisms in the ampulla and the isthmus? What are the roles of the ciliated and non-ciliated cells in tubal fluid formation?
3. How is tubal fluid formation controlled?
4. How is the control mechanism(s) modified in different hormonal states?
5. What is the function(s) of the oviduct-derived glycoproteins?
6. What physical interactions occur between the ciliated and non-ciliated cells and the sperm, cumulus mass and preimplantation embryos?
7. What is the basis of the advantage(s) to preimplantation embryo development conferred by co-culture?

Acknowledgements. The authors' studies reported in this review, were supported by The Medical Research Council and The Wellcome Trust.

References

1. Nakanashi H, Wood C. Effects of adrenergic blocking agents on human fallopian tube motility in vitro. J Reprod Fertil 1968; 16:21–6.
2. Cooke HJ. Neutral and humoral regulation of small intestinal electrolyte transport. In: Johnson LR, ed. Physiology of the gastrointestinal tract. New York: Raven Press, 1987; 963–81.
3. Leese HJ. Studies on the movement of glucose, pyruvate and lactate into the ampulla and isthmus of the rabbit oviduct. Q J Exp Physiol 1983; 68:89–96.
4. Donnez J, Casanas-Roux F, Caprasse J, Ferin J, Thomas K. Cyclic changes in ciliation, cell height, and mitotic activity in human tubal epithelium during reproductive life. Fertil Steril 1985; 43:554–9.
5. Jansen RPS. Endocrine response in the fallopian tube. Endocr Rev 1984; 5:525–51.
6. Brenner RM, Maslar IA. The primate oviduct and endometrium. In: Knobil E, Neill J, eds. The physiology of reproduction. New York: Raven Press, 1988; 303–29.
7. Verhage HG, Bareither ML, Jaffe RC, Akbar M. Cyclic changes in ciliation, secretion and cell height of the oviductal epithelium in women. Am J Anat 1979; 156:505–22.
8. Brenner RM, West NB, McClellan MC. Estrogen and progestin receptors in the reproductive tract of male and female primates. Biol Reprod 1990; 42:11–19.
9. Leese HJ. Formation and function of oviduct fluid. J Reprod Fertil 1988; 82:842–56.
10. Hamner CE. Oviducal fluid – composition and physiology. In: Greep RO, Astwood EB, eds. Handbook of physiology, section 7, Endocrinology II, Washington DC: American Physiology Society, 1973; 141–51.
11. Bellve AR, McDonald MF. Directional flow of Fallopian tube secretion in the ewe at onset of breeding season. J Reprod Fertil 1970; 22:147–9.
12. Oliphant G, Bowling A, Eng LA, Keen S, Randall PA. The permeability of rabbit oviduct to proteins present in the serum. Biol Reprod 1978; 18:516–20.
13. Rao LV, Farber M, Smith RM, Heyner S. The role of insulin in preimplantation mouse development. In: Heyner S, Wiley AM, eds. Early embryo development and paracrine relationships. UCLA Symposia on Molecular and Cellular Biology New Series, Volume 117, New York: Wiley Liss, 1990; 109–24.
14. Hunter RHF. Fallopian tubal fluid: the physiological medium for fertilization and early embryonic development. In: Hunter RHF, The fallopian tubes. Their role in fertility and infertility. Berlin: Springer-Verlag, 1988; 30–52.
15. Haas GG, Beer AE. Immunologic influences on reproductive biology: sperm gametogenesis and maturation in the male and female genital tracts. Fertil Steril 1986; 46:753–765.
16. Parr EL, Parr MB. Uptake of immunoglobulins and other proteins from serum into epithelial cells of the mouse uterus and oviduct. J Reprod Immunol 9:339–54.
17. Oliphant G, Randall P, Cabot CL. Immunological components of rabbit Fallopian tube fluid. Biol Reprod 1977; 16:463–7.
18. Yang S-L, Schumacher GFB, Broer KA, Holt JA. Specific antibodies and immunoglobulins in the oviductal fluid of the rhesus monkey. Fertil Steril 1983; 39:359.
19. Maguiness SD, Shrimanker K, Djahanbakkch O, Grudzinskas JG. Oviduct proteins. Contemp Rev Obstet Gynaecol 1992; 4:42–50.
20. Kapur RP, Johnson LV. An oviductal fluid glycoprotein associated with ovulated mouse ova and early embryos. Dev Biol 1985; 112:89–93.
21. Robitaille G, St-Jacques S, Potier M, Bleau G. Characterisation of an oviductal glycoprotein associated with the ovulated hamster oocyte. Biol Reprod 1988; 39:687–94.
22. Stone SI, Huckel WR, Oliphant G. Identification and hormonal control of reproductive-tract-specific antigens present in rabbit oviductal fluid. Gamete Res 1980; 3: 169–77.
23. Jansen PS, Bajpai VK. Oviduct acid mucus glycoproteins in the estrous rabbit: ultrastructure and histochemistry. Biol Reprod 1982; 26:155–68.
24. Hyde BA, Black DL. Synthesis and secretion of sulphated glycoproteins by rabbit oviduct explants in vitro. J Reprod Fertil 1986; 78:83–91.
25. Newton GR, Hansen PJ, Low BG. Characterisation of a high molecular weight glycoprotein secreted by the peri-implantation bovine conceptus. Biol Reprod 1988; 39:553–60.
26. Boice ML, Geisert RD, Blair RM, Verhage HG. Identification and characterization of bovine oviductal glycoproteins synthesized at estrus. Biol Reprod 1990; 43:457–65.
27. Gerena RL, Killian GJ. Electrophoretic characterization of proteins in oviduct fluid of cows during the estrous cycle. J Exp Zool 1990; 256:113–20.
28. Buhi WC, Alvarez IM, Sudhipong V, Dones-Smith MM. Identification and characterization

of de novo-synthesized porcine oviductal secretory proteins. Biol Reprod 1990; 43:929–38.
29. Sutton R, Nancarrow CD, Wallace ALC. Oestrogen and seasonal effects on the production of an oestrus-associated glycoprotein in oviductal fluid of sheep. J Reprod Fertil 1986; 77:645–53.
30. Verhage HG, Fazleabas AT. The in vitro synthesis of estrogen-dependent proteins by the baboon (*Papio anubis*) oviduct. Endocrinology 1988; 123:552–8.
31. Verhage HG, Fazleabas AT, Donnelly K. The in vitro synthesis and release of proteins by the human oviduct. Endocrinology 1988; 122:1639–45.
32. Kapur RP, Johnson LV. Selective sequestration of an oviductal fluid glycoprotein in the perivitelline space of mouse oocytes and embryos. J Exp Zool 1986; 238:249–60.
33. Fowler RE, Barratt E. The uptake of (^{3}H)glucosamine-labelled glycoconjugates into the perivitelline space of preimplantation mouse embryos. Hum Reprod 1989; 4:821–5.
34. Kan FWK, St-Jacaues S, Bleau G. Immunocytochemical evidence for the transfer of an oviductal antigen to the zona pellucida of hamster ova after ovulation. Biol Reprod 1989; 40:585–98.
35. Abe H, Oikawa T. Ultrastructural evidence for an association between an oviductal glycoprotein and the zona pellucida of the golden hamster egg. J Exp Zool 1990; 256:210–21.
36. Boice ML, McCarthy TJ, Mavrogianis PA, Fazleabas, Verhage HG. Localization of oviductal glycoproteins within the zona pellucida and perivitelline space of ovulated ova and early embryos in baboons (*Papio anubis*). Biol Reprod 1990; 43:340–6.
37. Kan FWK, Roux E, St-Jacquest S, Bleau G. Demonstration by lectin-gold cytochemistry of transfer of glycoconjugates of oviductal origin to the zone pellucida of oocytes after ovulation in hamsters. Anat Rec 1990; 226:37–47.
38. Wegner CC, Killian GJ. In vitro and in vivo association of an oviduct estrus-associated protein with bovine zona pellucida. Mol Reprod Dev 1991; 29:77–84.
39. Findlay JK, Salamonsen LA. Paracrine regulation of implantation and uterine function. Bailliere's Clin Obstet Gynaecol 1991; 5:117–31.
40. Glass LE. Transfer of native and foreign serum antigens to oviducal mouse eggs. Am Zool; 3:135–56.
41. Anderson E. The localization of acid phosphatase and the uptake of horseradish peroxidase in the oocyte and follicle cells of mammals. In: Biggers JD, Schuetz AW, eds. Oogenesis. Baltimore: University Park Press, 1970; 87–117.
42. Borland RM, Biggers JD, Lechene C, Taymor ML. Elemental composition of fluid in the human Fallopian tube. J Reprod Fertil 1980; 58:479–82.
43. Leese HJ, Gray SM. Vascular perfusion: a novel means of studying oviduct function. Am J Physiol 1985; 248:E624–E632.
44. Gott AL, Gray SM, James AF, Leese HJ. The mechanism and control of rabbit oviduct fluid formation. Biol Reprod 1988; 39:758–63.
45. Kimber SJ, Waterhouse R, Lindenberg MA. In vitro models for implantation of the mammalian embryo. In: Bavister BD, ed. Preimplantation embryo development. Serono Symposia USA. New York: Springer-Verlag. (in press)
46. Brewis IA, Winston RML, Leese HJ. Energy metabolism of the human Fallopian tube. J Reprod Fertil 1992; in press.
47. Leese HJ, Bewis IA, Edwards LJ, Skiera S, Winston RML. Biochemistry of tubal secretions. In: Grudzinskas JG, Chapman MG, Chard T, Djahanbakhach O, eds. The fallopian tube in diagnosis and surgical treatment. London: Springer-Verlag, (in press).
48. Leese HJ. Physiology of the fallopian tube: the provision of nutrients for oocytes and early embryos. In: Evers JHL, Heineman MJ, eds. From ovulation to implantation. Excerpta Medica. International Congress Series. No 917, Amsterdam: Elsevier Science Publishers, 1990; 121–6.
49. Leese HJ, Barton AM. Production of pyruvate by isolated cumulus cells. J Exp Zool 1985; 234:231–6.
50. Gott AI, Hardy K, Winston RML, Leese HJ. The nutrition and environment of the early human embryo. Proc Nutr Soc 1990; 49:2A.
51. Hardy K, Hooper MAK, Rutherford AJ, Winston RML, Leese HJ. Non-invasive measurement of glucose and pyruvate uptake by individual human oocytes and preimplantation embryos. Hum Reprod 1988; 4:188–91.
52. Overstreet JW, VandeVoor CA, Sperm transport in the female genital tract. In: Asch RH, Balmaceda JP, Johnston I, eds. Gamete physiology. Norwell Mass: Serono Symposia, 1990; 43–52.
53. Hunter RHF. Physiology of the Fallopian tubes, with special reference to gametes, embryos

and microenvironments. In: Evers JHL, Heineman MJ, eds. From ovulation to implantation. Excerpta Medica. International Congress series, No 917, Amsterdam: Elsevier, 1990; 101–19.
54. Pollard JW, Plante C, King WA, Hansen PJ, Betteridge KJ, Suarez SS. Fertilizing capacity of bovine sperm may be maintained by binding to oviductal epithelial cells. Biol Reprod 1991; 44:102–7.
55. Suarez SS, Katz DF, Owen DH, Andrew JB, Powell RL. Evidence for the function of hyperactivated motility in sperm. Biol Reprod 1991; 44:375–81.
56. Suarez S, Redfern K, Raynor P, Martin F, Phillips DM. Attachment of boar sperm to mucosal explants of oviduct in vitro: possible role in formation of a sperm reservoir. Biol Reprod 1991; 44:998–1004.
57. Ortiz ME, Croxatto HB. Human eggs in the female genital tract. In: Asch RH, Balmaceda JP, Johnston I, eds. Gamete physiology. Norwell Mass: Serono Symposia, 1990; 173–85.
58. Leese HJ. Metabolism of the preimplantation mammalian embryo. Oxford Rev Reprod Biol 1991; 13:35–72.
59. Kutteh WH, Hatch KD, Blackwell RE, Mestecky J. Secretory immune system of the female reproductive tract: I. Immunoglobulin and secretory component-containing cells. Obstet Gynecol 1988; 71:56–60.
60. Moor RM, Nagai T; Gandolfi F. Somatic cell interactions in early mammalian development. In: Evers JHL, Heineman MJ, eds. From ovulation to implantation. Excerpta Medica. International Congress Series. No 917, 1990; 177–91.
61. Bongso A, Ng S-Y, Fong C-Y, Ratnam S. Cocultures: a new lead in embryo quality improvement for assisted reproduction. Hum Reprod 1991; 56:179–91.

Chapter 10

Falloposcopy

J. F. Kerin

Introduction

Falloposcopy is defined as microendoscopy of the fallopian tube lumen from the uterotubal ostium to the fimbria using a non-incisional transvaginal approach [1–3]. To date the only section of the human female genital tract that has not been explored adequately by endoscopy resides within the fallopian tube and extends from the uterotubal ostium to the ampullary-isthmic junction. The inability to visualise this narrowest medial third of the tubal lumen has been a major drawback because it is vulnerable to the sequelae of ascending genital tract infections and surgical interventions. Falloposcopy may aid in the diagnosis of subclinical epithelial and vascular damage, stricture formation and intraluminal non-obstructive adhesions, and the detection of endotubal debris, which may play a significant role in infertility, or predispose women to tubal pregnancyy. Previous attempts to explore the tubal lumen endoscopically using the transvyaginal approach have been unsatisfactory because of the difficulty in negotiating the narrow medial tubal lumen and the limited resolution and illumination characteristics of the endoscopes used [4]. Modifications and refinements to current angioscopic instruments, and techniques that are currently used for intraoperative and percutaneous angioscopy of the vascular tree [5,6] have been incorporated into the falloposcope.

Brief History of Transvaginal Fallopian Tube Cannulation

A description and an illustration of the passage of a wire probe through the intramural segment of the human fallopian tube using a transvaginal approach were first published by Gardner in 1856 [7]. In 1954, Lisa studied the gross anatomy of the intramural tube from 300 hysterectomy specimens and noted that it was 2 cm in average length and varied from 0.2 to 0.4 mm in a diameter [8]. In 1963 Sweeney studied the course the intramural tube took through each lateral uterine horn to the uterotubal junction (UTJ) in 100 post-hysterectomy uterine specimens [9]. He noted that the intramural tubal length ranged from 1 to 3.5 cm and that the course of the tube through the uterine wall was variable in both calibre and direction. In 69% of cases the intramural tube was tortuous, in 23% straight, and in 8% curved. He concluded that it was not possible to pass probes from the uterine cavity along the tubal lumen and that such attempts may cause damage to its integrity and therefore be of little value. This sentiment was conveyed in a critical and challenging editorial by De Cherney as late at 1987 where he cautioned over-enthusiastic investigators about the complex, narrow and tortuous nature of the intramural tube and stated that too much harm may be done by persevering with retrograde cannulation procedures [10]. With this background in mind the functional in vivo anatomy of the intramural and isthmic tubal lumen, and how it differs from the above ex vivo studies, will be described.

The Importance of Understanding the Living Tubal Lumen

Unlike a blood vessel, the tubal lumen is collapsed and therefore considered as a potential space. Its primary longitudinal and secondary epithelial folds provide great surface area to its collapsed lumen, which contains mucus-like secretions throughout most of the menstrual cycle. The tubes are paired organs which vary from 7 to 14 cm in length and are enclosed in the mesosalpingeal borders of the broad ligament. The tubes' smooth muscle wall becomes thinner both in relation to its distance from the uterine horn and as its luminal diameter increases. Furthermore, the tubes may take a gentle C-shaped curve to a more tortuous path as they proceed from the uterine horn, anterolateral and then posterior to the ovary. These variations in tubal configuration may be further distorted by uterine fibroids, peritubal adhesions, infection, ectopic pregnancy, intratubal disease, and space-occupying tumours in the broad ligament, ovary and adjacent pelvic structures. An appreciation of the dimensions, curvatures, muscular contractions and variable epithelial appearances within different parts of the tubal lumen is essential for performing intelligent and successful cannulation procedures. One of the most important things that has been learnt from in vivo tubal cannulation experience [2] is that the dimensions and path of the intramural lumen bear little resemblance to observations made on ex vivo uterotubal specimens by Lisa [8], Hernstein [11] and Sweeney [9]. It was pointed out by Corfman [12] that observations made from in vivo studies such as those of Rubin [13], Rubin and Bendeick [14] and Kerin et al. [2], indicate that the tube tends to take a straight or gently curved course in its intramural segment. It is, therefore, likely that the tortuous course of the intramural tube in non-living ex vivo uterine specimens is due to artifactual changes secondary to myometrial contraction, resulting in distortion. In addition, the intramural lumen diameter of 0.2–0.4 mm described by Lisa [8] in ex vivo specimens is probably also due to collapse and contraction of the tubal wall. It has been observed during falloposcopy that the intramural tube is 0.8–1.2 mm in diameter and can accommodate a cannula of 1–1.2 mm in diameter without epithelial damage [2]. The healthy elastic tubal wall can probably tolerate minor extra distension provided the cannula is flexible and smooth walled and is advanced using a coaxial [2] or a linear eversion catheter (LEC) system [15]. As a prerequisite for minimising technical difficulties associated with falloposcopy procedures it is most important to understand the living cyclical anatomy and behaviour of each tubal segment undergoing cannulation. Like the ovary and endometrium, the tube is a dynamic organ which undergoes important cyclical secretory, epithelial, vascular and motility changes as it provides a functional environment for critical stages of sperm, oocyte and embryo development.

Delivery Systems Used for Falloposcope Tubal Placement

The first successful delivery system developed for placing the falloposcope successfully and safely in the fallopian tube was a coaxial delivery system [1–3]. The other system that has been developed specifically for falloposcopy using a non-hysteroscopic technique involves the use of a miniaturised linear eversion catheter (LEC) delivery system. Each system will be described in detail and the advantages of each system will be outlined. However, it must be emphasised that the technique of falloposcopy is in its formative stages and these techniques are constantly being refined and upgraded. Therefore the conclusions drawn at this stage should be regarded as preliminary.

Development and Description of a Falloposcope

Based on evaluations made with small flexible fibreoptic endoscopes, which were used to cannulate fallopian tubes from women who were undergoing hysterectomy and salpingectomy, the initial prototype falloposcopes were made by Olympus (Olympus PF-5X; Olympus Corp., Lake Success, NY), Intramed (Intramed Laboratories, San Diego, CA), Medical Dynamics (Medical Dynamics, Inc., Englewood, CO), Advanced Interventional Systems (AIS, Costa Mesa, CA) and Imagyn (Imagyn Medical Inc., San Clemente, LA). As a result, falloposcopes measuring 1.0–1.5 m in length and 0.5 mm in outside diameter (OD), containing an atraumatic leading end, and sufficiently flexible to minimise the risk of perforation or damage to the endosalpinx and myosalpinx, were designed and evaluated.

Falloposcopes with adequate optical qualities for defining endotubal surface anatomy contained 1800–2200 imaging fibres with a centre to centre cross-sectional fibre distance of 4 μm, a distal lens of 0.3 mm diameter, a depth of field from 2 mm to infinity, two-point discrimination down to 10 μm, 8–50×

magnification, an angle of end-on field of view in air of 72° and in water of 54°, and 8–12 illuminating fibre bundles each of 100–130 μm in diameter. A 300W xenon light-source provided illumination through a flexible cable attached to the falloposcope at its proximal end. A 35–70 mm video coupler relayed images from the falloposcope to a light-sensitive colour microchip video camera. During the procedure, the real-time images were displayed on a high-resolution colour monitor and were recorded permanently on tape using a high-quality video cassette recorder. Still photographs were obtained from the recorded video-tape using a computer graphics freeze-frame technique and a colour video printer for production of hard copy prints.

Development of a Flexible, Miniature Hysteroscope

Small flexible hysteroscopes were also evaluated as suitable for effectively and atraumatically guiding the falloposcope into the ostium of the fallopian tube under direct vision (Olympus Corp., Lake Success, NY; Intramed Laboratories, San Diego, CA). The 25–30 cm working fibre of these hysteroscopes was semirigid and the distal 5 cm flexible and steerable about a tight radius of 1.5 cm with a range of movement from neutral (straight) to 90°. The focus and steering were operated from the proximal end, next to the eyepiece. The working channels of these hysteroscopes ranged from 1.6 mm to 2.2 mm in diameter, and were adequate to accommodate the cannulation and falloposcope instrumentation. The small outer diameter (OD) (3.3–4.5 mm) of these hysteroscopes and their distal flexibility permit them to be introduced through the cervical canal without prior dilatation. The firm seal created between the hysteroscope and endocervical canal was important for producing adequate distention of the uterine cavity and tubal ostium for good visualisation to facilitate easy passage of the falloposcope, associated guide wires and cannulas into the tube under direct vision, and for an economy of the use of the fluid distention medium, because of minimal leak back into the vagina.

After refinement of the hysteroscope and falloposcope prototypes, approval was obtained from the Institutional Review Board of Cedars-Sinai Medical Centre, Los Angeles, California, in March 1987, for use of the falloposcope in the human female under general anaesthesia. The female subjects who were undergoing elective hysteroscopy or combination hysteroscopy–laparoscopy procedures were fully informed about the nature and risks of this experimental procedure and voluntarily signed an appropriately designed consent form, before the procedure.

The Coaxial, Hysteroscopy-Assisted Delivery System for Falloposcopy

A hysteroscope was gently introduced through the cervical canal and into the uterine cavity after uterine sounding but without cervical dilatation. The uterine cavity was then irrigated with lactated Ringer's solution at a rate of 10–40 ml min^{-1}, using a gravity-fed system. Under video monitoring, the distal tip of the hysteroscope was directed to within 3 mm of one of the tubal ostia so that a direct long-axis view of the intramural tubal lumen was obtained. The coaxial technique involved the passage of small "floppy", steerable, Teflon-coated or stainless-steel platinum-tip tapered guide wires (Target Therapeutics, Fremont, CA; Cook OB/GYN, Spencer, IN; Glidewire, Medi-Tech Incorporated, Watertown, MA) into the fallopian tube under video hysteroscopic monitoring. These flexible guide wires ranged from 0.3 mm to 0.8 mm OD and were able to negotiate the natural curvatures of the tubal lumen atraumatically. When the wire had been passed to a point of resistance, or 15 cm beyond the uterotubal ostium, flexible Teflon cannulas of 1.2–1.3 mm OD with polished leading ends (Target Therapeutics; Cook OB/GYN) were introduced over the guidewire and through the tubal ostium under direct hysteroscopic vision to a point of resistance or for approximately 15 cm to the region of the fimbrial segment of the fallopian tube and into the immediate peritoneal cavity.

The guidewire was then withdrawn while the Teflon cannula remained in place and the falloposcope introduced into the Teflon cannula. The Teflon cannula provided protection to the entire length of the falloposcope and also served as a conduit to carry fluid irrigation to the distal end of the falloposcope, to minimally distend the tube and permit a lens-fluid-epithelial interface for image facilitation. The direct apposition of the endothelium onto the lens otherwise resulted in a "white-out reflection". Second, by lifting the endothelium off the lens, the falloposcope could be advanced under direct vision, thus preventing epithelial damage. Furthermore, if the falloposcope met with an obstruction or tubal distortion, this limited hydrotubation facilitated negotiation of the falloposcope along a narrow intramural, isthmic or ampullary segment, through strictures or intraluminal adhesions, under video monitoring. It became very obvious when the leading end of the falloposcope passed beyond the fimbria, because the light carried by the falloposcope was insufficient to light adequately the larger peritoneal cavity. However, when the falloposcope lens came to within millimetres of bowel, mesentery or ovary, transient views of these moving organs were obtained [2]. When this procedure was carried out with laparoscopic peritoneal illumination, more light was present for a tunnel view of intraperitoneal structures through the falloposcope.

In all cases a better inspection of the tubal epithelium was obtained by retrograde withdrawal of the falloposcope from the fimbria toward the uterine ostium with continuous insufflation by irrigation. This was due to elimination of an "antegrade white-out reflection" associated with advancing the falloposcope lens against a curvature of the tubal lumen before it straightened out again, when moving toward the fimbria. Dual video monitoring was used simultaneously for both continuous hysteroscopic and falloposcopic recordings. A fluid-tight irrigation system was achieved by using a Tuohy-Borst type Y adaptor (Cook OB/GYN; Rotating Hemostatic Valve; Advanced Cardiovascular Systems; Temecula, CA) connection for introducing the falloposcope and fluid via the proximal end of the hysteroscope operating channel (Figure 10.1). The normal falloposcopic appearance and dimensions of the intramural, isthmic, ampullary and fimbrial tube as viewed falloposcopically have been described in detail [2].

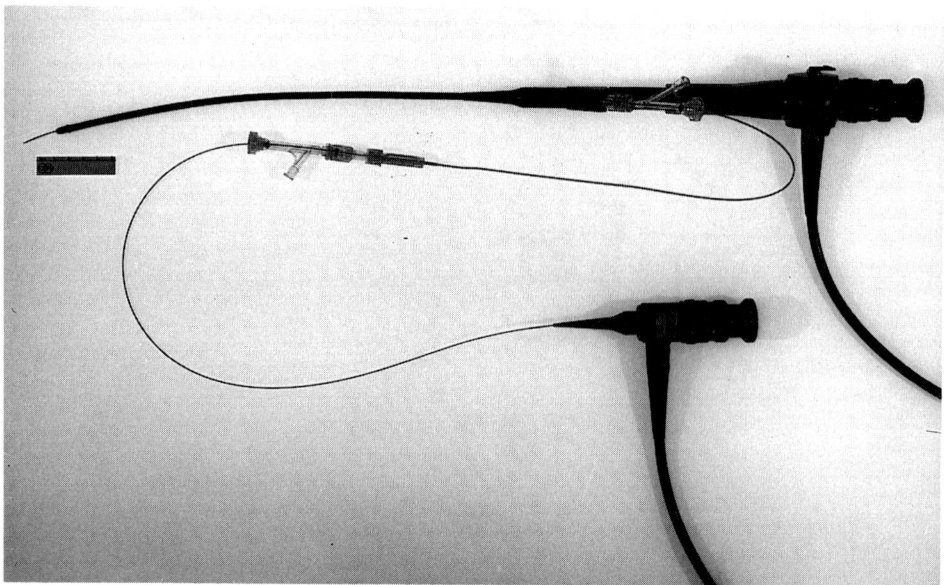

Fig. 10.1. A falloposcope ("Baby scope", bottom) with a fibre OD of 0.5 mm (Olympus Corp. PF-5X) inserted through a Y-connector assembly connected to a Teflon cannula which in turn passes through a second Y-Connector attached to a 1.7 mm operating channel of a steerable miniature hysteroscope ("mother scope", top) with an optical fibre OD of 4.5 mm (Olympus Corp). Irrigation for the tubal lumen and uterine cavity is run through the other arms of the respective Y-connectors.

The Linear Eversion Catheter Delivery System for Falloposcopy

Until recently the coaxial technique has been the primary method of introducing the falloposcope and endotuboplasty devices into the tube. A promising refinement in catheter technology has been the development of a miniaturised linear eversion catheter (LEC) system for the introduction of the falloposcope into the tube with or without the need of a hysteroscope (Imagyn Medical). Unlike most coaxial catheter systems where guide wires and catheters need to be pushed along the tubal lumen, the linear eversion catheter system advances the falloposcope using the gentle action of a linear everting catheter principle. Currently this LEC system has been approved by the Food and Drug Administration (FDA) for a multicentre clinical trial in selected medical centres throughout North America. A diagrammatic description of the LEC System is illustrated in Fig. 10.2. The inner lumen of the balloon is constructed of a material that is flexible enough to conform to the fallopian tube path and diameter variations but retains sufficient radial integrity to maintain an open internal lumen under balloon pressure. This open lumen facilitates irrigation and the free, independent movement of the falloposcope or an embryo/gamete

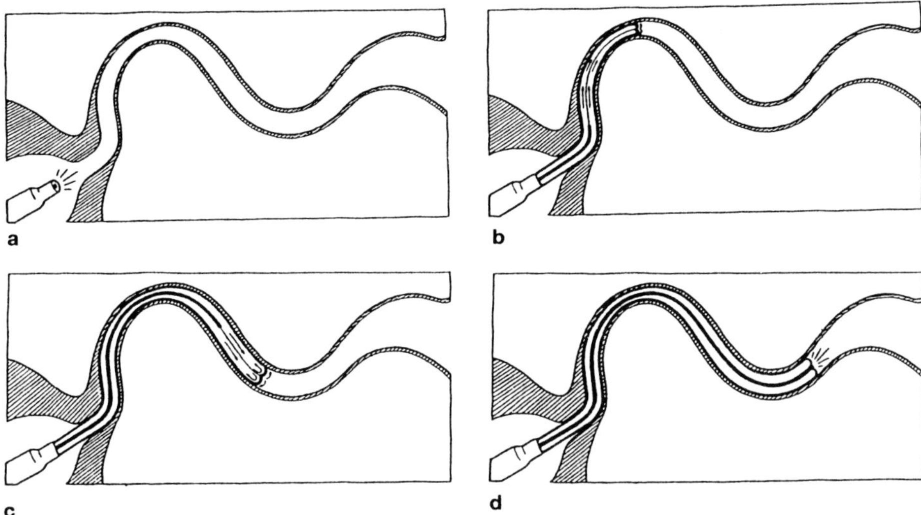

Fig. 10.2. **a** displays the relationship of a precurved transuterine guiding catheter and falloposcope for end-on viewing of the tubal ostium. **b** displays a close-up diagrammatic view of the construction of the LEC delivery system showing the relationship between the outer transuterine catheter tip, balloon, scope and irrigation channel. **c** shows the LEC delivery system carrying the falloposcope advanced through the intramural and isthmic segments of tubal lumen. **d** displays the LEC delivery system everted into the ampulla with the falloposcope flush with the balloon's end for optimal viewing. (With kind permission from Imagyn Medical, Inc.)

cannula along the length of the everted balloon. Once the tubal ostium is visualised and coaxial alignment is achieved the LEC is advanced into the tube. The actual LEC consists of a long miniature tubular balloon with an outside diameter of 1.2 mm, which is inverted in on itself. As hydraulic pressure is driven into the proximal end of the balloon its inner lumen is forced to roll outside itself to become its outer lumen, thereby advancing itself in a forward manner along the tubal lumen. Furthermore, any cannula system located within the inverted balloon lumen, such as a falloposcope, irrigation system or gamete transfer catheter, will be carried forward.

The advantages of the LEC system are fourfold. First, the LEC and its balloon automatically keep the falloposcope coaxially aligned along the tubal lumen. Second, its unrolling action effectively eliminates any shear forces between the balloon element and the inner walls of the tube. Third, the forward progression of this LEC balloon system concentrates its energy at its tip, making it a self-guiding delivery system that follows the tubal lumen and conforms to variations in tubal diameter. Fourth, the LEC system can be delivered via the transvaginal route to the uterotubal ostium using either a hysteroscopic or non-hysteroscopic approach [15]. The use of the LEC systems without the need for hysteroscopy enables falloposcopy or tubal cannulation procedures to be performed in the non-anaesthetised woman with little or no sedation. These properties may make the LEC system an attractive method for placing gametes or embryos at a known position in the oviduct lumen under visual monitoring, using the transvaginal approach. As the balloon catheter continues to roll out, it advances the falloposcope automatically, but at the same time the falloposcope can be moved independently for fine adjustments to keep it aligned with the leading end of the balloon catheter for optimal visualisation and prevention of epithelial trauma, particularly during antegrade manoeuvres.

Falloposcopic Classification, Scoring, and Localisation of Tubal Lumen Disease

A falloposcopic classification and scoring system for determining the localisation, nature and extent of tubal lumen disease was developed so that objective evaluation of the endoscopic findings could be recorded and graded [16]. A scoring system for a series of parameters including degree of patency, abnormal epithelial changes, abnormal vascular patterns, degree of adhesion formation, amount of dilatation, and abnormal intraluminal contents was developed. Scores of 1 (normal), 2 (mild to moderate disease), and 3 (severe disease) were ascribed to each of the four segments of the left and right tubes. A total minimum score of 20 for each tube reflected normality, a score between 20 and 30, mild to moderate endotubal disease, and a score of 30, severe endotubal disease (Fig. 10.3).

Such a scoring system could then be prospectively evaluated against the diagnostic value of falloposcopy in relation to tuboplasty treatment and pregnancy outcome. The classification and scoring table is divided into three sections. The top section covers the history, and the middle section contains the scoring chart. The third section contains a diagram of the tubes, uterus and ovaries, treatment summary, pregnancy prognosis, and a recommendation for future treatment. If the tube was considered to be falloposcopically normal throughout its intramural, isthmic, ampullary and fimbrial segments (normal score = 20 points), no further procedures were performed. Falloposcopic characterisation of each segment of the fallopian tube has been described and its position confirmed laparoscopically by transillumination from the leading end of the falloposcope through the wall as it progresses along the tubal lumen [2]. If the tubes contained intraluminal debris or filmy adhesions, simple tubal cannulation and "aquadissection" were performed [1,3,17]. If the tube contained polypoid structures, thicker criss-cross adhesions, or short segmental strictures, the passage of flexible dilatation wires with OD up to 0.8 mm (Cook OB/GYN, Spencer, IN; and Target Therapeutics, Fremont, CA) were used in an attempt to dilate stenoses or break down adhesions [1,3]. When there was evidence of thick intraluminal adhesions, rigid structures or obstructions small "Stealth" balloon catheters (Target Therapeutics, Fremont, CA) with OD of 1.0 mm in the deflated and up to 2.5 mm in the inflated state were used [1,3] (Fig. 10.4).

Operative falloposcopy procedures were performed under laparoscopic monitoring to observe if the tuboplasty technique was associated with excessive tubal distortion, partial or complete perforation, or bleeding. These tuboplasty techniques have been described in detail [1]. Before any falloposcopic-directed tuboplasty procedure, the nature, site and extent of the lesion was classified and scored. The lesion was falloposcopically reassessed immediately after the

FALLOPOSCOPIC CLASSIFICATION, SCORING AND LOCALIZATION OF TUBAL LUMEN DISEASE

Patient's Name ______ Date ______ Phone # ______
Age ______ G ______ P ______ SAB ______ TAB ______ Ectopic ______ Infertile: Yes ______ No ______
Other Significant History (i.e. surgery, infection, etc.) ______

HSG ______ Sonography ______ Photography ______ Laparoscopy ______ Laparotomy ______
Cycle Details ______
Semen Details ______

	RIGHT TUBE				LEFT TUBE			
SITE of DISEASE	INTRAMURAL	ISTHMIC	AMPULLARY	FIMBRIAL	INTRAMURAL	ISTHMIC	AMPULLARY	FIMBRIAL
PATENCY Patency — — — — 1 Stenosis — — — — 2 Fibrotic obstruction — — 3								
EPITHELIUM Normal — — — — — 1 Pale, Atropic — — — 2 Flat, featureless — — 3								
VASCULARITY Normal — — — — — 1 Intermediate — — — 2 Poor pallor — — — — 3								
ADHESIONS None — — — — — 1 Thin, weblike — — — 2 Thick — — — — — 3								
DILITATION None — — — — — 1 Minimal — — — — 2 Hydrosalpinx — — — 3								
* OTHER — — — 2-3								
CUMULATIVE SCORE								
TOTAL SCORE	RIGHT TUBE =			(NORMAL = 20)	LEFT TUBE =			(NORMAL = 20)

A cumulative score for each tube of: 20 = Normal Tubal Lumen; > 20 but < 30 = Moderate Endotubal Disease; > 30 = Severe Endotubal Disease.
* Mucus Plugs or Tubal Debris, Endotubal Polyps, Endometriosis, Salpingitis Isthmica Nodosa, Inflammatory, Infectious, Neoplastic conditions and absent tubal segments are each assigned a score of 2 to 3 depending on the significance of the lesion.

Treatment (Specify R & L Tube Surgical Procedures).

Nothing ______
Aquadissection ______
Guidewire Cannulation ______
Wire Dilitation ______
Direct Balloon Tuboplasty ______
Other ______

Prognosis for Conception:

______ Excellent (> 75%)
______ Good (50-75%)
______ Fair (25-50%)
______ Poor (< 25%)

Recommended Followup Treatment: ______

Surgeons ______

RIGHT
LEFT

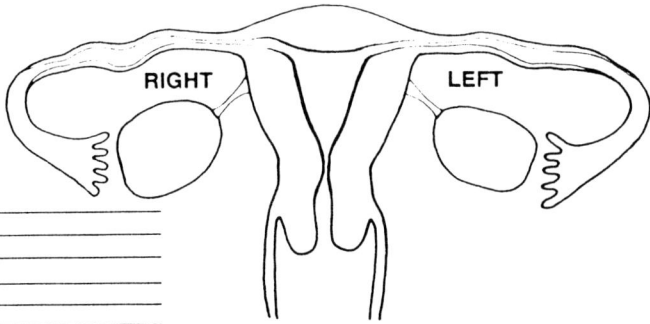

Fig. 10.3. Falloposcopic classification, scoring and localisation of tubal lumen disease.

tuboplasty procedure was completed to observe if there was improvement to the lesion of if evidence of tubal damage had been induced. Improvement was considered to be significant if a stenosis, adhesion, or obstruction had been visibly enlarged or broken down with a concurrent improved flow of fluid

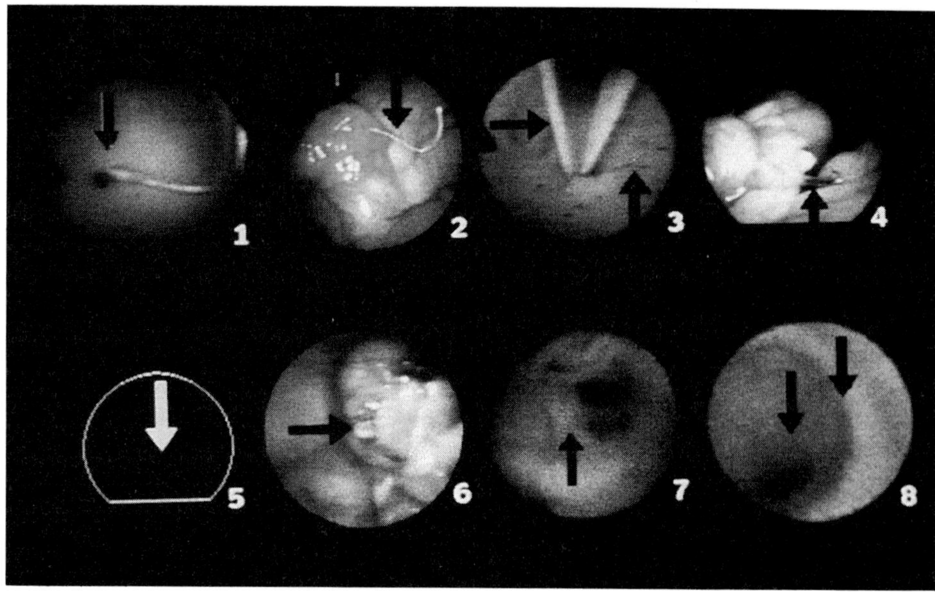

Fig. 10.4. 1, Flexible guidewire (Target Therapeutics. Fremont, CA) entering the ostium. 2, Same platinum-tipped guidewire exiting the fimbria. 3, Falloposcope (Olympus Corp.) enclosed in its Teflon cannula (Target Therapeutics) entering the ostium. The openings of individual proliferative endometrial glands can be seen. 4, Close-up laparoscopic view of the same transparent Teflon cannula enclosing the falloposcope, exiting the fimbria. 5, The proximal white end of a tubal plug occupying the isthmic lumen as contrasted against the Indigo Carmine solution instilled by a tubal cannula. 6, Mobilisation of the same plug as seen in 5, presenting to the fimbria, followed by the appearance of Indigo Carmine fluid, indicating the achievement of patency. 7, Stringy, non-obstructive weblike or fenestrated adhesions in a pale devascularised fibrotic and atrophic epithelium. 8, Stenosis in isthmic segment down to 0.2 mm diameter. Note the proximal distension of the lumen due to irrigation during the falloposcopy procedure.

through the tube and associated fimbrial spill. The patients were prophylactically given an intravenous antibiotic at the commencement of the falloposcopy procedure. They were reviewed within the first week of surgery and then at 3-month intervals for a minimum of 12 months. Patients with subsequent tubal patency were followed prospectively for 1 year to see if pregnancy occurred spontaneously.

Clinical Results Following Falloposcopy Procedures Using the Coaxial Delivery System

To date 75 women with a provisional diagnosis of endotubal disease have had a total of 121 tubes available for falloposcopy. Falloposcopy was performed successfully within 112 tubes. Technical failures, defined as an inability to advance the falloposcope within the tubal lumen in the absence of a detectable intraluminal obstruction, occurred in 9 cases (7%). Technical failures were because of the inability of the currently available coaxial cannulation systems to negotiate successfully the more tortuous tubal lumens that are encountered in any population of women [2]. Therefore, the falloposcopic evaluations and pregnancy rates (PRs) were confined to the 66 women who underwent the procedure successfully. There were no intraoperative or postoperative complications related to the falloposcopy procedure. However, when flexible wire cannulation or direct balloon tuboplasty procedures were attempted on 27 occasions in which severe endotubal disease was diagnosed, five instances of partial and one instance of complete tubal perforation were observed laparoscopically to occur just proximal to the obstructive lesion. These perforations were small and the same size as the dilator or balloon (<1 mm in diameter) and were not associated with external tubal bleeding as viewed laparoscopically, but were associated with minor, endotubal bleeding when viewed falloposcopically; this bleeding spontaneously resolved in less than 1 min on two of the five occasions in which it was noted. If the tube had undergone advanced fibrosis, the normal vascular epithelium was absent and bleeding was not likely to occur with perforation. Therefore a total absence of bleeding was considered to be a poor prognostic sign. In all instances, the nature of the intratubal disease was of a major degree, and the perforation was not considered sufficient to further compromise the patients' tubal status. There was no evidence of pyrexia, inflammation, or infection associated with either diagnostic or operative falloposcopy.

Of the 112 tubes examined falloposcopically 60 (54%) had evidence of disease, and the total number of lesions detected was 119. Therefore, each diseased tube had an average of two endotubal lesions present. The incidence of falloposcopically disease-free tubal lumens was 46%, despite suggestive evidence from prior hysterosalpingograms and hydrochromotubation procedures carried out under laparoscopic monitoring that tubal patency or spill could not be demonstrated. Of the 28 women who had falloposcopically normal tubes (falloposcopy score = 20), 6/28 (21%) conceived within a year of the procedure. However, the clinical pregnancy rate in women with mild to moderate disease (falloposcopy score 21–30) was poor during a one-year follow-up period where only 2/22 (9%) conceived. None of the 16 women with severe endotubal disease (falloposcopy score greater than 30) conceived during the one-year follow-up period [16].

Preliminary Evaluation of Falloposcopy Using the Non-Hysteroscopic Linear Eversion Catheter Delivery System

The LEC delivery system (Imagyn Medical) has been specifically designed to deliver a 0.5 mm OD falloposcope successfully and safely into the human fallopian tube lumen using the transvaginal approach and without the need for hysteroscopic guidance.

It is important to realise that the balloon or membrane employed in the linear everting catheter system is quite distinct from the conventional "balloon catheters" used for tuboplasty procedures. The linear everting catheter balloon unrolls or everts to achieve linear progression, whereas conventional balloons expand radially, without advancing in a forward direction.

Prospective FDA-approved clinical studies to assess the clinical efficacy and safety of the LEC system are currently in progress in North America and collaborating medical centres in the United Kingdom, Europe and Australia. Preliminary studies have indicated that the non-hysteroscopic LEC delivery system can access the tubal lumen with an acceptable degree of reliability. Furthermore, clinically useful diagnostic information could be obtained from the falloposcopic images to influence the course of infertility treatment [15]. In a series of observations made within five separate institutions following 45 falloposcopy procedures in women with suspected endotubal disease, using the LEC system, successful imaging within the tubal lumen occurred in 39 instances (87%). Identifiable lesions within the tube were found in 26/39 cases (62%) and were significant enough to alter the patient's counselled management in 18/26 instances (67%). In these procedures, falloposcopy findings assisted in selecting candidates for IVF versus endotuboplasty or conventional microsurgery. Furthermore, falloposcopy can provide an assessment of healthy tubal mucosa and tubal potency immediately prior to transvaginal intratubal transfer procedures. Thus falloposcopy with the LEC system provides an effective diagnostic tool which can significantly influence patient management [18].

Conclusion

A stepwise developmental outline of falloposcopic exploration of the human fallopian tube lumen has been described. The falloposcope's construction and necessary properties and the technique for safe and successful microendoscopy of the tube have been outlined. The two currently used falloposcope delivery systems and the differences between the coaxial and LEC system have been described. An attempt has been made to define normal and abnormal lesions as viewed falloposcopically and a classification and scoring system has been constructed in order to quantify the degree of severity of endotubal lesions. Preliminary results indicate that meaningful clinical information can be obtained from falloposcopy procedures which may influence the patient's management. The initial studies indicate that falloposcopy may add a useful diagnostic and therapeutic dimension to the clinician's armamentarium for treating tubal infertility and for providing a means of retrograde transfer of gametes and embryos into the fallopian tube.

Acknowledgements. Special thanks to Mrs Anna Manuel for typing the manuscript and to Barrie Schwortz for his expert production of the endoscopic photographs.

References

1. Kerin J, Daykhovsky L, Grundfest W, Surrey E. Falloposcopy: a microendoscopic transvaginal technique for diagnosing and treating endotubal disease incorporating guide wire cannulation and direct balloon tuboplasty. J Reprod Med 1990; 35:606–12.
2. Kerin J, Leon Daykhovsky L, Segalowitz J et al. Falloposcopy: a microendoscopic technique for visual exploration of the human fallopian tube from the uterotubal ostium to the fimbria using a transvaginal approach. Fertil Steril 1990; 54:390–400.
3. Kerin J, Surrey E, Daykhovsky L, Grundfest W. Development and application of a falloposcope for transvaginal endoscopy of the fallopian tube. J Laparendoscopic Surg 1990; 1:47–56.
4. Mohri T, Mohri C, Yamadori F. Tubaloscope: flexible glass fibre endoscope for intratubal observations. Endoscopy 1970; 2:226–9.
5. Grundfest WS, Litvack F, Sherman T et al. Delineation of peripheral and coronary detail by intraoperative angioscopy. J Am Surg 1985; 202:394–400.
6. Grundfest W, Litvack R, Sherman T et al. The current status of angioplasty and laser angioscopy. J Vasc Surg 1987; 5:667–72.
7. Gardner AK. The causes and curative treatment of sterility with a preliminary statement of the physiology of generation. New York: DeWitt and Davenport, 1856.
8. Lisa JR, Gioia JD, Rubin IC. Observation of the interstitial portion of the fallopian tube. Obstet Gynecol 1954; 99:159–60.
9. Sweeney WJ. The interstitial portion of the uterine tube: its gross anatomy, course, and length. Obstet Gynecol 1963; 19:3–8.
10. DeCherney AH. Anything you can do I can do better ... or differently! Fertil Steril 1987; 48:374–6.
11. Hernstein A, Neustadt B. Intramural portion of the fallopian tube. Z Geburtshilfe Gynakol 1924; 88:431–7.
12. Corfman RS. Falloposcopy: frontiers realized ... a fantastic voyage revisted. Fertil Steril 1990; 54:574–6.
13. Rubin IC. Observations on the intramural and isthmic portions of the fallopian tubes with special reference to so called "isthmospasm". Surg Gynecol Obstet 1928; 46:87–94.
14. Rubin IC, Bendeick AJ. Metrosalpingography with the aid of iodized oil. AJR 1928; 19:348–52.
15. Bauer O, Diedrick K, Bacich S et al. Transcervical access and intra-luminal imaging of the fallopian access technology. Hum Reprod in press.
16. Kerin JF, Williams DB, San Roman GA, Pearlstone AC, Grundfest WS, Surrey ES. Falloposcopic classification and treatment of fallopian tube lumen disease. Fertil Steril 1992; 57:731–41.
17. Kerin JS, Surrey ES, Williams DB, Daykhovsky L, Grundfest WS. Falloposcopic observations of endotubal mucus plugs as a cause of reversible obstruction and their histological characterisation. J Laparoendoscopic Surg 1990; 1:97–101.
18. Kerin J, Pearlstone A, Surrey E et al. Non-hysteroscopic transvaginal falloposcopy using a linear eversion catheter (LEC) delivery system. International Congress of Gynaecologic Endoscopy. September 1992; Chicago Illinois.

Discussion

Aitken: There is some evidence now, from our own work and others, that oxidative stress does play a role in arrested embryonic development. Has anybody looked to see whether radical scavenging enzymes like superoxide dismutase or glutathione peroxidase are secreted into the tubal lumen?

Leese: There is a paper showing superoxide dismutase (SOD) present throughout the epithelial cells. As for the secreted enzyme, I do not know.

Aitken: Would it be worth assessing the antioxidant activity of tubal fluid?

Leese: That has not been done as far as I know. And there would be the problem of collecting sufficient tubal fluid. I guess one would have to use a domestic animal to do it. Certainly SOD is present in the epithelial cells, and could be secreted I guess. But we do not know. The compound pyruvate will scavenge free radicals to some extent, and that is present in quite high concentrations at fertilisation. It could be an added function of pyruvate but we have no real evidence for that.

It is perhaps worth adding, the PO_2 in the tube is about 40 mmHg, which is equivalent to about 5% oxygen. So those clinics, which grow human embryos in 5% oxygen, that is probably physiological. Indeed people will claim that they get much better embryo development rates at that PO_2 whereas other people will claim different.

Templeton: Jansen when he was working in San Antonio suggested that the isthmus might secrete mucus which acted as a sperm reservoir in the human, possibly similar to other species. Is there evidence of mucus secretion at all times of the cycle?

Leese: It is quite copious, and clears at the time that the embryos need to go down to the uterus. That is an old finding and I guess I should like to see more recent data, but it looks very sound, as though there is copious mucus prior to fertilisation which then disappears as the embryos have to progress. In embryos such as the rabbit it is very conspicuous; the embryo has a very pronounced mucus coat. It could be argued that all embryos need a mucus coat to some extent.

Kerin: It is a fascinating area. I have seen mucus within the fallopian tube. We have taken about 10 women now around the time of ovulation and very gently put the falloposcope in the tube with very low irrigation rates. Our biggest frustration is that we probably flush out so much information. But we have seen white mucus and I have got it on videotape.

The other fascinating thing is that purely by chance when we were passing the falloposcope under laparoscopic control with fluid through the isthmus we could see very fine mucus strands, very springy, similar to pericervical mucus. It is interesting how we can extend it out beyond the fimbria with the falloposcope, and when we touch this mucus on to the pre-ovular follicle it sticks to it very firmly and we can then draw it back into the tube again.

Winston: Occasionally it can be encountered under the operating microscope.

Watson: How long do fallopian tube epithelial cells in culture remain differentiated as apparently good tubal cells?

Leese: We have not got beyond 17 days. This is primary culture and we have shown steady glucose consumption and lactate production over days 7 to 17; they become confluent in 2, 3 or 4 days. Other people have passaged human tubal epithelial cells but not grown on plastic. Bongso has passaged them many times, as he has recently described [1]. They then begin to dedifferentiate and the cilia are lost; that is the message there. These seem to be the ones that are difficult to maintain with the phenotype. But we have not passaged ours. They are primary culture and 17 days is our limit so far. But the obvious aim would be to passage them, freeze them, know that they are AIDS free, HIV free, and that would be the substrate on which to put human eggs and embryos.

Gosden: Going back to the question of mucins both in the oviduct and in the cumulus mass round the oocyte, one of the interesting properties of these materials is that they are very highly charged and by binding cation, they may profoundly alter the microenvironment. The sampling that we take from the tubes, may be very different from that in the cells' microenvironment. There could be dangers of modelling culture media precisely on the fluid rather than on the environment within the glycosaminoglycans.

Leese: I could not agree more. We are down to about a microlitre in the human. The expression that has recently been used about the human tubal lumen was "a potential space". Often it is no space at all and so potential space is a very apposite description. We are not down to the picolitre or nanolitre samples we would need to take for the microenvironment. We have produced a model looking at unstirred layers around embryos and this could well be important. And we have no analytical access to that compartment.

Winston: Certainly one of the problems with the tube is that drainage does not come through the uterotubal junction, and that is one of the reasons for the mechanism of the hydrosalpinx formation. Similarly we see this in the sterilised tube where one segment becomes distended, i.e., the uterine end, but the distal end does not become distended.

Braude: Dr Leese mentioned wories about the viral problem in connection with cocultured cells. Are there any other cells that have been successfully used.

Leese: The consensus in the coculture field is that oviduct is best, but it is not overwhelming. Some people find no advantage of coculture or of conditioned medium. But there were 100 abstracts in *Theriogenology* not so long ago and about 80 found some advantage from co-culture. It then comes down to different laboratories, different species, different methods. But I do not think it will go away, and I suspect it will be frozen human tubal lines or something like that. It is whether they de-differentiate too far to sustain the embryos, and we just do not known that yet.

Aitken: Does it have to be of the same species?

Leese: No. They can be mixed. One can put any embryo in a rabbit oviduct and it is quite happy. The HFEA might be interested if we were to do that! The literature is a mess at the moment. The people who culture other cells will say that what we are going through now, the mess of coculture, is something we have to go through, because coculture has been used in other cell types. It will resolve itself eventually, it just needs more research. But at the moment the literature is confusing and contradictory.

Cooke: The legislation prohibits doing an embryo transfer into another species, but as I recall it does not mention cell lines.

Leese: No. I think it will have to take that one on board. Roger Gosden and I are always interested in simple mixing. We keep saying this, but we have never done the work ourselves, which I guess we should do. Embryos will be stirred within the female tract, both by cilia and by the mesosalpinx, as will sperm, and we just keep saying this and not doing the research to show whether that simple mixing might confer some advantage.

Reference

1. Bongso A, Ng SC, Fong CY, Ratnam S. Cocultures: a new lead in embryo quality improvement for assisted reproduction. Fertil Steril 1991; 56:179–91.

Chapter 11

The Role of Reproductive Surgery

R. M. L. Winston and R. A. Margara

Introduction

The treatment of tubal disease has been reappraised in the last two decades. It used to carry a virtually hopeless prognosis except in a few cases. Microsurgical techniques and the development of in vitro fertilisation have changed this and the treatment of tubal infertility now has an excellent outlook.

Increasing use of IVF has changed the indications for surgery. The best tubal surgery will inevitably be conducted where the alternative of IVF is also available. In those centres, microsurgery will be increasingly used for the "best" cases, i.e., those with lesser damage and where the prognosis will be better with surgery than with assisted conception.

There can be little argument now that the clinical and anatomical results of operating using a microscope are superior to surgery done using the naked eye. It is sad that, in the 1990s, so many surgeons still cause unnecessary iatrogenic damage by using instruments and techniques that should have been abandoned years ago. The attitude seems to be "We do not do microsurgery, let's give the patient some hope whilst she is on a waiting list for IVF". In our view, that is a cruel deception. The results are unnecessary pain and waiting, avoidable National Health Service expenditure, and the risk that pelvic damage may worsen the prognosis when definitive treatment is finally conducted.

The Scope of Reproductive Surgery

In this chapter, most attention will be given to tubal surgery for inflammatory disease, such as adhesiolysis, salpingostomy, cornual damage and sterilisation reversal. Endometriosis is also an important indication for surgery, particularly when the ovaries are heavily involved. The scope is a good deal wider than this, however, and we frequently perform surgery for various uterine problems such as fibroids, Asherman's syndrome and congenital conditions. Conservative surgery for ectopic pregnancy also has an important place.

Pelvic Adhesions

Pelvic adhesions are the result of inflammation, endometriosis, or previous abdominal surgery. Occasionally congenital adhesions are a significant cause of infertility, sometimes in association with other genital tract anomalies. The clinician will often see infertile patients with pelvic adhesions of varying severity, but with patent tubes.

Basic Microsurgical Approach

We used to make a large incision giving wide exposure, limiting peritoneal handling and traction, and in the past have usually opened the abdomen with a big Pfannenstiel incision. An excellent alternative is the Czerny incision, which gives a bloodless approach with superb access. This incision can be kept small, allowing patients home after three nights. Haemostasis is best secured by use of blended microdiathermy; very often a retractor is not required, avoiding bruising and hence, postoperative discomfort. Whether loupes are used or the microscope, it is important to avoid handling the tissues more than absolutely necessary, to obtain complete haemostasis, to use sharp rather than blunt dissection, and to irrigate the whole field constantly with Ringer's lactate solution. Raw peritoneal areas should be repaired without tension; free peritoneal grafts are very useful in strategic sites. Although some authors favour the use of magnifying spectacles, we invariably use an operating microscope. The coaxial illumination, the freedom of movement and the variable magnification give unsurpassed advantages.

Laparoscopic Adhesiolysis

Laparoscopic surgery is finding increasing favour, because of the rapidity of recovery and the reduced pressure on hospital beds. Patients like it because it avoids a major wound. Most surgeons prefer videolaparoscopy, because keeping one's eye to the eyepiece for long periods of time is tiring. Moreover, the use of closed circuit television allows assistants a view of the procedure. However, the use of even the most sophisticated three-tube colour camera with the best light source does not give the same resolution achieved using direct vision. This is important because some subtle pathological changes may simply be missed using television and accurate dissection is more difficult.

The technique varies considerably from surgeon to surgeon. Excellent descriptions are published by Nezhat et al. [1] and Reich [2]. There is argument about the best method of dissection, and the use of the laser [3], diathermy or simple cold cutting [4] instruments is favoured by various authors. There is no statistical evidence that any of these instruments or methods is superior. Whichever method for dissection is used, all workers are agreed that multiple abdominal punctures are used, in addition to the portal for the telescope. Usually three further entry sites are needed; one for lavage, through which gas, or blood-contaminated fluid, can be aspirated, and two others (usually in each iliac fossa) for grasping or dissection of tissues. Instruments have now been designed for applying sutures during laparoscopic surgery, and these are usually placed with instruments situated in one flank.

Results of Adhesiolysis

Operations conducted to divide adhesions tend to have higher success rates [5] than other types of tuboplasty. Unfortunately, there is always the lurking suspicion that the patient might have conceived without any surgical procedure. There are no published reports of controlled studies on the value of salpingolysis when the tubes are patent. Suitable studies, using life-table analysis of results, are still much needed. There are suggestions in the literature that adhesiolysis can speed conception, as many patients become pregnant very soon after surgery of this kind. In the past we had low success rates after adhesiolysis, but since the advent of IVF with better patient selection, there has been substantial improvement.

One contentious area is whether laparoscopic surgery gives results which are as good, or better than open microsurgery. No clear picture emerges, because controlled randomised trials have not been done. Many of the surgeons doing excellent laparoscopic surgery are uncomfortable using an operating microscope; conversely most practitioners of microsurgery inherently feel that they get better anatomical results with open surgery. There is no doubt that laparoscopic adhesiolysis may be effective for pain [6]. There is also some evidence that open surgery is likely to be associated with recurrent adhesion formation as Lundorff et al. [7] showed in their randomised study of patients followed after removal of ectopic pregnancy. However, it is also clear that laparoscopic surgery is not free of risk of causing further adhesions [8].

Our own figures for adhesiolysis are shown in Table 11.1. To qualify for this study patients had to be under 42 years of age and have both tubes patent, with no evidence of serious cornual disease. This is not a randomised study, though in general, the patients on whom laparoscopic surgery was done had less extensive pathology (hence the laparoscopic approach). Consequently, the results of open surgery seem impressive. Nevertheless, a properly randomised, controlled study is seriously needed. The main prognostic factor seems to be ovarian fixity.

Salpingostomy

Microsurgical salpingostomy has been extensively described [9,10] and a full description of technique is inappropriate here. The most important aspects can be summarised:

Table 11.1. Comparison of pregnancy rates after microsurgery and laparoscopy

	Patients	Pregnant	>28 weeks	Ectopics
Microsurgery				
Free ovaries	47	26 (55%)	22[a]	2
Fixed ovaries	41	17 (41%)	15[a]	0
Total	88	43 (49%)	37	2
Laparoscopy				
Free ovaries	42	17 (40%)	15	4[b]
Fixed ovaries	33	9 (27%)	7	2
Total	75	29 (35%)	23	6

[a]Two of each of these patients had failed laparoscopic surgery.
[b]Three patients with tubal ectopics also had a term pregnancy.

1. The tube should be opened at its most terminal part. Any incision which is not terminal will, in effect, be linear. Linear incisions in the tube tend to heal over [11].
2. The reconstructed fimbria should be positioned so that the tubal ostium can move over the whole ovarian surface.
3. It is important to avoid cutting across mucosal folds. Transection of the folds may disturb the blood supply and also prevent the passage of the egg into the tube.
4. Reconstruction of the fimbria ovarica may help provide the main communication between ovary and tube.
5. Tubal mucosa should be everted sufficiently to achieve a stable tubal ostium. Too much eversion may devitalise the fimbrial "lip".
6. Too large an ostium should be avoided, otherwise there is a risk that mucosal folds will not gain contact with each other.
7. Tubal tissue should not be resected unless it is obviously redundant. Following salpingostomy, involution of some large hydrosalpinges occurs.
8. Damage to the fimbrial blood supply should be avoided.
9. No raw areas should be left around the margins of reconstructed fimbria otherwise there may be recurrent tubal closure.
10. The surface area of the ovary available for ovulation should be maximised.

Careful dissection can only, in our view, be achieved under good magnification. We feel less confident that good anatomical results are achieved using the laparoscope, particularly when the laser is used. Laser treatment to the end of the tube causes considerable damage and denudes serosal surfaces. Check laparoscopy after laser salpingostomy suggests that the effects of this damage are often considerable with fibrous adhesions and reclosure of the tube being frequent.

Results of Salpingostomy

Following microsurgical salpingostomy [12] in a series of 323 primary operations, we found that 74 patients had a term pregnancy. Over half the women who conceived went on to have at least one other term pregnancy; emphasising the return to natural fertility which can occur after tubal operations. The results were closely related to pre-existing pathology. Stage I disease (occluded tubes with a thin wall and relatively normal mucosa, no fibrosis, and few ovarian adhesions) resulted in 39% of patients having a term pregnancy. Operations on stage II disease (thick-walled tubes, or tubes with fibrous adhesions and poor mucosa) resulted in 18% of patients conceiving. In stage III and stage IV disease the results were worse with 13% and 8% of patients being successful. Only a minority (in our series, 23%) of patients have stage I disease, but in these patients surgery undoubtedly should be preferred to IVF. Our results were similar with salpingostomy done as a repeat procedure (secondary salpingostomy) in women whose tubes reoccluded after conventional surgery. Secondary salpingostomy, widely thought of as useless, was comparatively successful when pathology was limited. Our results for primary surgery are similar to those reported by others in smaller series [9,13–16]. These results are certainly no worse, and possibly better than, those obtained by authors undertaking laparoscopic salpingostomy or laser surgery [17–20].

Our own results for laparoscopic salpingostomy are worse than those obtained by open microsurgery. Our study (Table 11.2) is not randomised and there is a possibility that the cases on whom laparoscopic surgery was undertaken were those with more badly damaged tubes, it tending to be offered more frequently to women who already had failed open macrosurgery.

Surgery for Cornual Occlusion

Cornual anastomosis [21] is now the standard procedure for dealing with cornual block, or blockage in the proximal isthmus. The commonest cause of

Table 11.2. Comparison of pregnancy rates after laparoscopy and open surgery

	Patients	Term	>2 term	Intrauterine	Ectopics
Open surgery					
Primary	323	74	39	106 (33%)	32 (10%)
Repeat	65	12	7	16 (25%)	6 (9%)
Total	388	86	46	122 (31%)	38 (10%)
Laparoscopy					
Primary	33	4	0	6 (18%)	2
Repeat	45	2	0	4 (8%)	4
Total	78	6	0	10 (13%)	6

such blocks is infection, following an abortion or the insertion of an intrauterine contraceptive device, or salpingitis isthmica nodosa. Cornual occlusion following adenomyosis or as a result of congenital atresia of the intramural portion can also be treated by cornual anastamosis. Some patients requesting reversal of sterilisation are also best with tubocornual anastomosis if there has been injury to the cornua. Tubal implantation used to be the standard until microsurgical cornual anastomosis was introduced. Implantation should only be contemplated when all the intramural portion is damaged. Occasionally this part of the tube is so fibrotic, either as a result of severe infection or extensive high-frequency diathermy during sterilisation, that implantation only is possible. Tubal implantation has several disadvantages. It causes undue shortening of the tube, destroys the uterotubal junction and is associated with excessive bleeding and damage to tubal blood supply. Moreover, there is a risk of rupture of the uterus in the event of subsequent pregnancy and therefore most authorities agree that delivery by elective caesarean section is best. For these reasons tubocornual anastomosis should be done whenever feasible.

Results of Cornual Anastomosis after Inflammation

The results of tubal anastomosis following resection of an inflammatory block are less good than the results of sterilisation reversal. There is no doubt that inflammation, particularly salpingitis isthmica nodosa and especially tuberculosis, causes diffuse muscular and mucosal damage which may extend along the whole length of the tube. A high proportion of patients with salpingitis isthmica nodosa have hydrosalpinges or major adhesions. In spite of this, the use of microsurgery has improved the results of surgical treatment. Previously, isthmic and cornual obstruction was treated by various methods of tubal implantation [22] and, with the exception of a very few series, only 20% of patients had live births. Rocker [23] pointed out that the intramural portion of the tube was seldom obstructed after inflammation, demonstrating that so-called cornual block usually occurred where the isthmic portion of the tube joins the uterine muscle externally. This observation led Shirodkar [24] and Ehrler [25] to try to join the tube directly onto the patent intramural segment over an indwelling splint. However, neither of these surgeons produced superior results to those following uterotubal implantation. This is probably because the lumen of the tube is so narrow at this point that a conventional macrosurgical approach is inadequate to get proper apposition. The microsurgical approach to anastomosis improved results.

Results of tubocornual anastomosis depend on the proficiency of the surgeon as the technique demands experience in identifying pathological tissue requiring excision and manual dexterity. A frequent mistake is to resect too little rather than too much tube. High power magnification (16–25×) is required to identify healthy mucosa and muscle. Inadequate resection of tissue (one cause of recurrent blockage early in our series) may result in a continuing inflammatory process.

Pathology of Cornual Obstruction

Fortier and Haney [26] found that the commonest causes of cornual block were obliterative fibrosis (38.1%), salpingitis isthmica nodosa (23.8%), chronic inflammation (21.4%) and endometriosis (14.3%). Wiedemann et al. [27] felt that the degree of histological damage was an important prognostic factor; they also reported that careful serial sections revealed that some tubes, apparently blocked at the time of laparotomy, showed no definite occlusion. Jansen [28] pointed out that, whatever the pathology, wide excision of abnormal tissue was important for best results. Pathological findings vary greatly from one country to another. For example, Wang [29] found non-specific infection in about half his cases and tuberculosis in one-third. Endometriosis was rare in his experience. The surgical results were poor, suggesting that he probably was operating on severely damaged tubes.

Results of Cornual Anastomosis

Our results following surgery for inflammatory or endometriotic cornual obstruction are reported in Table 11.3. When long lengths of tube require excision, or after tuberculosis, the results are less good. Even when patients do not conceive, tubes joined microsurgically remain patent in 85% of cases. A limiting factor is recrudescence of disease or extension of the original inflammation seen at check laparoscopy one year after tubal reconstruction. Gomel [30] reported similar results, 53% of his patients having at least one term pregnancy after this approach. Gillett and Herbison [31] found little correlation between degree of pathology or adhesion formation, and subsequent pregnancy rate. A poor sperm count or poor ovulation were more significant. Donnez and Casanas-Roux [32] reported that 44% of patients achieved a term pregnancy and that good tubal length, absence of chronic inflammation and lack of endometriosis all favourably influenced prognosis.

Ectopic pregnancy has not been a large problem in any major series; this is surprising as many of these women have diffusely damaged but patent tubes. One advantage of cornual anastomosis is that it does not weaken the uterine muscle and caesarean section is not mandatory afterwards.

Table 11.3. Cornual anastomosis for inflammatory disease

	Patients	Delivered	Ectopic	%
<1 cm isthmus resected	107	65	4	59
>1 cm isthmus resected	83	39	3	46
Ampullocornual	53	21	4	40
Uteroisthmic	25	8	0	32
Uteroampullary	32	14	1	45
Implantation (macro)	23	6	1	26
Total	323	153	13	47

Alternatives to Sutured Microsurgical Cornual Anastomosis

Some authors have used tissue adhesives or thrombin instead of sutures. Rouselle et al. [33] pointed out that thrombin, an adhesive, appeared to reduce implantation rate in the rabbit. Dargenio et al. [34] found no effect on nidation rate when they used fibrin glue, and demonstrated good patency rates in animals; however, they did not translate this success into clinical practice. Swolin [35] proposed the use of a single suture to hold the tube in apposition; he reported a single pregnancy, since when, like so many "advances" in tubal surgery, no further success has been published. Another approach was suggested by Ngheim Toan [36] who threaded an absorbable filament past the cut ends of tube and allowed the tube to heal over this. Although he claimed 100% patency rates, there is no clinical advantage in this approach.

Two other techniques have received widespread attention. The laser has been used to resect the damaged cornu by various authors; most have not reported any clinical improvement. However, in a small uncontrolled study, it was claimed that 100% of the cornua treated with CO_2 laser excision remained patent and that 71% of patients became pregnant [37]. These results are surprising as no other worker using the laser has achieved such a success rate. Moreover, it is difficult to understand how the laser could improve the quality of the join at the cornu.

Mechanical cornual canalisation has also had some adherents. Daniell and Miller [38] appear to have been first in establishing a pregnancy after cannulating the intramural portion from below. They used a ureteric stent and reported a term pregnancy. Thurmond and Rosch [39] performed transcervical fallopian tube catheterisation under fluoroscopic control in 100 patients with blocked cornua; 26 intrauterine pregnancies resulted. It is unclear whether all these women were fully evaluated and laparoscopic confirmation of tubal pathology was not confirmed in most cases. In a more recent report, Thurmond [40] states that 10% of his patients suffered ectopic pregnancy. In a more cautious report, Deaton et al. [41] suggested the alternative of hysteroscopically guided wires passed into the cornu. They felt that this was therapeutic, six out of eleven of their patients conceiving. Further studies are needed to evaluate a technique which may have a definite place when the cornua show only limited damage.

Reversal of Sterilisation

Reversal of sterilisation was the first procedure to be widely undertaken using microsurgery [21]. The microsurgical approach has stood the test of time and the best results are obtained when sterilisation has been done using clips in the isthmic region. Results are less good when the ampulla is damaged, or when the lumen of the two joined segments show wide disparity in diameter. Our results are shown in Table 11.4.

Table 11.4. Sterilisation reversal

	Patients	Term	Ectopics	%
Isthmoisthmic	21	17	0	81
Isthmoampullary	32	20	3	62
Cornualisthmic	27	22	0	81
Cornualampullary	33	18	2	55
Ampullary–ampullary	19	9	0	47
Clip removal	25	24	0	95
Miscellaneous[a]	30	13	2	43
Fimbriectomy	8	1	1	12
Total	195	124	8	64

Site of anastomosis classified according to "better" tube. [a]Miscellaneous refers to unclassifiable, e.g. where many adhesions, large amounts of the tube missing unequally on both sides or double-clipping or ligation.

Endometriosis

The surgical treatment of endometriosis is controversial. Many infertile women have minor endometriotic lesions in the peritoneum which are generally easy to ablate. There has been a vogue for treating these with diathermy, or with the laser. There is no controlled evidence that this improves fertility, though some authors claim that such treatment prevents the development of more serious lesions later. Conversely, severe endometriosis represents a very difficult technical problem. There is no clear consensus about how useful surgical management really is. Certainly, surgery is helpful for women in pain, but restoration of fertility is unpredictable. One problem is that none of the published series are truly comparable with each other. This is because most authors rely on the American Fertility Society (AFS) classification system which is difficult to score and may not be accurately related to prognosis. A second problem is that, in the case of severe endometriosis, there are no controlled studies published, making proper evaluation impossible.

We are uncertain about the place of surgery for endometriosis. It seems to be a significant advantage in preparing the pelvis for IVF and in our hands, superovulation and egg collection has been more successful after all endometriotic cysts have been resected with careful ovarian repair. Our results are reported in Table 11.5.

There has recently been increasing use of laparoscopic surgery for endometriosis. The results seem to differ from centre to centre, possibly in part due to poor standardisation of material, variable reporting and indifferent follow-up. Kojima et al. [42] reported that only 37% of patients conceived after surgery, though most obtained pain relief. Nezhat et al. [43] claimed an astonishing success rate in a large series. In their report, 168 patients out of 243 (69.1%) operated for endometriosis conceived. No figures are given for overall length of infertility before treatment. No further fertility treatment was

Table 11.5. Ovarian cystectomy for endometriosis

	No tubal pathology	Tubal damage
Single cyst		
Patients	45	32
Pregnant	18 (40%)	11 (34%)
Bilateral cysts		
Patients	32	27
Pregnant	10 (31%)	5 (18%)

Rather than use the vagaries of the AFS classification, only patients with a cyst of greater than 3.5 cm diameter are included in these results. Bilateral involvement, or concomitant tubal damage, gave the worse results.

given after surgical intervention. In this series, there was no significant difference at all in success rate between those patients who had the mildest disease, those with moderate involvement, and those with severe damage (AFS stage IV). These results are difficult to understand as in almost all other reported series a relationship has been reported between pathology and prognosis. This mystery is compounded by an earlier report by the same authors [44] in which 102 patients were treated for endometriosis, some without cysts; 75% with mild endometriosis conceived; only 42.1% with stage III disease conceived.

It is not possible to comprehend why different centres should achieve such widely varying results. These paradoxes can only be solved by properly conducted randomised studies which are carefully evaluated by life-table analysis.

Uterine Surgery

The commonest indication for uterine surgery in infertile women is leiomyomata. Several authors [45–47] have tried to evaluate the success of myomectomy in restoring fertility. A problem remains that the indications for myomectomy are diffuse and there is no evidence that all fibroids cause infertility. In general, reports on myomectomy do not give the length of time patients were infertile before surgery and do not analyse cumulative conception rates afterwards. This makes judgement of the effect of surgery impossible.

Babaknia et al. [48] reported a term pregnancy rate of 38% after myomectomy for primary infertility, 50% for secondary infertility. They could not find evidence that endometrial cavity distortion before surgery correlated with subsequent prognosis. Berkeley et al. [49] concluded that myomectomy may actually result in decreased fertility, possibly due to adhesions. However, in their series conventional surgical techniques were used. Garcia and Tureck [50] removed submucosal myomata larger than 5 cm from the uterine wall; 53% of their patients attempting a pregnancy conceived within ten months of surgery.

They concluded that myomectomy was beneficial. Rosenfeld [51] removed intramural or subserous leiomyomata only; 65% of his patients conceived within one year. No correlation was found between success rate and the patient's age, duration of fertility, size of fibroids, or the presence of menorrhagia. Egwatu [52] reported reduced miscarriage rates after myomectomy in a large series from Nigeria.

Various modifications of traditional myomectomy technique have been reported, of which the most popular seems to be by laser. In Starks' [53] series, 59% had a term pregnancy. He found no evidence that the size or positon of tumours made any difference to outcome, but presumably most of his patients would have had relatively superficial myomata for laser myomectomy to be possible.

In our own series (Table 11.6) patients have been classified according to the position of the myoma. Intrauterine fibroids occupying the cavity have a very significant effect on fertility and their removal carries a marked increase ($P < 0.001$) in the chance of conception. Unlike other authors, we found that age apparently carried an effect and that older women did less well after myomectomy.

We have had no experience in laser myomectomy; we are reluctant to remove submucosal myomata by transcervical resection because we are concerned that the endometrial cavity may be irreparably damaged by this approach.

Table 11.6. Results of myomectomy from different sites

Position	Mean age (years)	Patients	Pregnant	MCI[a]
Polyp >1 cm diam[b]	37.3	22	16	4.2
Submucosal	34.3	31	15	7.1
Subserous	37.1	34	14	8.3
Cornual	31.1	9	3	5.2
Others all >35		41	17	11
Others all <34		21	11	7.8

All patients had been infertile for a minimum of 18 months. [a]MCI, mean conception interval in months after surgery. [b]Myomectomy done by hysterectomy, except in 10 cases. The recorded miscarriage rate in this series was 13.8%.

Age and Female Reproductive Surgery

Age appears to have an important influence on the outcome of microsurgery. We recently analysed the outcome of sterilisation reversal by age in 504 patients (Fig. 11.1). As nearly all these women had been very fertile before sterilisation and all were checked for ovulation and male fertility before reversal, and were generally sexually active, we question whether tubal surgery generally is particularly helpful in women much over 40 years old. By comparison, we have found that IVF can have surprisingly good success rates in older women (Fig.

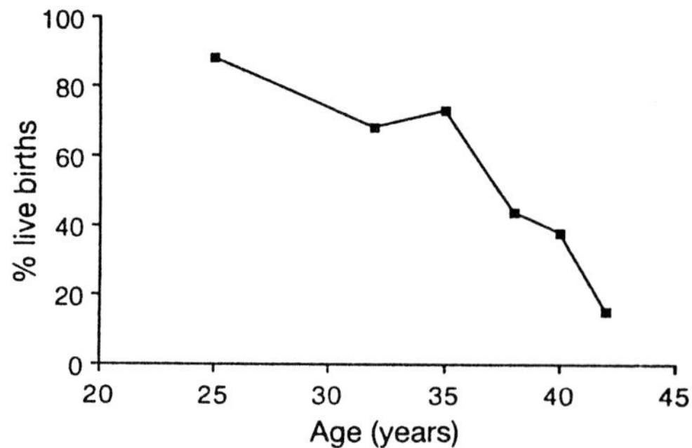

Fig. 11.1. The outcome of sterilisation reversal in 504 patients.

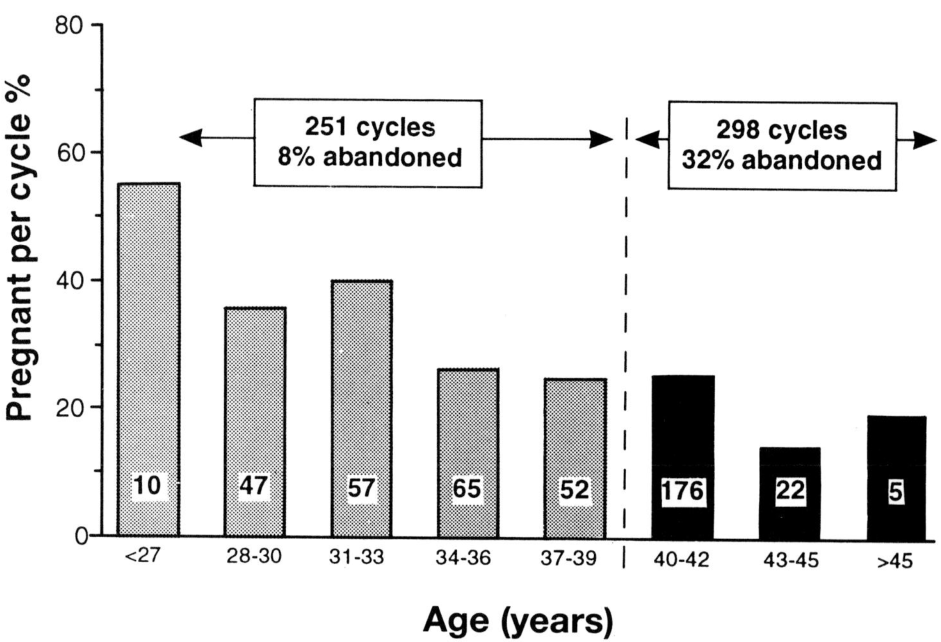

Fig. 11.2. Analysis of 298 treatment cycles in women over 40 years old compared with 251 cycles in a matched cohort of women under 40 years, treated during the same period by IVF. Most had two embryos transferred. Of women over 40 years getting to oocyte collection 19% had a pregnancy, but in a high proportion of cycles (32%) superovulation was inadequate and treatment was abandoned.

11.2). IVF cycles, with the transfer of more than one embryo, carry conception rates which are as good as, or better, than those achieved naturally. In view of this apparent advantage, we tend to recommend that older women undergo IVF rather than surgery as a first option. Perhaps fertility surgery is better reserved for those women who have more time to conceive naturally following operation.

References

1. Nezhat C, Hood J, Winer W, Nezhat F, Crowgey SR, Garrison CP. Videolaparoscopy and laser laparoscopy in gynaecology. Br J Hosp Med 1987; 38:219–24.
2. Reich H. Laparoscopic treatment of extensive pelvic adhesions, including hydrosalpinx. J Reprod Med 1987; 32:736–42.
3. Donnez J, Nisolle M. CO_2 laser laparoscopic surgery. Adhesiolysis, salpingostomy, laser uterine nerve ablation and tubal pregnancy. Bailleres Clin Obstet Gynecol 1989; 3:525–43.
4. Serour GI, Badroui MH, el Agizi HM, Hamed AF, Abdel-Aziz F. Laparoscopic adhesiolysis for infertile patients with pelvic adhesive disease. Int J Gynaecol Obstet 1989; 30:249–52.
5. Herschlag A, Diamond MP, DeCherney AH. Adhesiolysis. Clin Obstet Gynecol 1991; 34:395–402.
6. Kolmorgen K, Schulz AM. Ergebnisse nach per laparoscopiam ausgefuhrten Adhäsiolysen bei Patientinnem mit chronischen Unterbauchbeschschwerden. Zentralbl Gynäcol 1991; 113:291–5.
7. Lundorff P, Hahlin M, Kallfelt B, Thorburn J, Lindblom B. Adhesion formation after laparoscopic surgery in tubal pregnancy: a randomised trial versus laparotomy. Fertil Steril 1991; 55:911–15.
8. Operative Laparoscopy Study Group. Postoperative adhesion development after operative laparoscopy: evaluation at early second-look procedures. Fertil Steril 1991; 55:700–4.
9. Gomel V. Salpingostomy by microsurgery. Fertil Steril 1978; 29:380–6.
10. Winston RML. The future of microsurgery in infertility. Clin Obstet Gynaecol 1978; 5:607–25.
11. Israel SL. Total linear salpingostomy. Fertil Steril 1952; 2:505–9.
12. Winston RML, Margara RA. Microsurgical salpingostomy is not an obsolete procedure. Br J Obstet Gynaecol 1991; 98:637–42.
13. Donnez J, Casanas-Roux F. Prognostic factors of fimbrial microsurgery. Fertil Steril 1986; 46:200–4.
14. Fioretti P, Teti G, Maffei S, Murru MS. Utilita e risultati microchirurgica nel trattamento della sterilita femminile. Ann Obstet Ginecol Perinat 1990; 111:239–44.
15. Kitchin JD, Nunley WC, Bateman BG. Surgical management of distal tubal occlusion. Am J Obstet Gynecol 1986; 155:296–300.
16. Singhal V, Li TC, Cooke ID. An analysis of factors influencing the outcome of 232 consecutive tubal microsurgery cases. Br J Obstet Gynaecol 1991; 98:628–36.
17. Tulandi T. Adhesion reformation after reproductive surgery with and without the carbon dioxide laser. Fertil Steril 1987; 47:704–6.
18. Mage G, Pouly JL, Canis M, Bruhat MA. CO_2 laser microsurgery with long term results. Microsurgery 1987; 8:89–91.
19. Dubuisson JB, Bouquet de Joliniere J, Aubriot FX et al. Terminal tuboplasties by laparoscopy: 65 consecutive cases. Fertil Steril 1990; 54:401–3.
20. McComb P, Paleologou A. The intussusception technique for the therapy of distal oviductal occlusion at laparoscopy. Obstet Gynecol 1991; 78:443–7.
21. Winston RMI. Microsurgical tubocornual anastomosis for reversal of sterilisation. Lancet 1977; i:284–5.
22. Green-Armytage VB. Tubo-uterine implantation. J Obstet Gynecol Br Emp 1957; 64:47–53.
23. Rocker I. The anatomy of the utero-tubal area. Proc R Soc Med 1964; 57:707–12.
24. Shirodkar VN. Plastic surgery of the fallopian tubes. West J Surg 1961; 69:253–64.
25. Ehrler P. Die intramurale Tubenanastomose, ein Beitrag zur Uberwindung der tubaren Sterilität. Zentralbl Gynäkol 1963; 85:93–8.
26. Fortier KJ, Haney AF. The pathologic spectrum of fallopian tube obstruction. Obstet Gynecol 1985; 65:93–8.
27. Wiedemann R, Schiedel P, Wiesinger H, Hepp H. Die Pathologie des proximalen Tubenverschluss-Morphologische AuswertungenErgebnisse nach mikrochirurgischer Anastomose. Geburtshilfe Frauenheilkd 1987; 47:96–100.
28. Jansen RP. Tubal resection and anastomosis. Aust NZ J Obstet Gynaecol 1986; 26:300–4.
29. Wang XY. Female infertility due to tubal obstruction – management and clinicopathologic study of 66 cases. Chung Hua Fu Chan Ko Tsa Chih 1989; 24:201–2.
30. Gomel V, McComb P. Microsurgery in gynaecology. In: Silber SJ, ed. Microsurgery. Baltimore: Williams and Wilkins, 1978; 162.
31. Gillett W, Herbison GP. Tubocornual anastomosis: surgical considerations and coexistent infertility factors in determining the prognosis. Fertil Steril 1989; 51:241–6.
32. Donnez J, Casanas-Roux F. Prognostic factors influencing the pregnancy rate after microsurgical anastomosis. Fertil Steril 1986; 46:1089–92.

33. Rouselle D, Jarrel J, Belbeck L, Andryjowicz E, Hibbert P, McMahon A. Thrombin as an adjuvant to reanastomosis of the fallopian tubes reduces implantation in the rabbit. Can J Surg 1988; 31:66–8.
34. Dargenio R, Ranelletti FO, Cimino C, Ragusa G, Panetta V, Garcea N. Fibrin glue vs nylon in the anastomosis of rabbit fallopian tubes. J Reprod Med 1986; 31:361–5.
35. Swolin K. Tubal anastomosis. Hum Reprod 1988; 3:177–8.
36. Nighiem Toan N. Anastomose tubaire par cicatrisation dirigée. Rev Fr Gynecol Obstet 1991; 86:247–8.
37. Vilos GA. Intramural-isthmic fallopian tube anastomosis facilitated by the carbon dioxide laser. Fertil Steril 1991; 56:571–3.
38. Daniell JF, Miller W. Hysteroscopic correction of cornual occlusion with resultant term pregnancy. Fertil Steril 1987; 48:490–2.
39. Thurmond AS, Rosch J. Nonsurgical fallopian tube recanalization for treatment of infertility. Radiology 1990; 177:371–4.
40. Thurmond AS. Selective salpingography and fallopian tube recanalization. AJR 1991; 157:415.
41. Deaton JL, Gibson M, Riddick DH, Brumsted JR. Diagnosis and treatment of cornual obstruction using a flexible tip guidewire. Fertil Steril 1990; 53:232–6.
42. Kojima E, Morita M, Otaka K, Yano Y. Nd:YAG laser laparoscopy for ovarian endometriomas. J Reprod Med 1990; 35:592–6.
43. Nezhat C, Crowgey S, Nezhat F. Videolaproscopy for the treatment of endometriosis associated with infertility. Fertil Steril 1989; 51:237–40.
44. Nezhat C, Crowgey SR, Garrison CP. Surgical treatment of endometriosis via laser laparoscopy and videolaparoscopy. Contrib Gynecol Obstet 1987; 16:303–12.
45. Wallach EE. The uterine factor in infertility. Fertil Steril 1972; 23:138–58.
46. Cooke ID. The current status of infertility surgery. Clin Obstet Gynecol 1978; 591–606.
47. Markham S. Cervico-utero-tubal factors in infertility. Curr Opin Obstet Gynecol 1991; 3:191–6.
48. Babaknia A, Rock JA, Jones HW Jr. Pregnancy success following abdominal myomectomy for infertility. Fertil Steril 1978; 30:644–7.
49. Berkeley AS, de Cherney AH, Polan ML. Abdominal myomectomy and subsequent fertility. Surg Gynecol Obstet 1983; 156:319–22.
50. Garcia CR, Tureck RW. Submucosal leiomyoma and infertility. Fertil Steril 1984; 42:16–19.
51. Rosenfeld DL. Abdominal myomectomy for otherwise unexplained fertility. Fertil Steril 1986; 46:328–30.
52. Egwatu VE. Fertility and fetal salvage among women with uterine leiomyomas in a Nigerian Teaching Hospital. Int J Fertil 1989; 34:341–6.
53. Starks GC. CO_2 laser myomectomy in an infertile population. J Reprod Med 1988; 33:184–6.

Chapter 12

Advanced Videolaparoscopy and Videolaseroscopy

C. Nezhat and F. Nezhat

Introduction

The first attempt at endoscopy was by Philip Bozzini of Italy in 1805, using a tube and a candle. Since that time, the instrumentation for laparoscopy has developed at an acceptable rate although the applications have lagged. Laparoscopy was moderately successful in Europe during the first half of the 20th century. After Jacobaeus of Sweden first induced pneumoperitoneum and placed a Nitze cystoscope into the peritoneal cavity (1910), the technique was applied to diagnostic and simple sterilisation procedures by Kalk (Germany), Ruddock and Hope (United States) in the 1930s. Although the results which were reported were promising, the procedure was not accepted in the United States. The next significant developments occurred in the 1950s. These were cold light (Fourestier, Gladu and Valmiere) and fibreoptics (Kampany and Hopkins). In the late 1940s, Raoul Palmer of France was the main promoter of the use of laparoscopy in gynaecology. He reported the first human tubal fulguration in 1962.

With the groundwork for operative laparoscopy established, the next logical progression would seem to have been a gradual increase in the application to various types of pelvic and abdominal surgery. Instead, the laparoscope received limited use as a diagnostic tool.

The advantages of this instrument were diminished by three serious drawbacks. First, the surgeon had to work crouched over the patient and had to peer with one eye through the scope; visibility was limited, the position was uncomfortable, and the surgeon's back was easily fatigued. Second, the rest of the surgical team was unable to view the procedure and as a result, were prevented from anticipating the surgeon's needs. Third, the auxiliary instruments were not available to perform procedures more complicated than tubal ligation.

Despite these limitations, Gomel [1] began reporting the merits and safety of operative laparoscopy in the early 1970s. He successfully performed a number of procedures, including salpingo-ovariolysis, fimbrioplasty and management of ectopic pregnancy. In 1977, he reported his experience with salpingostomy in a series of patients.

At the same time, Kurt Semm headed the German-based Kiel School's development of instruments for and promotion of operative laparoscopy, particularly fertility-enhancing procedures [2].

Gomel and Semm significantly contributed to this new technique during the 1970s, but operative laparoscopy was not integrated into the operating room. It was performed and promoted by a limited group of surgeons, and failed to be accepted by the medical community at large.

As laparoscopy stagnated, another medical breakthrough was evolving. The CO_2 laser, developed by Patel in 1964, was used in experimental surgery the following year. The "high-tech" concept of the laser gained public attention. Through lay publications, attention was focused on this new surgical tool, and its development continued until in the late 1970s, it was coupled with the laparoscope. Unfortunately, this combination was still subject to the earlier-mentioned limitations and was not frequently used.

Finally, in 1978, we attached a video camera to the eyepiece of the laparoscope while working in an animal laboratory [3]. The camera magnified the image and projected it onto monitors in the operating room. The major disadvantages of the laparoscope were eliminated, but the early cameras were cumbersome and the first videolaparoscopic operations on humans were difficult due to the cameras' weight, inadequate light sources, and poor resolution. In the early 1980s, equipment companies began to recognise the potential market for a miniature video camera and produced lighter versions with higher resolution and better light sources. With these final modifications, all the elements were in place by 1984–85 for a revolution in abdominal and pelvic surgery [4].

Using the more refined cameras and the CO_2 laser via the operative channel of the laparoscope, we successfully performed laparoscopic procedures on over 1000 patients including those with extensive endometriosis (Fig. 12.1, Tables 12.1–12.5) [5]. Through the publication of case reports and detailed studies, as well as through national and international meetings, we have promulgated the use of videolaparoscopy (VL) in gynaecology (for benign and malignant disease), lower gastroenterology and urology [3,4]. The demand for the technique was created when patients recommended the procedure to acquaintances. Media attention [6–8] and word of mouth made operative laparoscopy one of the first consumer-driven medical advances.

We are currently witnessing a rapid increase in both the applications of operative laparoscopy and the number of surgeons who are learning this technique. VL offers several benefits which are not possible with an open procedure. First, pelvic and abdominal anatomy is magnified by the video camera and laparoscope, allowing the surgeon to perform microsurgical procedures. Second, the pressure created by pneumoperitoneum decreases bleeding and provides a cleaner operating field. Third, areas like the upper abdomen, pouch of Douglas and posterior aspect of the broad ligaments may be more thoroughly evaluated. Fourth, operative laparoscopy produces fewer

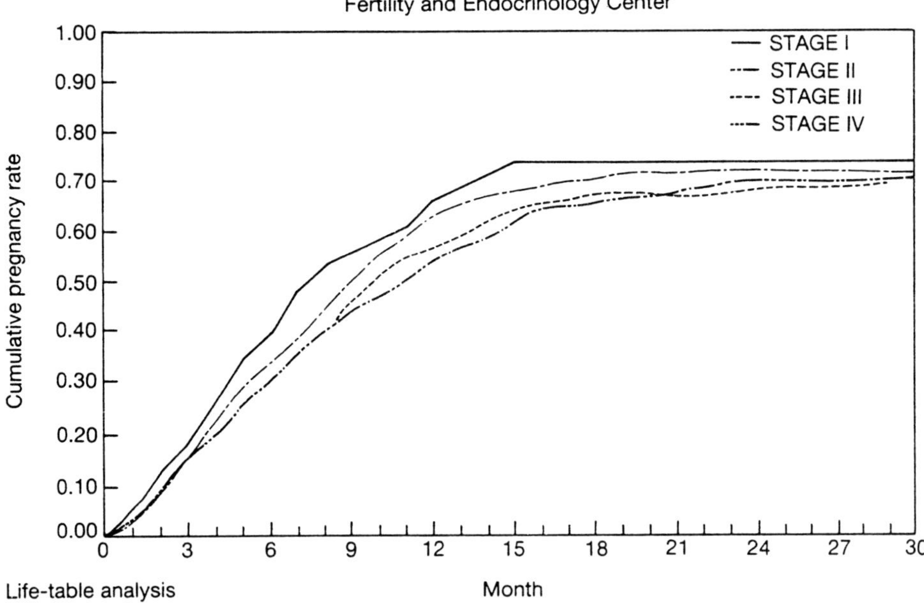

Fig. 12.1. Life table analysis.

Table 12.1. Total of 102 patients with endometriosis

Stage		No. of patients	No. of pregnancies	%	Spontaneous abortion	%
I.	Mild	24	18	75.1	4	22.0
II.	Moderate	51	32	62.7	4	12.5
III.	Severe	19	8	42.1	2	25.0
IV.	Extensive	8	4	50.0	0	
Total		102	62		10	

From Nezhat et al. [5], with permission.

Table 12.2. Pregnancy rate in 8 to 12 and 18 months in 102 patients with endometriosis

No. of months	No. of pregnancies	%
0–8	41	66.1
8–12	18	29.0
12–18	3	4.8
Total	62	

From Nezhat et al. [5], with permission.

Table 12.3. Distribution of pregnancy in 102 patients with endometriosis

Age	No. of patients	No. of pregnancies	%
20–25	19	12	24.1
25–30	31	20	32.0
30–35	43	27	43.0
35–41	9	3	4.8
Total	102	62	

From Nezhat et al. [5], with permission.

Table 12.4. Duration of infertility in 102 patients with endometriosis

No. of months	No. of patients	%
12–23	14	13.1
24–49	48	47
59 or more	40	39.2

Table 12.5. Endometriosis patients by stage of disease and pregnancy rates

	All patients	Stage I	Stage II	Stage III	Stage IV[a]
No.	243	39	86	67	51
Age[b]	29.4 ± 4.1	26.8 ± 2.2	27.6 ± 2.4	30.5 ± 1.9	31.5 ± 4.1
Years infertile[b]	3.6 ± 1.9	3.3 ± 1.9	3.6 ± 2.1	3.8 ± 1.2	3.6 ± 2.0
Pregnant	168 (69.1%)	28 (71.8%)	60 (69.8%)	45 (67.2%)	35 (68.6%)

[a]The American Fertility Society: Classification of Endometriosis.
[b]95% confidence limit.
From Nezhat et al. Fertil Steril 1989; 51:239.

adhesions [9,10] and requires a shorter recovery period. Finally, with technological advances such as the development of the CO_2 laser, disease may be treated more effectively and possibly with a greater margin of safety.

We would like to emphasise that the laser is not a panacea [5]. As with any instrument, its efficacy is primarily determined by the surgeon's skill. Further, most of the following procedures could be performed with different types of lasers or electrocoagulators. Our use of the CO_2 laser combined with bipolar electrocoagulator is a personal preference. It is important to note that the CO_2 laser may be used with a greater margin of safety than electrocoagulation.

For the past several years, we have avoided over 98% of all laparotomies which would have been performed to treat benign gynaecological disease. It is our belief that this same percentage of all pelvic and abdominal operations will be performed by laparoscopy by the early 21st century.

To prepare a patient for videolaparoscopy, the surgeon should follow a protocol similar to that for laparotomy, including thorough clinical and laboratory evaluation. Additionally, the procedure is explained to the patient and proper consent obtained. The evening before surgery, patients with more advanced disease are given a bowel preparation consisting of 4 l polyethylene glycol-3350 (Go-LYTELY, Braintree Laboratories, Braintree, MA) and take 1 g metronidazole (Flagyl, G.D. Searle, Chicago, IL) at 11:00 p.m. Cefoxitin (Mefoxin, Merck, Sharp and Dohme, West Point, PA) is administered prophylactically, both preoperatively and postoperatively.

The room set-up and trocar placement have been described in detail [11]. VL is performed under general endotracheal anaesthesia, with the patient placed in a modified dorsolithotomy and Trendelenburg position.

We recommend identifying the courses of sensitive structures in the pelvic cavity, such as the bowel, ureters or vessels, from their entrance to exit during the initial diagnostic portion of the procedure to help decrease the possibility of injury. When the procedure is complete, the pelvic cavity should be inspected as thoroughly as possible, with careful attention given to all surgical sites to assure complete haemostasis [12].

Tubal Adhesiolysis and Hydrosalpinges

Using VL to lyse and remove peritubal adhesions has proven effective in preserving fertility, providing the anatomy of the lumen, including major and minor folds of the mucosa and cilia, has not been destroyed by disease [13, 14]. Fimbrioscopy [15] can be used to further evaluate the tubes during the operation. This is done by suspending the fimbria in fluid. The pouch of Douglas is filled with lactated Ringer's solution (Baxter), and the 3 mm salpingoscope with video camera is introduced through one of the suprapubic portals or the operative channel of the laparoscope. The smaller salpingoscope magnifies folds of the fimbria, and agglutination or adhesions can be readily identified and incised with the laser.

When tubal anatomy has been destroyed, the probability of conception is poor [14,16]. Although an attempt should be made to repair even those tubes during VLS, these patients may eventually become candidates for in vitro fertilisation and embryo transfer (IVF-ET). For these women, exposure of ovaries for subsequent ovum retrieval by future laparoscopy or ultrasound aspiration is as important as tubal repair. Ovarian adhesiolysis is carried out using VLS techniques described above.

If a hydrosalpinx is encountered, results of repair are comparable to microsurgical laparotomy repair [16]. The distal end of the tube is distended with indigo carmine chromotubation, and stabilised with two grasping forceps. Stellate incisions are made in the end of the tube using focused, high-power laser, while carefully preserving all remaining fimbria. The edges of the neosalpingostomy must be everted to preserve tubal patency; low-power laser applied circumferentially to the distal serosal surface causes serosal contraction and eversion of the mucosa. Sutures should be avoided, unless absolutely necessary to keep the tube open, as they may increase adhesion formation.

In a study of 62 women who underwent laparoscopic treatment for unilateral or bilateral hydrosalpinges (Group 1) and 69 who were treated for peritubal and/or periovarian adhesions (Group 2), the pregnancy rates were evaluated. The rates for Group 1 were highly affected by the condition of the tubal lining. Of the 64 women who were followed for 12 months, 42 had undamaged lining, and 15 achieved intrauterine pregnancy. The remaining 22 had sustained damage to the lining, and only three conceived. Two of these pregnancies were ectopic. In Group 2, 34 of the 45 patients followed for 12 months achieved intrauterine pregnancy.

When compared to reported pregnancy rates following laparotomy [17], the results by laparoscopy are more encouraging. A total of 42 patients underwent videolaseroscopy for hydrosalpinges and peritubal and periovarian adhesions. Nine were aged 20–25 years, 15 aged 26–30 years, 12 aged 31–35 years and six were aged 36–39 years. Potency of at least one tube was achieved in 38 patients (90%), intrauterine pregnancy in 13 (30%) and ectopic pregnancy in two (5%). Forty-five patients underwent videolaseroscopy for peritubal and periovarian adhesion. Three were aged 20–25 years, 21 aged 26–30 years, 10 aged 31–35 years and 11 aged 36–42 years. Intrauterine pregnancy was achieved in 34 patients (75%) and two patients (4%) had ectopic pregnancies.

Ectopic Pregnancy Management

The chance of managing unruptured ectopic pregnancy laparoscopically has increased with rapid serum human chorionic gonadotropic (HCG) assays and high-resolution vaginal ultrasounds [18–20]. New medical management techniques using ultrasonically guided injection of methotrexate, potassium or prostaglandin $F_{2\alpha}$, may some day replace primary laparoscopic management of ectopic pregnancy.

When an unruptured tubal pregnancy is confirmed laparoscopically, surgical management should be instituted at that procedure. The surgeon should talk to the patient preoperatively to ascertain her desire for fertility preservation and to determine whether or not her history is significant for previous ectopic pregnancies. The patient must understand that salpingectomy or laparotomy may become necessary. However, an experienced endoscopic surgeon can laparoscopically manage ectopic pregnancies, regardless of size and independent of location [21]. Finally, the need for careful follow-up with conservative management must be explained to the patient.

In patients who prefer permanent sterilisation, coagulation using bipolar Kleppinger forceps over a small ectopic pregnancy will destroy the tubal pregnancy and sterilise the patient. For larger ectopics (greater than 6 cm), cases of spontaneous tubal rupture or more than one recurrent ipsilateral ectopic pregnancy, salpingectomy may be indictated [12]. This procedure is accomplished by placing the fallopian tube under traction using grasping forceps. The isthmus and mesosalpinx are serially coagulated with bipolar electrocoagulator and cut with high-power laser. An alternative to salpingectomy involves using a GIA-type endoclip applicator which is placed along the mesosalpinx, parallel to the fallopian tube and triggered, simultaneously clamping and cutting the pedicle.

Salpingectomy has been shown to produce higher subsequent pregnancy rates and lower recurrence rates of ectopic pregnancy than salpingectomy [19]. Linear salpingotomy should be performed in patients who want to preserve the affected tube and are haemodynamically stable. Dilute vasopressin is injected into the mesosalpinx using a 22 gauge aspirating needle inserted through a 5 mm portal, or using a 22 spinal needle inserted directly through the abdominal wall; direct vascular injection of vasopressin must be avoided. The fallopian tube is grasped and extended, and a 1.5 cm linear incision is made along the antimesenteric surface of the tube. The products of conception are allowed to extrude through the newly created salpingotomy, or are flushed out using the Nezhat–Dorsey probe (Karl Storz) inserted into the tubal opening, then grasped and removed from the abdomen. Copious irrigation within the tube and the pelvis will ensure removal of the tissue and adequate haemostasis. Chromotubation with indigo carmine will indicate tubal patency and also flush remaining products from the tube. Vigorously grasping tissue which remains adherent to the mucosa can cause haemorrhage and must be avoided. Weekly HCG levels must be followed until they decrease to non-pregnant levels. Consistent or rising levels require further medical or surgical management.

Segmental resection of the tube has a better prognosis in the narrow isthmic portion of the tube, possibly because pregnancies in this site tend to infiltrate deeper into tubal tissue layers and are smaller, and therefore less well-defined for surgical manipulation [19]. Segmental resection may also be used in cases of spontaneous rupture without active bleeding, and persistent tubal pregnancies. The segment to be excised is coagulated at each end, using the bipolar electrocoagulator or sterilisation clips, and transected using the laser, scissors or electrocoagulator [12–16,18–22]. The mesosalpinx of the segment is then coagulated and incised. As much fallopian tube as possible should be preserved so that reanastomosis with microsurgical technique can be accomplished at a later date.

We have now managed 112 ectopic pregnancies laparoscopically, of whom 76 desired to achieve pregnancy, and 57 have been unsuccessful. Nine women (12%) experienced recurrent ectopic pregnancy.

Peritoneal Endometriosis

To facilitate safe use of the CO_2 laser, we have developed the technique of hydrodissection [23]. As the CO_2 laser does not penetrate water, a fluid backstop allows the surgeon to work on selected tissue with a more comfortable margin than would otherwise be available.

To treat endometriosis of the bladder, for example, an aspiration needle is used to inject 20–30 ml of lactated Ringer's (Baxter) subperitoneally in an avascular area approximately 2 cm from the endometriotic lesion. This elevates the peritoneum and backs it with a fluid bed. A 0.5 cm incision is made with the laser on this elevation, through which 100–200 ml of lactated Ringer's (Baxter) are injected subperitoneally.

The incision may then be vaporised or excised using the CO_2 laser in the ultrapulse mode [23]. For excision, a circular line is made with a radius of 1–2 cm from the lesion. The peritoneum is then grasped and pulled away with the help of the CO_2 laser and the tip of the Nezhat–Dorsey probe (Karl Storz).

When the endometriosis forms scarring to the subperitoneal connective tissue, creating openings as described above and injecting fluid on the lesion's lateral sides allows water to "tunnel" under the lesion. This often separates scarring and the implant can then be vaporised or removed. Irrigation and washing should follow to remove all byproducts and ensure complete treatment of the disease. More extensive endometriosis may be treated after removing the peritoneum, again followed by irrigation and washing.

As shown in Table 12.6, the pregnancy rate achieved following laparoscopic treatment of endometriosis is lower when the AFS score (revised American Fertility Society classification, 1985) is higher, and when adhesions are more dense and/or more prevalent.

Proper use of hydrodissection allows experienced laparoscopic surgeons to treat mild to extensive endometriosis with the CO_2 laser not only more thoroughly, but also more safely than was previously possible [24].

Myomectomy

Patients with indications for laparoscopic myomectomy are managed with GnRH analogues (Lupron, Tapp Pharmaceuticals, North Chicago, IL or Synarel, Syntex, Palo Alto, CA) for up to 3 months preoperatively. Reducing tumour size, improving operative handling and reducing intraoperative blood loss are theoretical advantages. Three months of amenorrhoea does improve preoperative haematocrits, but, patients are also given the option of autologous blood donation [25].

Table 12.6. Pregnancy rate after laparoscopic treatment of endometriosis in relation to AFS score

AFS Score	No. of women	No. of pregnancies	%	Other contributing factors
10–40	19	14	73	No tubal involvement
40–110	30	10	33	Tubal adhesions 14 patients with 1 or more endometriomas
>116	12	1	8	Tubal adhesions Previous laparotomies 9 patients with 1 or more endometriomas
Total	61	25		

Laparoscopic myomectomy has two stages: the first – removing the tumour from the uterus – is generally straightforward. However, the second – removing the tumour from the abdomen – can be long and tedious.

A pedunculated myoma is simply excised at the stalk using the high-power laser and bleeding is controlled with the bipolar electrocoagulator. For intramural or subserol myomata, 5–10 ml of dilute vasopressin (20 U in 100 ml of sterile saline) are injected under the capsule. The capsule is incised with the CO_2 laser (between 60–100 W) and gradually dissected using a combination of the Nezhat–Dorsey probe, hydrodissection and laser. When CO_2 gas is used for the pneumoperitoneum, the high power CO_2 laser (except the latest Ultrapulse 500L, Coherent, Palo Alto, CA) creates a large spot between 2 and 3 mm, which is haemostatic for vessels with diameters of 2–3 mm. Traction on the myomas can be produced with a small hook or claw forceps. Once the myoma is removed, the base is thoroughly irrigated, and haemostasis generally requires a bipolar electrocoagulator rather than a low power laser. To close the myometrial defect, 4-0 PDS endosutures (Ethicon) are probably the most appropriate, but clips (Ethicon) might also be used. Intraligamentous myomata are approached by incising the anterior or posterior leaf of the broad ligament (depending on the location of the myoma) with the laser, after identifying the location of large vessels, ureter and bladder. Excision is then accomplished as described for subserol myomata.

The basic surgical principles for myomectomy are identifying and removing the tumour, permanent haemostasis and eliminating the dead space. Laparoscopic microsurgical repair of the uterus using 6-0 or 7-0 sutures at the present time (1992) is not possible. Our experience suggests possible vascular adhesion formation following laparoscopic repair of the uterus [25]. Another problem in laparoscopic removal of myoma remains its extraction from the abdominal cavity, although this has been simplified by the availability of an 18 mm trocar (Ethicon).

Posterior colpotomy, when feasible, is the most technically acceptable route for the removal of large (>5 cm) and multiple leiomyomata. Morcellation in the pouch of Douglas by the vaginal assistant becomes an additional option. However, this procedure lengthens the duration of the surgery and increases the risk of morbidity including infection, and rectal and ureteral injuries and should be performed cautiously.

The strength of the uterus following laparoscopic myomectomy is unknown, and must be determined to predict the ability of the organ to withstand labour and delivery. While uterine healing appears adequate after the removal of small myomata, indentations have been noted upon removal of larger lesions without suturing, which might represent structural defects. Even when endosutures are applied, the meticulous reapproximation of layers available by microsurgical laparotomy is not possible during laparoscopy.

Finally, adhesion formation is of great concern when future fertility is desired. The data obtained after second-look procedures reveal an increase in the number and density of adhesions when the suturing is undertaken (Table 12.7) [25].

In a group of 154 women who underwent laparoscopic myomectomy, the projected pregnancy rates are similar to those following myomectomy by laparotomy. Of 38 infertility patients, 32 were available for follow-up (1 to 3 years). Ten had only leiomyomas; 18 also had mild to severe endometriosis;

Table 12.7. Incidence of adhesion formation and grade after laparoscopic myomectomy at 56 sites during second-look laparoscopy or laparotomy

Size of leiomyoma	No suture used				Suture used			
	0	1	2	3	0	1	2	3
<3 cm	18/21	3/21	0/21	0/21	N/A	N/A	N/A	N/A
>3 cm	10/16	5/16	1/16	0/16	0/19	4/19	10/19	5/19

Grades of adhesions: 0 = no adhesions; 1 = filmy and non-vascular; 2 = thick and non-vascular; 3 = thick, vascular, and bowel.
From Nezhat et al. [25], with permission.

two had hydrosalpinges and adhesions; and two had pelvic adhesions. Of the 10 with only leiomyomas and 13 with mild to moderate endometriosis, nine achieved pregnancy. Three of the five patients with severe endometriosis conceived. Neither of the women with hydrosalpinges achieved pregnancy, but both women with adhesions conceived. Of the 14 pregnant women, one miscarried, then later conceived and delivered. One other woman had an abortion.

Four women underwent myomectomy due to miscarriage. Two conceived and delivered. The third miscarried, but is currently pregnant. The fourth experienced a blighted ovum and is no longer trying to conceive.

Oophorectomy and Adnexal Mass Management

Although operative laparoscopy has proven to be a safe and effective diagnostic and therapeutic tool in the hands of experienced laparoscopists, doubts remain about the laparoscopist's ability to diagnose and properly manage early ovarian cancer if the adnexal mass in question is found to be malignant [26]. One of the chief concerns is that the spillage of a cancer confined to the ovary may worsen prognosis. However, when a competent surgeon follows proper protocol, adnexal masses may be safely evaluated laparoscopically.

Patients should be evaluated clinically with a pelvic examination and vaginal ultrasound, along with a review of previous intraoperative records. Simple (unilocular) cysts are initially managed with hormonal suppressive therapy using oral contraceptive pills containing 50 μg of oestrogen, depo-provera or danazol. A serum CA 125 level or other tumour marker, if indicated, should also be obtained. Cystic, complex or solid masses up to 25 cm have been laparoscopically managed by the authors [27].

Informed consent should include a statement to the patient that laparoscopic diagnosis and treatment of adnexal mass are not standard medical practice at this time (1992). Patients are further informed that if a cancer is found, intraoperative cancer cell spillage could occur, and could influence the chance of survival. In addition, the patients must understand that a second operation, specifically a laparotomy, might be required if the findings at laparoscopy could not be properly managed laparoscopically.

Intraoperative management of all patients with masses is carefully standardised, and includes inspecting the pelvis, ovaries, upper abdomen and diaphragmatic surfaces for any vegetation or other sign of malignancy. Peritoneal washings are obtained for cytology. If a strong suspicion of malignancy exists based on intraoperative findings, an attempt is made to obtain frozen section biopsies without rupturing the cyst. If that is not possible, the laparoscopic procedure is terminated, and the patient will undergo a laparotomy.

The management of the cystic mass itself includes aspirating the fluid and sending it for cytology, followed by opening the cyst and inspecting the wall for excrescences or irregular thickening. Frozen section biopsies are obtained if the surgeon feels any surfaces are suspicious. Finally, depending on the patient's age and pertinent clinical history, an ovarian cystectomy or oophorectomy may be performed [28]. The CO_2 laser is set at 30–80 W in ultrapulse mode to cut, and the bipolar electrocoagulator is used to control bleeding.

After exploring the pelvis and abdomen, oophorectomy is carried out as follows: any existing adhesions between the ovary and adjacent organs, pelvic walls, and broad ligament are lysed with the CO_2 laser. A Nezhat–Dorsey probe is introduced suprapubically, and used both as a backstop and to provide constant suction, irrigation, and smoke evacuation. When the ovary is severely adherent to the lateral pelvic wall, and in cases with previous hysterectomy, hydrodissection [23] is used to open the peritoneum, beginning at the pelvic brim, and the courses of the ureters and major blood vessels are identified. Using the same techniques, the descending colon and rectosigmoid colon are dissected from the left infundibulopelvic ligament and ovary. Ovarian cysts are aspirated, allowing easier handling of the deflated cyst and smaller ovary. Once the ovary is completely mobilised, it is held under tension with grasping forceps. The ovarian ligament is desiccated using bipolar cautery and transected with the laser at its junction to the uterus. Using 20–25W, the bipolar electrocoagulator is applied briefly to blanch and desiccate the tissue, care being taken not to over-desiccate, in order to reduce the blood flow to the pedicle to be transected. The mesovarium is then serially blanched, coagulated and transected at 1–2 cm increments, working medial to lateral, until the ovary is removed. When the ipsilateral fallopian tube is also removed, the isthmic portion of the tube is severed and incised along with the ovarian ligament; after identification of the ureter, the infundibulopelvic ligament is coagulated and transected at 1–2 cm increments, working from lateral to medial, until the adnexa is removed as described above. Alternatives to coagulating and blanching include the Endo GIA Stapler and Endoloop sutures (Ethicon). It is imperative that the ovary is free of any attachments, however, as the stapler has been associated with haematoma, and the Endoloop may trap ovarian tissue, resulting in ovarian remnant syndrome.

In most cases, the ovary is removed from the peritoneal cavity through a 10 mm trocar sleeve placed into one of the suprapubic puncture sites; the ovary is held with forceps, and the forceps and sleeve are removed together, delivering the ovarian tissue to the abdominal wall, where it is then grasped by a Kelly clamp and removed. Alternatively, the tissue may be removed by posterior colpotomy [29], tissue removal bag (Endopouch, Ethicon) or an 18 mm trocar

(Ethicon). It is then submitted for standard histological examination. If at all possible, the tissue should be removed without coming into contact with the edges of the incisions or abdominal wall.

Thorough irrigation is necessary to clean the incision. After the cyst has been removed, the abdominal and pelvic cavities are thoroughly washed with copious amounts of irrigation, especially in cases of dermoid cysts or mucinous cystadenomas.

Before the procedure is terminated, any associated pelvic pathology such as endometriosis or other adhesions is treated, and haemostasis is assured. Six weeks and 6 months postoperatively, patients with benign ovarian neoplasms (mucinous or serous cystadenomas or cystic teratomas) are followed by a biannual pelvic examination and an ultrasound examination to look for possible recurrence.

We believe that laparoscopic evaluation and management of a benign adnexal mass performed by experienced operative laparoscopists is safe. However, no substitute exists for sound clinical judgement. A surgeon should perform the techniques he/she is comfortable with, and should conduct careful preoperative patient screening [27].

Endometriosis

Superficial endometriosis of the ovaries can be treated by vaporisation. Because small endometriomas (<2 cm) tend to be fibrotic and difficult to remove, they can be biopsied and then vaporised. Larger endometriomas, however, must be removed completely, including the capsule, to reduce the risk of recurrence [30]. Simple aspiration or fenestration of endometriomas can lead to an unacceptably high level of recurrence, as demonstrated by Nezhat et al. [30] and Hasson [31].

Management of endometriomas must be approached in the same way as all other adnexal masses, keeping in mind that endometrioid carcinoma can co-exist with endometriosis, and is indistinguishable at surgery until histology is reported. We have encountered this type of case, which appeared at laparoscopy to be typical bilateral endometriomas. The smaller cyst (3 cm diameter), however, proved to be endometrioid adenocarcinoma in situ. Thus, histological examination of the cyst wall is mandatory in even the most apparently typical case [27].

When endometriomas are suspected on the basis of ultrasound appearance or previous operative reports, hormonal suppressive therapy (danazol or GnRH analogues) given for six to eight weeks preoperatively will reduce vascularity and suppress ovarian activity, as Buttram suggests in his review of preoperative danazol [32]. Consequently, intraoperative haemorrhage will be reduced, surgical manipulation of follicular or corpus luteum cysts will be avoided, and more of the normal ovarian tissue will be preserved. Similarly, six weeks of postoperative suppressive therapy in cases of endometrioma removal will, in our opinion, facilitate better healing. For patients who are not interested in achieving pregnancy, we suppress ovulation with oral contraceptives indefinitely.

The laparoscopic approach to an endometrioma is as follows. The cyst is first aspirated and drained through an 18 gauge aspiration needle inserted through one of the suprapubic portals. After copious irrigation of the cyst and pelvis, and cyst wall is opened further and inspected. The capsule is then stripped from the ovarian stroma using two grasping forceps, or excised using the laser and submitted for histological examination. Hydrodissection can be used for easier removal of the ovarian cyst capsule [33]. The laser can then be used at low power (10–20 W, continuous) to seal blood vessels at the base of the capsule, and at higher power to vaporise any small capsule remnants. The ovarian defect is left to heal without suturing. If the edges of the ovarian capsule do not spontaneously approximate, the low power laser or the bipolar coagulator can be used to invert them by treating the inner surface of the defect, causing the surface so treated to contact and invert. In the event that the edges still will not approximate, one or two 4-0 polydioxanone sutures may be used. It is important to keep in mind that fewer suturer will result in fewer adhesions [34].

In a prospective study of 216 haemorrhagic, or "chocolate", ovarian cysts, four types of endometriomas were identified, both clinically and pathologically [35]. They were classified based on gross appearance, size, content, ease of removal of the capsule and pathological findings. Superficial endometriomas of less than 2 cm (type I) were difficult to remove. All type I cysts contained a histologically confirmed endometrial gland and stroma. In endometriomas larger than 2 cm (types II, III and IV, based on the proximity of the endometrial implant to the cyst's capsule), 0, 50% and 85% contained an endometrial lining (Table 12.8). We have postulated that large ovarian endometriomas represent secondary involvement of functional cysts with superficial endometriosis.

To treat these cysts, aspiration and irrigation of only selected haemorrhagic or "chocolate" cysts without adjacent endometrial implants (essentially old corpora lutea) may be adequate. However, neoplastic cysts could be missed with this method. During the past several years, we have encountered one early endometrioid carcinoma and several other epithelial-type cysts. These all clinically appeared to be "chocolate" cysts.

We do not advocate vaporising large endometriomas as the considerable thickness of the capsule renders this method time-consuming [5]. In addition, it produces a significant amount of plume and substantially increases the possibility of missing any neoplasm.

Table 12.8. Characteristics of presumed endometriosis

Type	No.	Size (mean)	Luteal lining	Endometrial lining	No diagnostic lining	Adhesions dense/ filmy	Haemosiderin and/or fibrosis
I	15	1–2 cm (1.67)	0	15	0	5/3	15
II	57	2–6 cm (3.9)	46	0	9[a]	5/11	6
III	46	3–12 cm (5.4)	14	23	9	22/13	35
IV	98	3–20 cm (7.0)	10	84	4	88/2	88

[a]The remaining two cysts were haemorrhagic cystic teratomas.

When the histological appearance was compared with the clinical classification, small superficial cysts were always endometriomas (Type I), large cysts with walls that were easily removed were usually luteal cysts (Type II) and large cysts that had adhesions or were associated with superficial endometriotic implants were often endometriomas, but also had histological characteristics of corpus luteum cysts (Types III and IV). Thus, cysts of pure types (endometrioma or luteal) as well as mixed types were encountered and were distinguishable, based on adherence of the capsule to the ovarian cortex. The presence of endometriosis on the ovarian surface, particularly in the absence of dense adhesions, was associated with the absence of tissue planes between the capsule and the ovarian stroma, giving the impression that the surface endometriosis may have invaded the functional ovarian cyst. On histological examination, variable degrees of "invasion" of the cyst wall by the surface endometriosis could be observed, thus forming the basis of the differentiation between Type III and IV cysts, Type III cysts showing early signs of "invasion" and Type IV demonstrating more advanced association between the surface endometriosis and the cyst wall.

Our observations, which distil the classification into four types of endometriomas, suggest a natural progression of endometrioma development. At one end of the spectrum are "pure" endometriomas which develop from surface endometriosis and resemble endometriomas at other pelvic sites in organisation and aetiology. At the other end are the haemorrhagic corpus luteum cysts which may coexist with surface ovarian endometriosis. Based on the microscopic findings in this study, it seems reasonable to assume that large ovarian endometriomas result from secondary involvement of haemorrhagic non-endometrial cysts by endometriosis. Large primary ovarian endometriomas may be exceptional because of adjacent scarring and adhesions which would limit their size.

Several mechanisms of development of endometrioma are possible based on our observations. Rupture of a follicle at the time of ovulation regularly disrupts the continuity of the ovarian surface epithelium. If this occurs in proximity to a surface implant of endometriosis, this may allow endometriosis to enter the ovary resulting in endometrioma formation. Alternatively, rupture of a functional cyst (haemorrhagic or otherwise) could also allow an infolding of surface endometriosis. Small endometriomas (Type I) may coalesce with functional cysts of the ovary and result in endometriomas. Finally, if a functional cyst persists for a sufficient length of time, surface endometriosis may invade the thin ovarian cortex which separates it from the cyst capsule resulting in endometrioma formation.

Finally, our findings and theory are supported by the observations by Sampson [36], who 70 years ago noted

At operation the cyst or ovary is found adherent and in freeing it the chocolate contents escape because a previous perforation reopened or the cyst is torn. The histologic findings in these cysts vary in different portions of the same cyst. A portion (usually deeper) ... is lined by a luteal membrane. The rest of the cyst, usually toward the perforation, is apparently being relined by the invasion of epithelium, through the perforation, from epithelium situated in the periphery of the ovary at the site of rupture. With the advance of the epithelial invasion, the luteal membrane retrogresses and eventually the entire cyst may be relined by this epithelial tissue.

Our observations have extended those of Sampson to include additional clinical observations which may aid in preliminary diagnosis of the cyst type at the time of surgery.

These observations may explain the variable response of presumed ovarian endometriomas to medical therapy, the rarity of large cystic endometriomas in sites other than the ovary, the infrequency of torsion of endometriomas because of associated adhesions, and the lack of development of endometriomas in patients on ovulation suppression therapy despite development of small surface implants while under this therapy. The ability to classify endometriomas may enhance further study into the aetiology and pathophysiology and provide possible methods to prevent and treat endometriosis.

References

1. Gomel V. Operative laparoscopy: time for acceptance. Fertil Steril 1989; 52:1–11.
2. Mettler L, Giesel H, Semm K. Treatment of female infertility due to tubal obstruction by operative laparoscopy. Fertil Steril 1979; 32:384–8.
3. Nezhat C. Videolaseroscopy: a new modality for the treatment of endometriosis and other diseases of reproductive organs. Colposc Gynecol Laser Surg 1986; 2:221–4.
4. Anonymous. Obstet Gynecol News 1986 (15–30 November); 41–42.
5. Nezhat C, Crowgey SR, Garrison CP. Surgical treatment of endometriosis via laser laparoscopy. Fertil Steril 1986; 45:778–83.
6. Wallis C. The career woman's disease? Time 1986; 28 April:62.
7. Clark M, Carroll G. Conquering endometriosis. Newsweek 1986; 13 October:95.
8. Cowley G. Hanging up the knife. Newsweek 1990; 12 February:58–9.
9. Nezhat C, Nezhat F, Silfen SL. Videolaseroscopy: the CO_2 laser for advanced operative laparoscopy. Obstet Gynecol Clin North Am 1991; 18:585–604.
10. Nezhat C, Nezhat F, Metzger DA, Luciano AA. Adhesion development after operative laparoscopy: evaluation at early second-look procedures. Fertil Steril 1991; 55:700–4.
11. Operative laparoscopy study group. Postoperative adhesion formation after reproductive surgery by videolaparoscopy. Fertil Steril 1990, 53:1008–11.
12. Nezhat C, Nezhat F, Winer W. Salpingectomy via laparoscopy: a new surgical approach. J Laparosc Surg 1991; 1:91–5.
13. Nezhat C, Nezhat SL, Nezhat F, Martin D. Surgery for endometriosis. Curr Opin Obstet Gynecol 1991; 3:385–93.
14. Nezhat C, Winer WK, Nezhat F, Nezhat CH. Videolaparoscopy and videolaseroscopy: alternatives to surgery? The Female Patient 1988; 13:46–56.
15. Nezhat F, Winer WK, Nezhat C. Fimbrioscopy and salpingoscopy in patients with minimal to moderate pelvic endometriosis. Obstet Gynecol 1990; 75:15–17.
16. De Bruyne F, Putteman P, Boecks W, Brosens I. The clinical value of salpingoscopy in tubal infertility. Fertil Steril 1989; 51:339–40.
17. Gomel V. Salpingostomy by laparoscopy. J Reprod Med 1977; 18:265.
18. Bruhat MA, Mage G, Manhes H, Pouly JL. Treatment of ectopic pregnancy by means of laparoscopy. Fertil Steril 1980; 33:411–8.
19. DeCherney A, Diamond MP. Laparoscopic salpingostomy for ectopic pregnancy. Obstet Gynecol 1987; 70:948–50.
20. Stangel JJ, Gomel V. Techniques in conservative surgery for tubal gestation. Clin Obstet Gynecol 1980; 23:1221–8.
21. Nezhat F. Conservative management of ectopic gestation, letter-to-the-editor. Fertil Steril 1990; 53:382–3.
22. Nezhat C, Nezhat F, Gordon S, Wilkins E. Laparoscopic versus abdominal hysterectomy. J Reprod Med 1992; 37:247–50.
23. Nezhat C, Nezhat F. Safe laser excision or vaporization of peritoneal endometriosis. Fertil Steril 1989; 52:49–51.
24. Nezhat C, Videolaseroscopy for the treatment of endometriosis. In: Studd J, ed. Progress in obstetrics and gynaecology, 7th edn. Edinburgh: Churchill Livingstone, 1989; 293–303.

25. Nezhat C, Nezhat F, Silfen SL, Schaffer N, Evans D. Laparoscopic myomectomy. Int J Fertil 1991, 36:275–80.
26. Maimon M, Seltzer V, Boyce J. Laparoscopic excision of ovarian neoplasms subsequently found to be malignant. Obstet Gynecol 1991; 77:563–5.
27. Nezhat C, Nezhat F, Welander CE, Benigno B. Four ovarian cancers diagnosed during laparoscopic management of 1011 adnexal masses. Am J Obstet Gynecol 1992, in press.
28. Nezhat F, Nezhat C, Silfen SL. Videolaseroscopy for oophorectomy. Am J Obstet Gynecol 1991; 165:1323–30.
29. Davis GD, Hruby PH. Transabdominal laser colpotomy. J Reprod Med 1989; 34:438–40.
30. Nezhat C, Winer WK, Nezhat F. Is endoscopic treatment of endometriosis and endometrioma associated with better results than laparotomy? Am J Gynecol Health 1988; 2:10–16.
31. Hasson HM. Laparoscopic management of ovarian cysts. J Reprod Med 1990; 25:863–7.
32. Buttram VC. Use of danazol in conservative surgery. J Reprod Med 1990; 35:82–6.
33. Nezhat C, Winer WK, Cooper JD, Nezhat F, Nezhat C. Endoscopic infertility surgery. J Reprod Med 1989; 34:127–34.
34. Nezhat C, Nezhat F. Postoperative adhesion formation after ovarian cystectomy with and without ovarian reconstruction. Presented at the 75th annual meeting of the American Fertility Society, Orlando FL, 19–24 October, 1991.
35. Nezhat C, Nezhat F. Comparison of different treatment methods of endometriomas by laparoscopy, letter to the editor. Obstet Gynecol 1992; 79:315.
36. Sampson JA. Perforating hemorrhagic (chocolate) cysts of the ovary. Arch Surg 1921; 3:245–323.

Chapter 13

Surgery and the Polycystic Ovarian Syndrome

R. W. Shaw and A. A. Gadir

Introduction

The polycystic ovarian syndrome (PCOS) is a heterogeneous group of conditions ranging from a mere ultrasonic ovarian finding to the full clinical picture described by Stein and Leventhal [1]. It is estimated that 3% of all women in their reproductive life and more than 50% of all patients with menstrual dysfunction exhibit this problem, making it the most frequently diagnosed female endocrinopathy [2]. Other modes of presentation include hyperandrogenisation, infertility, obesity and recurrent miscarriages. However, the exact figures representing the spectrum of clinical presentation vary in different series depending on the diagnostic criteria used. Moreover, women with PCOS are more prone to develop hypertension, diabetes, hyperlipidaemia, cardiovascular and gall bladder diseases and to have a significant degree of insulin resistance not related to obesity [3]. Some of these conditions may be related to the hyperandrogenic state and the pattern of circulating lipoproteins.

Goldzieher and Green [4] described PCOS as a non-tumorous dysfunctional condition of the ovaries characterised by luteinising hormone (LH) dependent hypersecretion of androgens from hyperplastic theca and stroma cells.

The histological picture seen in an individual case of PCOS depends on the relative number of the follicular and atretic cysts, the degree of stromal hyperplasia and the number of primordial follicles present. At one end of the spectrum, ovaries contain more follicular cysts with few atretic ones, minimal stromal hyperplasia and abundant primordial follicles. At the other end, there are numerous atretic follicles with a small number of follicular cysts and marked stromal hyperplasia with islands of luteinised stromal cells. Accordingly, one histological picture does not portray the whole histological spectrum seen in women with PCOS. Furthermore, the mode of clinical presentation [5] and the response to the different treatment modalities may reflect the prevalent histological picture in the individual patient. This last point is portrayed by the fact that patients with larger ovaries, and presumably more recruitable antral and preantral follicles, had a better clinical response to induction of ovulation with human menopausal gonadotrophins (hMG) [6].

Biochemically PCOS is characterised by one or more of the following criteria: high LH basal values or a high LH:FSH ratio, elevated androgens, deficient acyclic oestradiol (E_2) production with disturbed E_2: oestrone (E_1) ratio. Hormone pulse studies showed women with PCOS to have high LH and testosterone pulse amplitude values with or without high pulse frequency [5, 7]. Nevertheless, we showed that there was no universal LH pulse pattern common to all women with PCOS [8]. Some patients had high and others had low LH pulse pattern components and few had inverted LH:FSH ratio despite our stringent criteria for patient recruitment. This reflected a non-persistent or variable endocrine milieu at different times in the same women, despite synchronised blood sampling relative to the menstrual cycle. Moreover, using Pearson's products, moment statistics, we could not find a significant correlation between LH and testosterone values or ovarian volume in patients with PCOS [9]. This discrepancy may be due to the fact that once PCOS develops, the ovary assumes a primary role in androgen hypersecretion [4,10]. This may reflect the increased sensitivity of theca cells to LH as shown by increased production of androstenedione after LH stimulation in patients with PCOS in comparison with normal women [11]. It may, in addition, be a reflection of the build up of the secondary interstitial tissue in the ovarian stroma following the progressive follicular atresia. These points may explain the favourable endocrine and clinical changes that followed local manipulation of the ovaries during the different surgical modalities used in the treatment of patients with PCOS.

Historical Considerations

Various surgical procedures have been used for the treatment of patients with PCOS. Ovarian wedge resection [1] extroversion of the ovaries [12] decortication [13] and medullectomy [14] have been described. One report described resumption of menstrual function following unilateral oophorectomy without manipulation of the other ovary but this report has not been confirmed in three subsequent cases [15]. However, ovarian wedge resection remained the most popular surgical treatment for patients with PCOS till clomiphene became available [16], when surgical treatment was reserved for clomiphene non-responders. This may be a reflection of the fact that the initial reports [17] were not confirmed by other investigators [18–20] and there was a high recurrence rate following wedge resection [21]. Furthermore, there is a high incidence of peritoneal adhesions [20,22] and an eightfold increase in the incidence of ectopic pregnancy [23] following surgery. With the introduction of human menopausal gonadotrophins for the treatment of patients with anovulatory infertility there has been a reduced need for ovarian wedge resection in treating patients with PCOS resistant to clomiphene [24]. In 1983, Hjortrup et al [25] studied the effect of ovarian wedge resection on anovulation, hirsutism and obesity by following 29 patients for 2.3–9.5 years (mean 5.7 years). They showed that 90% of their patients had regular menstruation following surgery with a low recurrence rate. Eight of nine obese patients who had normal menstrual cycles postoperatively showed major weight loss after wedge resection. On the other hand, hirsutism was not cured and none of the patients who remained oligomenorrhoeic following surgery lost any weight. One may argue that persistent oligomenorrhoea might be a reflection of persistent obesity, rather than the reverse.

Much attention was focused on the mode of action of ovarian wedge resection. Suggestions included removal of the capsule with its presumed mechanical barrier effect to the act of ovulation, removal of androgen-producing stromal tissues and oestrogen-producing atretic follicles and the increased blood flow that follows surgery resulting in increased delivery of gonadotrophins into a smaller volume of ovarian tissue. This diversity of opinion reflected the lack of uniformity of opinion on the aetiology of PCOS. Interest shifted to the endocrine effects of ovarian wedge resection, but this was hampered by the fact that both gonadotrophins and steroids might be lower following the stress of anaesthesia and surgery. Furthermore, plasma gonadotrophins were reported to be lower in the early postoperative period even in patients undergoing non-gynaecological surgery [26]. However, the sustained effect after ovarian wedge resection was rather specific and most investigations reported a reduction in circulating luteinising hormone and androgen basal levels [26–28]. This was further supported by the finding that LH pulse amplitude and frequency were reduced following ovarian wedge resection [29].

Current Practice

Anovulatory patients with PCOS are usually prescribed antioestrogens as the first line of treatment for induction of ovulation. However, 15% of all patients treated with clomiphene citrate remain anovulatory [30] and will need alternative treatment modalities. This may be in the form of injectable gonadotrophins, the use of which has always been hampered with the need for serial endocrine and ultrasonic monitoring and by the risk of hyperstimulation. Difficulties in providing this service have led some investigators to describe different surgical methods for the treatment of patients with PCOS including ovarian electrocautery [31], multiple ovarian punch biopsies [32] and laser vaporisation of the ovarian capsule and atretic follicles [33–35]. These modalities were introduced to circumvent the need for ovarian wedge resection and the inevitable formation of adhesions that follows this procedure with the ensuing tubal infertility. All these procedures can be done through the laparoscope during an initial diagnostic infertility work-up though laser vaporisation needs special equipment which may not be available in all units. All reports published so far concentrate on the treatment of anovulatory patients not responsive to clomiphene citrate therapy and different success rates are reported. The results are, however, not comparable due to the differences in the criteria used for diagnosing PCOS and the different methods used for documenting the results e.g. cumulative pregnancy rate, pregnancy rate per cycle or percentage of total group. However there are no reports yet available which described the efficacy of these procedures in treating hirsutism, obesity or recurrent miscarriages. The endocrine effect of all procedures was similar to that following ovarian wedge resection, namely reduction in LH and androgen basal levels. The risk of adhesion formation was reported to be slight after ovarian electrocautery [36] and laser vaporisation [35]. Keckstein et al. [35] showed that non-contact neodymium (Nd:YAG) laser treatment caused a more drastic decline in the level of androgens and less adhesions than CO_2 laser treatment. Moreover, there are no quantitative studies published yet comparing the effect of these laparoscopic procedures with each other or with ovarian wedge resection. This chapter concentrates on laparoscopic ovarian electrocautery as it is a simple procedure not requiring expensive equipment.

Ovarian Electrocautery

Ovarian electrocautery was first described by Gjonnaess in 1984 [31] and there have been a number of reports describing its use in women with PCOS resistant to clomiphene citrate therapy [37–39]. Ovarian electrocautery has been shown to be as effective as human menopausal gonadotrophin (hMG) and pure follicle stimulating hormone (FSH) in treating patients with PCOS [6]. Successful induction of ovulation, pregnancy and abortion rates were similar in patients treated with ovarian electrocautery or injectable gonadotrophins. All pregnancies following ovarian electrocautery were singleton whereas 20% of pregnancies following gonadotrophin therapy were multiple. This favourable clinical response may be a reflection of the reduced androgen and LH levels following ovarian electrocautery described by many authors [37,39–41]. We studied the 6-hour hormone pulse patterns in patients with PCOS before and after ovarian electrocautery [42]. All patients showed a reduction in the LH and testosterone 6h mean values and pulse amplitude. This effect was quantitatively assessed against 8 weeks of intranasal buserelin (Hoechst, Germany) therapy in a dose of 800 μg/day. The magnitude of change in the 6h mean values and pulse amplitude of both hormones was equivalent in the two groups. However, the significant hypo-oestrogenic state following buserelin medication was not matched by a similar change in oestradiol (E_2) levels after ovarian electrocautery. Paradoxically there was an increase in the level of serum insulin after ovarian electrocautery not matched by any significant change in the level of cortisol, prolactin or blood glucose.

Since PCOS is a heterogeneous problem and different diagnostic criteria have been used by different workers, we studied the biophysical and endocrine attributes of responders and non-responders to determine the factors that control clinical response [43]. Patients with PCOS and high LH values (>12 IU/l) responded better to ovarian electrocautery than patients with high LH:FSH ratio and normal LH levels. However, serum testosterone, ovarian volume and body mass index were not useful in predicting the clinical outcome. Both responders and non-responders had a significant increase in FSH and a decline in LH and testosterone (6h mean values) one month after surgery. However, the magnitude of change in LH was significantly higher in responders compared with non-responders. There was no difference in the corresponding changes in FSH, testosterone or E_2 (Fig. 13.1). This improved endocrine milieu even in non-responders may explain the improved response to clomiphene citrate therapy after electrocautery [6,38,39]. Following patients with serial endocrine examinations during the first week after surgery may identify those who need supplementary clomiphene therapy. Despite equivalent reduction in testosterone values in responders and non-responders during the first week, there were significant differences in the pattern of FSH and LH between the two groups (Fig. 13.2). Patients who do not show a rise in FSH and a decline in LH during the first week after surgery are better supplemented with clomiphene therapy. We further assessed the clinical effect of ovarian electrocautery versus a luteinising hormone releasing hormone agonist (LHRH-A) (buserelin, Hoechst, Germany) on the response of patients with PCOS to hMG therapy [44]. There was no difference in the ovulation or pregnancy rates between the two groups. However, women in the LHRH-A + hMG group needed more hMG ampoules for longer induction periods than women pretreated with ovarian electrocautery. Furthermore, patients pretreated with LHRH-A had more cycles with multiple dominant follicles and accordingly a higher potential for hyperstimulation. They also had higher luteal phase serum testosterone levels and a miscarriage rate three times that of patients pretreated with ovarian electrocautery. This shows that despite a favourable endocrine milieu during the follicular phase following LHRH-A medication, there were higher testosterone values at the time of ovulation, fertilisation and nidation

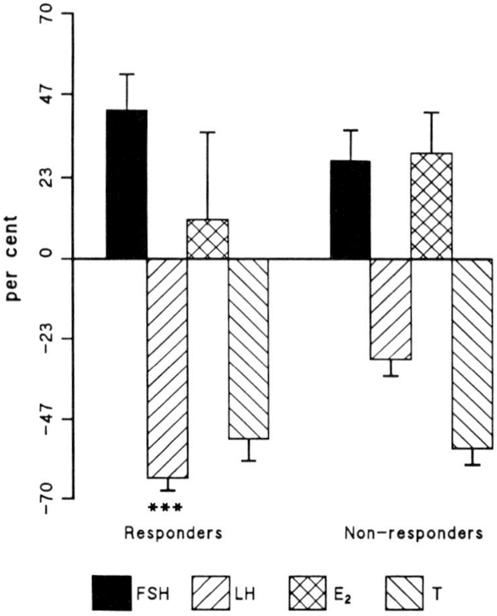

Fig. 13.1. The magnitude of change (%) in the 6-h mean value of hormones following ovarian electrocautery in responders and non-responders was significantly different only for LH. The error bars represent the standard error of the mean. The probability of a chance difference:***, $P < 0.001$.

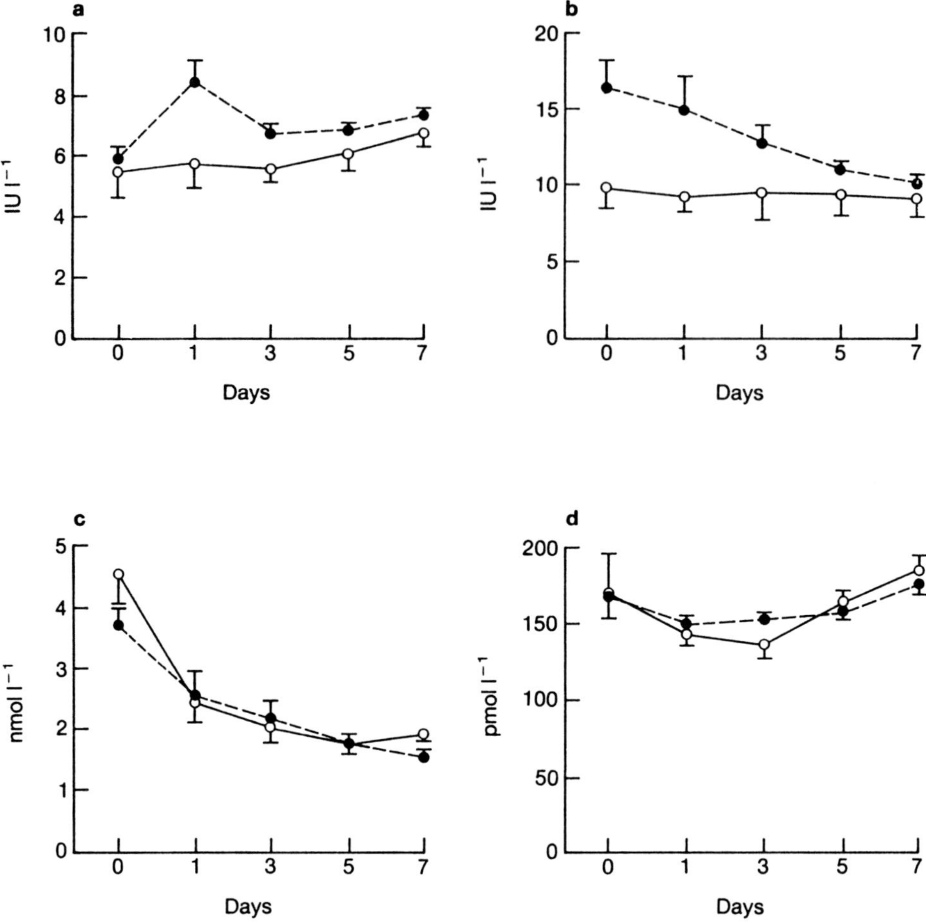

Fig. 13.2. The patterns of (**a**) FSH and (**b**) LH during the first week after ovarian electrocautery were significantly different between responders (●—●) and non-responders (○—○). FSH increased ($P < 0.01$) and LH values decreased ($P < 0.01$) significantly during the same period in responders only. There was no difference in the pattern of testosterone (**c**) or oestradiol (**d**) during the same period between the two groups. The error bars represent the standard error of the mean.

than in patients pretreated with ovarian electrocautery. The importance of this finding is amplified by the fact that patients with non-viable pregnancies had higher testosterone values during the luteal phase than patients with viable pregnancies.

Like ovarian wedge resection, the exact mode of action of laparoscopic ovarian surgery is not yet settled. The improved response to clomiphene therapy, with its central effect, and to hMG medication, with its local ovarian action, suggests that both central and ovarian mechanisms are involved. Most investigators reported a reduction in LH pulse amplitude which is suggestive of a modified pituitary response to LHRH stimulation [32,43]. However, Ohkouchi et al. [29] reported reduced LH pulse frequency after ovarian wedge resection, suggestive of an alteration in the hypothalamic electrical activity necessary to generate LHRH pulses. This central effect may be due to the reduction in androgen concentration without oestrogen intervention. This is especially so since the early decline in the level of LH following surgery was not matched by a significant decrease in the concentration of E_2 [42] or E_1 [45]. This central effect is coupled by a reduction in the intraovarian androgen concentration which facilitated the response of patients with PCOS to hMG medication following ovarian electrocautery. This is shown by an attenuated testosterone response after hCG administration in these patients [44]. Furthermore, Sakata et al. [45] reported a decrease in the concentration of the bioactive LH component following ovarian electrocautery [45]. The whole issue addressing the mode of action of ovarian surgery in the treatment of patients with PCOS was reviewed by Vaughan Williams [46] who showed that the common factor in ovarian wedge resection, laser vaporisation and electrocautery was the drainage of the androgen and inhibin rich contents of the atretic ovarian cysts rather than removal of part of the interstitial tissue or a breaching of the integrity of the ovarian tunica.

Unfortunately both electrocautery and LHRH-A therapy are temporary remedies. A total 15 patients with PCOS were followed after discontinuation of a three-month course of Nafarelin (Syntex Pharmaceuticals Limited, Maidenhead, Berkshire) [47]. The circulating gonadotrophin and androgen levels started to rise once LHRH-A medication was suspended. Seven women remained amenorrhoeic, four returned to an oligomenorrhoeic pattern but the remaining four had cycles of less than or equal to 42 days in length. One patient conceived spontaneously while being amenorrhoeic. In contrast, there was a 71.4% successful ovulation rate after ovarian electrocautery when 29 patients were followed for six months [43]; 14 women conceived and seven others were still menstruating. Six women had another pregnancy within 12 months following the first one. During the follow-up period there was a progressive rise in the level of LH during anovulatory cycles whereas testosterone values were not significantly different between ovulatory and anovulatory cycles (Fig. 13.3). More prolonged follow-up showed that 50% of patients reversed to chronic anovulation within two years [31]. Accordingly, one may argue that this treatment should be reserved for patients who wish to get pregnant. On the other hand, the more liberal use of ovarian electrocautery for the treatment of patients with other manifestations of PCOS, repeating the procedure every two years if necessary, sounds an attractive idea. This may reduce the need for the hormonal or surgical treatment needed by patients with dysfunctional uterine bleeding and may reduce the duration and the dose of antiandrogens necessary to suppress hyperandrogenism. This is especially so since the risk of adhesion formation after surgery appears to be minimal [36]. However, we advocate the use of a sharp needle for electrocautery and the application of the needle at a right angle to prevent slit cauterisation and to reduce the damaged area on the surface of the ovary. This procedure should be supplemented with the creation of artificial ascites with dextran to keep tissue afloat in the immediate postoperative period. These points tend to be in agreement with current microsurgical techniques to reduce adhesion formation.

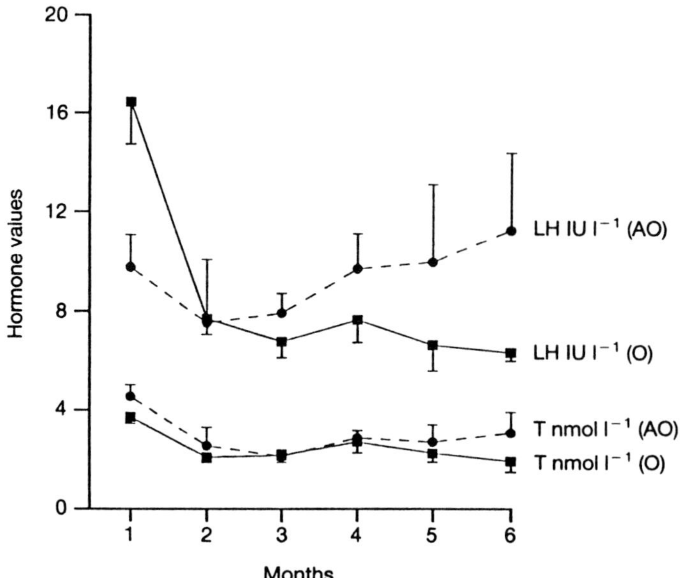

Fig. 13.3. Basal hormone levels on day 5 of the cycle after ovarian electrocautery during ovulatory (O) and anovulatory (AO) cycles. The error bars represent the standard error of the mean.

Conclusion

It is evident that 15% of all patients with PCOS treated with clomiphene citrate remain anovulatory. Medical treatment with hMG results in successful induction of ovulation in 73%–95% of patients in this group. However, the pregnancy rate is much lower and the multiple pregnancy rate and hyperstimulation risk are high following hMG medication, despite thorough endocrine and ultrasonic monitoring. Furthermore, the expenses and expertise involved in this medication may be beyond the grasp of some centres involved in infertility management. The introduction of LHRH-A in the management of patients with PCOS did not result in a reduction in the risk of multiple pregnancy or hyperstimulation. On the other hand, ovarian electrocautery done at the initial diagnostic work-up may reduce the need for any medication to induce ovulation and eliminate the need for regular endocrine and ultrasonic monitoring. The potential role for laparoscopic ovarian surgery in the management of patients with PCOS and related problems needs further evaluation.

References

1. Stein IF, Leventhal ML. Amenorrhoea associated with bilateral polycystic ovaries. Am J Obstet Gynecol 1935; 29:181–91.

2. Vaitukaitis JL. Polycystic-ovary syndrome. What is it? N Engl J Med 1983; 309:1245–6.
3. Chang RJ, Nakamura RM, Judd HL, Kaplan SA. Insulin resistance in non-obese patients with polycystic ovarian disease. J Clin Endocrinol Metab 1983; 57:356–9.
4. Goldzieher JW, Green JA. The polycystic ovary: clinical and histological features. J Clin Endocrinol Metab 1962; 22:325–38.
5. Yen SSC. Chronic anovulation caused by peripheral endocrine disorders. In: Yen SSC, Jaffe MD, eds. Reproductive endocrinology, physiology, pathophysiology and clinical management. Philadelphia: W.B. Saunders, 1986; 441–99.
6. Abdel Gadir A, Mowafi RS, Alnaser HMI, Alrashid AH, Alonezi OM, Shaw RW. Ovarian electrocautery versus human menopausal gonadotrophins and pure follicle stimulating hormone therapy in the treatment of patients with polycystic ovarian disease. Clin Endocrinol 1990; 33:585–92.
7. Waldstreicher J, Santoro NF, Hall JE, Fillicori M, Crowly WF. Hyperfunction of the hypothalamo-pituitary axis in women with polycystic ovarian disease: indirect evidence for partial gonadotrophin desensitization. J Clin Endocrinol Metab 1988; 66:165–72.
8. Abdel Gadir A, Khatim MS, Mowafi RS, Alnaser HMI Alzaid HGN, Shaw RW. Polycystic ovaries: do these represent a specific endocrinopathy? Br J Obstet Gynaecol 1991; 98:300–5.
9. Abdel Gadir A, Khatim M, Mowafi R, Alnaser HMI, Muharib NS, Shaw RW. Implications of ultrasonically diagnosed polycystic ovaries (1)-Correlations with basal hormonal profiles. Hum Reprod 1992; 7:453–7.
10. Fisher ER, Gregorio R, Stephan T, Nolan S, Donowski TS. Ovarian changes in women with morbid obesity. Obstet Gynecol 1974; 44:839–44.
11. McNatty KP, Moor-Smith D, Makris A et al. The intraovarian sites of androgens and oestrogen formation in women with normal and hyperandrogenic ovaries as judged by in vitro experiments. J Clin Endocrinol Metab 1980; 50:755–63.
12. Bailey KV. The operation of extroversion of the ovaries for functional amenorrhoea especially of the secondary type. J Obstet Gynaecol Br Emp 1937; 44:637–49.
13. Reycraft JL. Surgical treatment of ovarian dysfunction. Am J Obstet Gynecol 1938; 35:505–12.
14. Allen WM, Woolf RB. Medullary resection of the ovaries in the Stein Leventhal Syndrome. Am J Obstet Gynecol 1959; 77:826–37.
15. Goldzieher JW. Polycystic ovarian disease. In: Wallach EE, Kempers RD, eds. Modern trends in infertility and conception control, Vol. 2. Philadelphia: Harper and Row, 1982; 65–88.
16. Greenblatt RB, Barfield WE, Jungck EC, Ray AW. Induction of ovulation with MRL/41. JAMA 1961; 178:101–4.
17. Stein IF, Cohen MR. Surgical treatment of bilateral polycystic ovaries, amenorrhoea and sterility. Am J Obstet Gynecol 1939; 38:465–80.
18. Goldzieher JW, Axelrod LR. Clinical and biochemical features of polycystic ovarian disease. Fertil Steril 1963; 14:631–53.
19. Zarate A, Henandez-Ayup S, Rios-Montiel A. Treatment of anovulation in the Stein Leventhal Syndrome. Analysis of 90 cases. Fertil Steril 1971; 22:188–93.
20. Buttram VC, Vaquero C. Post-ovarian wedge resection adhesive disease. Fertil Steril 1975; 26:874–6.
21. Starup J. Treatment of patients with polycystic ovarian syndrome. Ugeskr Laeger 1976; 138:2866–70.
22. Kistner RW. Peri-tubal and peri-ovarian adhesions subsequent to wedge resection of the ovaries. Fertil Steril 1969; 20:35–42.
23. Adashi EY, Rock JA, Guzick D, Wentz AC, Jones GS, Jones HW. Fertility following bilateral ovarian wedge resection: a critical analysis of 90 consecutive cases of the polycystic ovary syndrome. Fertil Steril 1981; 36:320–5.
24. Lunenfeld B. Treatment of anovulation by human gonadotrophins. Int J Gynecol Obstet 1963; 1:153–67.
25. Hjortrup A, Kehlet H, Lockwood K, Hasner E. Long-term clinical effects of ovarian wedge resection in polycystic ovarian syndrome. Acta Obstet Gynecol Scand 1983; 62:55–7.
26. Katz M, Carr PJ, Cohen BM, Millar RP. Hormonal effects of wedge resection of polycystic ovaries. Obstet Gynecol 1978; 51:437–44.
27. Lloyd CW, Lobotsky J, Seare EJ, Kobayaski T, Taymor ML, Batt RE. Plasma testosterone and urinary 17-ketosteroids in women with hirsuitism and polycystic ovaries. J Clin Endocrinol Metab 1966; 26:314–24.
28. Judd HL, Rigg LA, Anderson DC et al. The effect of ovarian wedge resection on circulating gonadotrophins and ovarian steroid levels in patients with polycystic ovaries. J Clin Endocrinol Metab 1976; 43:347–55.

29. Ohkouchi T, Tanaka T, Okawa M, Sakuragi N, Fujimoto S, Ichinoe K. Effect of ovarian wedge resection and spironolactone administration on pulsatile LH release and steroid hormone levels in women with polycystic ovarian disease. Acta Obstet Gynaecol Jpn 1987; 39:626–32.
30. Franks S, Adams J, Mason H, Polson D. Ovulatory disorders in women with polycystic ovary syndrome. Baillieres Clin Obstet Gynaecol 1985; 12:605–32.
31. Gjonnaess H. Polycystic ovarian disease treated by ovarian electrocautery through the laparoscope. Fertil Steril 1984; 41:20–5.
32. Sumioki H, Utsunomyiya T, Matsuoka K, Korenata M, Kadota T. The effect of laparoscopic multiple punch resection of the ovary on hypothalamo-pituitary axis in polycystic ovary syndrome. Fertil Steril 1988; 50:567–72.
33. Huber J, Hosmann J, Spona J. Polycystic ovarian syndrome treated by laser through the laparoscope. Lancet 1988; ii:215.
34. Daniel JF, Miller W. Polycystic ovaries treated by laparoscopic laser vaporisation. Fertil Steril 1989; 51:232–6.
35. Keckstein G, Rossmanith W, Spatzier V, Kirstin B, Steiner R. The effect of laparoscopic treatment of polycystic ovarian disease by CO_2-laser or Nd:YAG laser. Surg Endosc 1990; 4:103–7.
36. Dabirashrafi H, Mohamad K, Behjatnia Y, Moghadami N. Adhesions formation after ovarian electrocauterization on patients with polycystic ovarian syndrome. Fertil Steril 1991; 55:1200–1.
37. Greenblatt E, Casper RF. Endocrine changes after laparoscopic ovarian cautery in polycystic ovarian syndrome. Am J Obstet Gynecol 1987; 156:279–85.
38. Van Der Weiden RM, Alberda AT. Laparoscopic ovarian electrocautery in patients with polycystic ovarian disease resistant to clomiphene citrate. Surg Endosc 1987; 1:217–19.
39. Armar N, McGarrigle HHG, Honour J, Holownia P, Jacobs HS, Lachelin GCL. Laparoscopic ovarian diathermy in the management of anovulatory infertility in women with polycystic ovaries: endocrine changes and clinical outcome. Fertil Steril 1990; 53:45–9.
40. Aakvaag A, Gjonnaess H. Hormonal response to electrocautery of the ovary in patients with polycystic ovarian disease. Br J Obstet Gynecol 1985; 92:1258–64.
41. Gjonnaess H, Norman N. Endocrine effects of ovarian electrocautery in patients with polycystic ovarian disease. Br J Obstet Gynaecol 1987; 94:779–83.
42. Abdel Gadir A, Khatim MS, Mowafi RS, Alnaser HMI, Alzaid HGN, Shaw RW. Hormonal changes in patients with polycystic ovarian disease after ovarian electrocautery or pituitary desensitization. Clin Endocrinol 1990; 32:749–54.
43. Abdel Gadir A, Khatim MS, Mowafi RS, Alnaser HMI, Alzaid HGN, Shaw RW. Hormonal changes following laparoscopic ovarian electrocautery in patients with polycystic ovarian syndrome. In: Shaw RW, ed. Advances in reproductive endocrinology, Vol. 3, Carnforth, UK: Parthenon Publishing, 1991; 135–47.
44. Abdel Gadir A, Alnaser HMI, Mowafi R, Shaw RW. The response of patients with polycystic ovarian disease to human menopausal gonadotropin therapy after ovarian electrocautery or a luteinizing hormone-releasing hormone agonist. Fertil Steril 1992; 57:309–13.
45. Sakata M, Tasaka K, Kurachi H, Terakawa N, Miyake A, Tanizawa O. Changes of bioactive luteinizing hormone after laparoscopic ovarian cautery in patients with polycystic ovarian syndrome. Fertil Steril 1990; 53:610–13.
46. Vaughan Williams CA. Ovarian electrocautery or hormone therapy in the treatment of polycystic ovary syndrome. Clin Endocrinol 1990; 33:569–72.
47. Shaw RW. Use of Nafarelin to investigate the pathophysiology of the polycystic ovary syndrome. J Reprod Med 1989; 34 suppl:1039–43.

Discussion

Cooke: Are there histological data on the damage inflicted on an ovary by electrocautery?

Shaw: We have not yet had to go in to remove any of these ovaries and at the moment these studies are up to three years ongoing. We have done repeat laparoscopies on 15 of the original 30 and there was only one where there were minor adhesions. I cannot speculate. Obviously there must be damage and destruction of surrounding primordial follicles following the area of electrocautery. Whether this may induce premature ovarian failure or something in the long term I guess has to be looked at if this procedure is to be done repeatedly.

Nezhat: I should like to say something about PCO and the role of electrocautery or laser. We feel that this should definitely be a secondary role, we should not attempt to cauterise the ovary as a first-line therapy. In our own data, not yet published but in press, we did notice a significant amount of adhesion formation following laparoscopic cauterisation of ovaries in patients with PCO.

Cooke: Dr Nezhat said that he did not believe that following laser there would be a substantial adhesion formation, and yet I thought the multicentre study in the US showed that there was a substantial, i.e., 70%, recurrence rate of adhesions treated by laser.

Nezhat: I seem to have been misunderstood. I mentioned that all the studies have shown that there is no significant difference. None of the lasers have proven to be associated with less adhesion formation than electrocautery. Laser is not a panacea. The laser is a long knife that makes the operative channel of the laparoscope an operating tool. Adhesion formation is the same as with electrocautery.

Franks: I shall be discussing our results with low-dose gonadotrophins later. The patients who do not do well in our series are those who have persistently elevated LH levels, which is very reminiscent of the women that Professor Shaw is looking at who are the non-responders. That remains the most difficult group of patients to treat, and that is the group where we see either no conception or early miscarriage. That is the most challenging group and Professors Shaw's data would suggest that surgery for the ovaries is not the ideal method of treatment in that group.

Shaw: I think that is true, because if they do not get that fall in LH they are the non-responders.

Braude: Can I clarify something about LH levels and when they are high and when the bloods should be taken?

Franks: From a diagnostic point of view it does not matter when we take it. A high LH level is not particularly useful in diagnosis. From a prognostic point of view, we found that if we do serial LH measurements through the follicular phase and the LH levels stay high, those are the ones that do not do well.

Braude: Why is it not useful diagnostically? I seem to have missed something. An LH of 15 or 16 in the follicular phase of the cycle does not mean anything?

Franks: Not if it comes down.

Abdalla: Turning to Professor Winston's chapter on tubal surgery. The problem we are facing in this country is a loss of expertise. There are very few people who are good at tubal surgery, a lot of us are going over to IVF and no one is being trained and there are no comparative results. Several recently published papers have condemned tubal surgery – I wonder if this could be one of the recommendations from this Group, that training should be made available not only in IVF but also in tubal surgery. Furthermore, there are problems in assessing the value of laparoscopic surgery.

Templeton: There are clearly difficulties in doing large prospective randomised clinical trials. We have used a variety of study designs to overcome some of these difficulties. However, I have difficulty in trying to visualise the sort of trial that is being suggested comparing, for example, laser laparoscopy with laparotomy, in the management of fimbrial block.

Winston: I am sure a properly randomised prospective study on adhesiolysis could be conducted and it would be one obvious way to start. Nothing would be lost by conducting the adhesiolysis at the time of investigative laparoscopy and patients could be randomised to open surgery without any difficulty. But it is more difficult with the other issues.

Cooke: The logistics relate to the main problem in doing a significant adhesiolysis on the routine laparoscopy list. It is very disorganising.

Winston: That is certainly one problem. But with regard to endometrioma, it could be done. With ultrasound it is possible to get a pretty good idea of what one will be looking at when it comes to doing diagnostic laparoscopy. I suspect it could be done with the tools that we have. But it is not easy.

Baird: Professor Winston did not really comment on why the fertility rate in the reversal of sterilisation group dropped off so rapidly compared with IVF rates in the older age groups.

Winston: I have analysed those figures. There is a very loose trend downwards in the number of follicles we can recruit with age; there is a very loose trend upwards in the amount of gonadotrophin we need to give to stimulate the ovaries. I have analysed, in those same patients, the number of eggs produced by age and there is a downward trend, but what is interesting is that, although there is a downward trend in number of eggs, the number of embryos that each patient has produced has been maintained up to age 45 years and then it drops off.

Baird: Were those pregnancy rates for treatments started, or for oocyte pick up?

Winston: All the numbers were patients started, so the 298 cycles were cycles started, of which just over 30% were abandoned and the remainder went on to egg collection.

Baird: If we take account of the treatments started, which is what people who have had reversal of sterilisation have to do, and also take into account that the pregnancy possibility is doubled by putting in two embryos instead of one, that probably comes some way towards explaining the results.

Winston: Yes. I am not arguing with the explanation. I am making a simple point, that there is a kind of knowledge around that women over 40 years are not worth treating by in vitro fertilisation. We have enough data to dispute that very clearly.

Franks: It is very likely that the difference that is seen between the two groups, which is reflected in the high cancellation rate of cycles, is to do with FSH levels. In that respect we do have a marker of response. We know that if a woman has a high basal FSH, response to superovulation will be poor, and it is quite likely that that is one of the major factors that determines whether or not she conceives. So we have a way of marking it.

Winston: We do not have the FSH values in all those patients before their treatment. We are now collecting those data.

Asch: I was also surprised at the success rates for IVF in women aged over 40 years. We have been working in that field for a number of years, have reviewed the literature, have visited many countries and seen the registries, and in general the results are dreadful. Life-table rates in general are between 3% and 5% from centres that do good ART procedures and have good pregnancy rates overall. But I really do not understand how one centre can have levels so much higher than practically everywhere else.

Kerin: I was equally impressed with that 19% of women aged over 40 and the quite large numbers. The bottom line is that most people experience a high abortion rate. Are there figures on the take-home baby rate and the abortion rate?

Winston: I do not have the figures, but the abortion rate has not been substantially different from the patients aged less than 40 years.

Chapter 14

Manipulation of the Ovarian Cycle

D. T. Baird

Introduction

Disorders of ovulation are the commonest treatable causes of infertility in developed countries. If the diagnosis has been made accurately, virtually normal rates of pregnancy can be obtained by appropriate treatment [1]. Antioestrogens, dopamine agonists, gonadotrophins and gonadotrophin releasing hormone (GnRH) all have their part to play in treatment of anovulatory infertility. The use of pituitary gonadotrophins to induce ovulation in hypophysectomised women by Carl Gemzell in 1958, was an example of classical replacement therapy of a deficiency of hormones [2]. Gonadotrophins still have a role in the treatment of anovulation (Chapter 15). The major challenge is to devise a regimen which avoids the risk of multiple birth.

In the last 30 years our understanding of the basic mechanisms underlying the central control of ovarian activity has increased and led to the development of methods of regulating ovarian function which have become clinically useful. The isolation and characterisation of the structure of gonadotrophin releasing hormone (GnRH), the hypothalamic peptide which stimulates the release of FSH and LH from the anterior pituitary was a landmark in the neuroendocrine control of reproduction [3,4]. Attempts to stimulate gonadotrophin secretion by continuous infusion or injection of large quantities of the synthetic decapeptide in monkeys rendered hypogonadotrophic by lesions in the hypothalamus were unsuccessful. In a series of elegant experiments, it was demonstrated that ovarian function could be restored only if the GnRH was administered in the form of small injections every hour, simulating the physiological situation [5]. This knowledge was rapidly applied to induction of ovulation in women with Kalmann's syndrome and other hypogonadotrophic conditions [6,7].

It was soon realised that alteration of one or more of the amino acids of the natural decapeptide might result in analogues of GnRH which could be used to stimulate or suppress gonadotrophins [8]. A series of potent agonists of GnRH were produced by substituting one or more amino acids with synthetic acids which were relatively resistant to breakdown by peptidase enzymes. Injection of these analogues results in a prolonged stimulation of the anterior pituitary so that the level of LH and FSH remains elevated for several hours. However, repeated injections results in desensitisation and within 10 days the levels of gonadotrophins are suppressed so that a hypogonadal state exists. GnRH analogues have been found useful in the treatment of a range of hormone sensitive conditions including endometriosis, fibroids, menorrhagia, premenstrual syndrome, contraception and breast cancer. Of particular interest in infertility is their use prior to treatment with gonadotrophins [9].

The development of antagonists of GnRH for clinical use has taken much longer than that of agonists. Antagonists are much more difficult to synthesise; require a much greater dose; and are much more expensive to make [10]. Moreover, until recently, virtually all the potent antagonists provoked the release of histamine and, hence, were unsuitable for routine clinical use. Antagonists block the release of gonadotrophins by binding to the GnRH receptor on pituitary cells and competing with the native GnRH. Unlike analogues, the suppression of gonadotrophins occurs very rapidly (within hours) and normal secretion of LH and FSH occurs as soon as the antagonist is cleared from the blood. Thus, to achieve prolonged suppression, large amounts of the antagonist must be given continuously.

This chapter will summarise the present use and future potential of manipulating the menstrual cycle with agonists and antagonists of GnRH in the treatment of infertility.

Stimulation of Follicular Development

All methods of stimulating follicular development depend on raising the levels of FSH above those found in the follicular phase of the cycle. It is possible to use native GnRH or its analogues to induce ovulation in anovulatory women [11] or in combination with gonadotrophins to produce hyperstimulation of the ovary for in vitro fertilisation–embryo transfer (IVF-ET) [12]. In the former the object is to produce a single ovulation by simulating the normal cycle and, hence, avoid the risks of hyperstimulation of the ovaries and multiple births. In contrast, in the latter, the treatment is designed to stimulate the development of many follicles so that an adequate number of ova can be recovered. Thus, in IVF, the normal feedback system which operates between the hypothalamus and ovary is overridden by the administration of exogenous FSH (Fig. 14.1). Alternatively, the secretion of endogenous FSH can be stimulated by administration of antioestrogens such as clomiphene or by the injection of GnRH agonist such as buserelin or nafarelin. Both these regimens result in the concomitant stimulation of LH release which may be deleterious to follicular growth and oocyte quality.

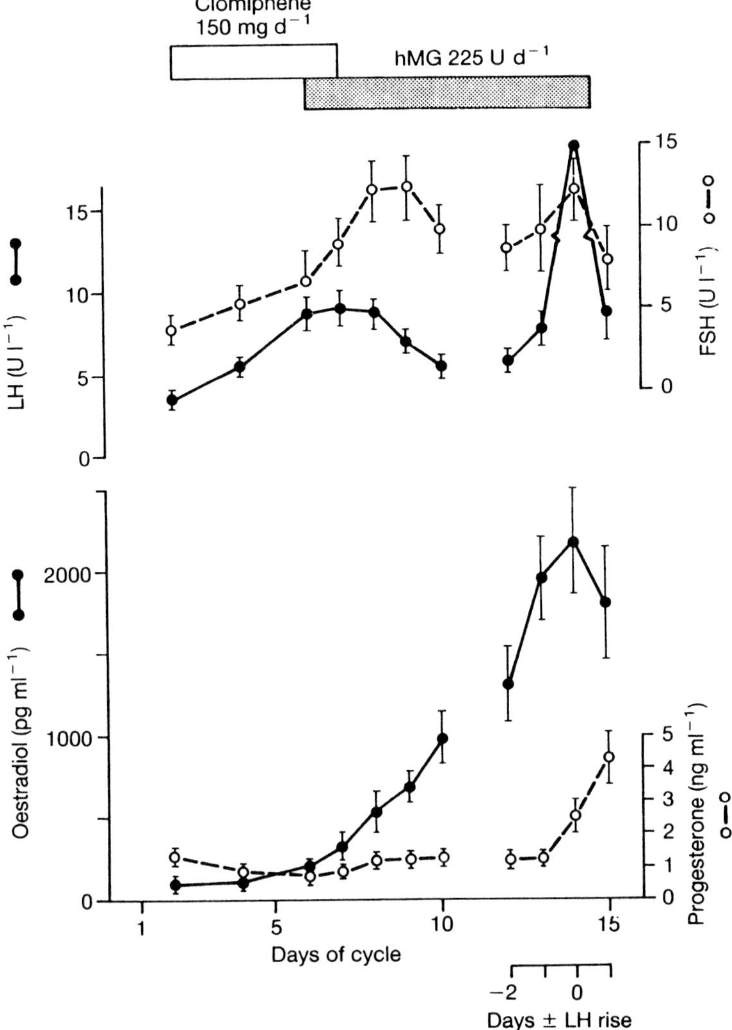

Fig. 14.1. Concentration of gonadotrophins and ovarian steroids in women treated with clomiphene and hMG to induce multiple follicular development. Note the rise in FSH concentration induced by clomiphene which is sustained by injections of hMG. An attenuated LH surge occurred about day 14 in all subjects. (From reference [27].)

Native GnRH

As indicated in the preceding section, the preferred method of inducing ovulation in the majority of women with anovulatory infertility due to hypogonadotrophic hypogonadism, is the injection of small boluses of GnRH at intervals of 60–90 min [11]. The treatment aims to reproduce the pattern of GnRH secretion which occurs in the follicular phase of the ovarian cycle. Intravenous injection of 2.5–5.0 μg GnRH induces follicular development in the majority of women who have a pituitary which is potentially normal. The cumulative pregnancy rate may be as high as 90% with a relatively low incidence of multiple pregnancies (<10%).

This treatment is much less effective in women with polycystic ovarian syndrome (PCOS) in whom gonadotrophin secretion is disturbed rather than absent. A novel approach has been pioneered by Filicori et al. who have treated these women with GnRH agonist for several weeks prior to treatment with Buserelin 300 μg s.c. bd. [13]. Pulsatile GnRH therapy is started within hours of stopping the GnRH agonist therapy when the anterior pituitary is down-regulated. The exogenous GnRH pulses override the endogenous secretion so that a normal pattern of follicular development and ovulation occurs. Although the percentage of ovulatory cycles is lower than in women with hypogonadotrophic hypogonadism, 28% of women with PCOS who were resistant to treatment with clomiphene, became pregnant after this regimen (Table 14.1). There were no multiple births but the abortion rate remained disappointingly high.

Programming Ovulation

Probably the most important application of GnRH agonists has been in combination with gonadotrophins. After a period of treatment with agonist to down-regulate the hypothalamic–pituitary system, gonadotrophin therapy (hMG or FSH) is given in combination with GnRH agonist. In women with PCOS the level of LH is often raised and results in stimulation of secretion of androgen by the ovaries [14]. The majority of women with PCOS ovulate in response to treatment with clomiphene. However, clomiphene stimulates a further rise in the concentration of LH as well as FSH which may result in luteinisation of granulosa cells before the follicles are fully mature. Treatment with hMG or

Table 14.1. Induction of ovulation with pulsatile GnRH (2.5 or 5.0 μg/60 min)

	Hypogonadotrophic hypogonadism $n = 66$	PCOS $n = 51$	
		Pre-GnRH^	Post-GnRH^
Cycles	95	42	50
Ovulatory	91%	43%	76%
Pregnant	27%	12%	28%
Multiple	11%	20%	0%
Abortion	7%	40%	43%

From Filicori et al. [13].

even FSH often results in similar premature discharge of LH. LH levels are suppressed following down-regulation with GnRH agonists with associated decline in androgen secretion [15]. Another study demonstrated that after suppression of ovarian activity with GnRH agonists, ovulation could be induced with exogenous gonadotrophin and the injection of the ovulating dose of hCG deferred until at least one mature follicle was present [16]. In their hands the pregnancy rate per cycle (28%) approaches that found in women with hypogonadotrophic hypergonadism and the cumulative pregnancy rate after six cycles was greater than 60%. This regime is attractive to women with PCOS who are resistant to clomiphene but the risk of multiple ovulation, high rates of spontaneous abortion and multiple pregnancy remains.

The techniques of down-regulation and GnRH^A originally developed by Fleming and Coutts for ovulation induction, has been widely applied in IVF and GIFT. The advantages of this combination treatment are twofold. First, the stimulation of follicular development is more predictable than in unsuppressed women. Second, because the anterior pituitary is suppressed, the levels of LH are lower and the risk of premature LH surge is virtually eliminated. It is unnecessary to make frequent detailed measurement of LH and/or progesterone to detect the presence of an endogenous LH surge and the number of unscheduled oocyte recoveries can be reduced to a minimum. Moreover, the number of cycles which are cancelled due to premature LH surge and ovulation are significantly reduced. In a controlled trial comparing treatment with hMG alone with hMG in combination with GnRH^A, no difference was noted in the number of eggs recovered or fertilised [17]. However, because the number of cancelled cycles was significantly reduced in the GnRH^A group (14% vs 29%; $P < 0.02$) the pregnancy rate per cycle started was much higher (21.2% vs 12.1%).

Most centres have adopted the so-called "long" regime which involves administration of the agonist over a two-week period during which follicular development is suppressed. In the "short" regime, the analogue is started at the beginning of the menstrual cycle and hMG is added after only 3 or 4 days. By utilising the initial stimulation of gonadotrophins ("flare") the period of treatment is shortened and the number of ampoules of gonadotrophin (and, hence, the cost) is reduced. Although this system inhibits the generation of an LH surge, the secretion of LH as well as FSH is stimulated during the period of "flare". There is a risk, therefore, that the developing follicles may be exposed to an inappropriately high level of LH which may have deleterious effects on their development and the orderly maturation of the oocyte.

Another advantage of the use of GnRH agonists is that it is possible to "programme" the period of treatment so that oocyte pick-up can be scheduled within a day or so several weeks in advance. Gonadotrophin treatment can be started on a given number of days (usually about 12) before anticipated day of oocyte recovery which can be scheduled for the convenience of both clinic and patient. Moreover, because there is no danger of a spontaneous LH surge, the injection of hCG can be deferred for up to 48 h after an adquate number of follicles have been recruited without compromising clinical efficacy. Although "programmed" IVF can be arranged by pretreatment with gestogens or combined oestrogen–gestogen mixtures as in oral contraceptives, the risk of premature LH release and luteinisation is still present and results in cancellation of approximately 20% of cycles [18,19].

In summary, the use of GnRH^{A} as a means of suppressing ovarian activity and inhibiting release of LH prematurely prior to treatment with gonadotrophins for ovulation induction or IVF has been a significant advance and has been largely responsible for the significant increase in the pregnancy rates per cycle started in most IVF clinics in the last five years.

GnRH Antagonists

GnRH antagonists inhibit the release of LH and FSH by competitive inhibition of GnRH at receptor sites on the pituitary gonadotroph [10,20]. Because of the decrease in the level of steroid feedback, the secretion of GnRH is actually increased but in physiological amounts is unable to bind with GnRH receptors which are blocked by the antagonist. The mode of action, therefore, of GnRH antagonists is quite different from that of agonists which down-regulate the postreceptor mechanisms. During treatment with GnRH antagonist it is still possible to provoke a release of LH after injecting a relatively large amount of GnRH [21]. It has been proposed that this could be the basis of a regimen to induce ovulation by pulsatile GnRH in women who had been rendered hypogonadotrophic by treatment with GnRH antagonist [22].

The initial GnRH antagonists required large quantities to suppress gonadotrophins and were relatively toxic [20]. The second generation, e.g. Nal-Glu, caused an immediate suppression of LH in doses of 1–5 mg/kg. They have been used for experimental studies in men and women although they still retain the property of histamine release and their use is associated with erythema and itching at the site of injection in a significant proportion of subjects. Nal-Glu has been shown to arrest follicular development when injected at a dose of 10 mg in the mid-follicular phase of the cycle and to cause luteal regression in the luteal phase [23,24]. At mid-cycle the effect is inconsistent with the LH surge ovulation being inhibited in less than 50% of women. It would appear that there comes a point in the preovulatory period when it is not possible to arrest the oestrogen-induced LH surge even in the presence of large amounts of antagonist. The most likely explanation is that after appropriate priming with GnRH, the anterior pituitary will release LH in response to oestrogen alone.

The most recent series of antagonists of GnRH, e.g. Antide, are apparently virtually devoid of ability to release histamine. Surprisingly, there is prolonged suppression of gonadotrophins after injection of Antide into rhesus monkeys lasting several weeks [25]. In intact animals, the suppression is variable and occurs after intravenous as well as subcutaneous injection. It seems likely, therefore, that the extreme hydrophobic nature of the substituted amino acids made the compound very resistant to enzymatic breakdown. Until further information about the pharmacokinetics have been obtained, it will not be possible to explore the possible clinical uses of the third generation of antagonists.

Because antagonists result in an immediate suppression of LH, it has been suggested that they might be used in combination with FSH for ovarian stimulation prior to IVF [26]. However, it would be necessary to ensure that the antagonist was given early enough in the treatment cycle to ensure inhibition of a premature LH surge. In addition, luteotrophic support would be required to avoid premature regression of the corpus luteum.

In an attempt to prevent a spontaneous LH surge in the "natural cycle", Frydman et al. [26] injected a 5 mg GnRH antagonist (Nal-Glu) when the concentration of oestradiol had reached $\geqslant$125 pg ml^{-1}. A second injection of GnRH antagonist was given 48 h later. In order to maintain follicular growth it was necessary to inject hMG (225 IU) simultaneously. Ovulation was induced by injection of 5000 IU of hCG when serum oestradiol $\geqslant$410 pg ml^{-1} and follicular diameter as measured by ultrasound $\geqslant$15 mm. Ovulation of a single follicle occurred in all women, suggesting that this may be a useful method of programming ovulation in spontaneous cycles.

Summary

A clearer understanding of the physiological mechanisms by which the hypothalamic–pituitary unit functions has led to the development of better methods of manipulating ovarian functions. GnRH and its analogues already have made a major impact on the management of the infertile woman. Induction of a hypo-oestrogenic state by down-regulation of the anterior pituitary has had wide application in reproductive conditions which are hormone sensitive, e.g. endometriosis, fibroids, breast cancer. Further research into the paracrine control of follicular development and formation and regression of the corpus luteum should lead to ways of manipulating ovarian function both for the treatment of infertility and for contraception.

References

1. Baird DT. Endocrinology of female infertility. Br Med Bull 1979; 35:193–8.
2. Gemzell CA. Induction of ovulation with human gonadotrophins. Rec Prog Horm Res 1965; 21:179–98.
3. Schally AV, Arimura A, Kastin AJ. Gonadotropin-releasing hormone: one polypeptide regulates secretion of luteinizing and follicle stimulating hormones. Science 1971; 173:1036–7.
4. Burgus R, Butcher M, Ling N, et al. Structure moleculair du facteur hypothalamique (LRF) d'origine ovine controlant la secretion de l'hormone gonadotrope hypophysaire luteinisation (LH). C R Acad Sci 1971; 273:1611–13.
5. Belcheltz PE, Plant TM, Nakai Y, Keogh EJ, Knobil E. Hypophysial responses to continuous and intermittent delivery of hypothalamic gonadotrophin-releasing hormone. Science 1978; 202:631–3.
6. Crowley WF Jr, McArthur JW. Stimulation of the normal menstrual cycle in Kallman's syndrome by pulsatile administration of luteinizing hormone-releasing hormone (LH-RH). J Clin Endocrinol Metab 1980; 51:173–5.
7. Leyendecker G, Wildt L, Hansmann M. Pregnancies following chronic intermittent (pulsatile) administration of Gn-RH by means of a portable pump (Zyklomat) – a new approach to the treatment of infertility in hypothalamic amenorrhoea. J Clin Endocrinol Metab 1980; 51:1214–16.
8. Corbin A, Bex FJ. Luteinizing hormone releasing hormone and analogues. Conceptive and contraceptive potential. In: Briggs M, Corbin A, eds. Progress in hormone biochemistry and pharmacology, Vol 1, Lancaster. MTP, 1980; 227–97.
9. Fraser HM, Baird DT. Clinical application of LHRH analogues. Bailliere's Clin Endocrinol Metab 1987; 1:43–70.

10. Conn PM, Crowley WF. Gonadotropin-releasing hormone and its analogues. N Engl J Med 1991; 324:93–103.
11. Leyendecker G, Wildt L. Induction of ovulation with chronic intermittent (pulsatile) administration of Gn-RH in women with hypothalamic amenorrhoea. J Reprod Fertil 1983; 69:397–409.
12. Fleming R, Haxton MJ, Hamilton MPR, McCune GS, Black WP, Coutts JRT. Successful treatment of infertile women with oligomenorrhoea using a combination of an LHRH agonist and exogenous gonadotrophins. Br J Obstet Gynaecol 1985; 92:369–79.
13. Filicori M, Flamigni C, Meriggiola MC et al. Endocrine response determines the clinical outcome of pulsatile gonadotropin-releasing hormone ovulation induction in different disorders. J Clin Endocrinol Metab 1991; 72:965–72.
14. Baird DT, Corker CS, Davidson DW, Hunter WM, Michie EA, Van Look PFA. Pituitary–ovarian relationships in polycystic ovarian syndrome. J Clin Endocrinol Metab 1977; 45:798–809.
15. Chang RJ, Laufer LR, Meldrum DR et al. Steroid secretion in polycystic ovarian disease after ovarian suppression by a long-acting gonadotropin-releasing hormone agonist. J Clin Endocrinol Metab 1983; 56:897–903.
16. Coutts JRT, Fleming R, Hamilton MPR, MacNaughton MC. Ovulation induction in women with polycystic ovarian syndrome: the use of combined gonadotrophin-releasing hormone analogue and exogenous gonadotrophin regimens. In: Shaw RW, ed. Advances in reproductive endocrinology. Polycystic ovaries. Lancs and New Jersey: Parthenon, 1991; 179–93.
17. Antoine JM, Salat-Baroux J, Alvarez S, Cornet D, Tibi Ch, Mandelbaum J, Plachot M. Ovarian stimulation using human menopausal gonadotrophins with or without LHRH analogues in a long protocol for in vitro fertilisation: a prospective randomised comparison. Hum Reprod 1990; 5:565–9.
18. Templeton AA, Van Look PFA, Lumsden MA, Angell R, Aitken J, Duncan AW, Baird DT. The recovery of pre-ovulatory oocytes using a fixed schedule of ovulation induction and follicle aspiration. Br J Obstet Gynaecol 1984; 91:148–54.
19. Frydman R, Forman RE, Rainhorn JD, Belaisch-Allart J, Hazout A, Testart J. A new approach to follicular stimulation for in vitro fertilization: programed oocyte retrieval. Fertil Steril 1986; 46:657–62.
20. Karten MJ, Rivier JE. Gonadotrophin-releasing hormone analogue design. Structure–function studies towards the development of agonists and antagonists. Rationale and perspectives. Endocrinol Rev 1986; 7:44–52.
21. Marshall GL, Akhtar FB, Weinbauer GF, Neischlag E. Gonadotrophin-releasing hormone (GnRH) overcomes GnRH antagonist-induced suppression of LH secretion in primates. J Endocrinol 1986; 110:145–50.
22. Gordon K, Danforth DR, Williams RF, Hodgen GD. Primate studies on GnRH antagonists. Gynaecol Endocrinol 4, Suppl 2, 1990; International Symposium on GnRH Analogues in Cancer and Human Reproduction. Parthenon Press. Abstract 33.
23. Flucker MR, Marshall LA, Monroe SE, Jaffe RB. Variable ovarian response to gonadotropin-releasing hormone antagonist-induced deprivation during different phases of the menstrual cycle. J Clin Endocrinol Metab 1991; 72:912–19.
24. Hall JE, Bhatta N, Adams JM, Rivier JE, Vale WW, Crowley WF Jr. Variable tolerance of the developing follicle and corpus luteum to gonadotropin-releasing hormone antagonist-induced gonadotropin withdrawal in the human. J Clin Endocrinol Metab 1991; 72:993–1000.
25. Leal JA, Williams RF, Danforth DR, Gordon K, Hodgen GD. Prolonged duration of gonadotropin inhibition by a third generation GnRH antagonist. J Clin Endocrinol Metab 1988; 67:1325–7.
26. Frydman R, Cornel C, de Ziegler D, Spitz JM, Bouchard P. Prevention of spontaneous LH surge in the natural cycle with an antagonist of GnRH (Nal-Glu). Gynaecol Endocrinol 4, Suppl 2, 1990; International Symposium on GnRH Analogues in Cancer and Human Reproduction. Parthenon Press. Abstract 26.
27. Messinis IE, Templeton AA, Baird DT. Endogenous luteinizing hormone surge during superovulation induction with sequential use of clomiphene citrate and pulsatile human menopausal gonadotropin. J Clin Endocrinol Metab 1985; 61:1076–80.

Chapter 15

Induction of Ovulation

S. Franks

The aim of induction of ovulation is to obtain normal fertility, using methods which are designed to restore ovarian function in as physiological a manner as possible. Current methods of diagnosis and treatment result in normal fertility rates in most patients with anovulation, with the notable exceptions of women with primary ovarian failure, and those with clomiphene-resistant anovulation associated with polycystic ovary syndrome (PCOS).

Causes of Anovulation

Disorders of ovulation can be classified in functional terms as illustrated in Table 15.1. Primary ovarian failure accounts for approximately 50% of cases of primary amenorrhoea, and 10%–15% of women presenting with secondary amenorrhoea [1]. In the former, genetic disorders are highly prevalent, the obvious example being gonadal dysgenesis associated with a 45XO karyotype. Evidence of autoimmune endocrinopathy can be found in up to 50% of women with primary ovarian failure, where the predominant abnormality is the finding of thyroid autoantibodies [2]. Ovarian autoantibodies are less commonly detected, but this may be largely due to the methodological problems involved in measuring these. The inference, therefore, is that many women with primary ovarian failure do have functionally significant ovarian autoantibodies.

Although the terms "premature menopause" and "resistant ovary syndrome" are frequently used as subcategories of primary ovarian failure, it is doubtful whether it is useful to distinguish between these entities. Resistant ovary syndrome is a presumptive diagnosis resting on the finding of antral follicles in women with high levels of follicle stimulating hormone. In practice, this may

Table 15.1. Causes of anovulation

A. Primary ovarian failure
Genetic (e.g. Turner's syndrome)
Autoimmune
Others (e.g. abdominal radiotherapy, chemotherapy)
B. Secondary ovarian dysfunction
Disorders of gonadotrophin regulation
(a) Specific
Hyperprolactinaemia
Kallmann's syndrome
Others (e.g. suprapituitary tumour)
(b) Functional
Weight loss
Exercise
Idiopathic
Gonadotrophin deficiency
Destructive lesion of pituitary (e.g. large tumour)
Pituitary ablative therapy
C. Polycystic ovary syndrome

simply reflect episodes of follicular activity in an ovary which has either a reduced number of primordial follicles and/or a major disorder of folliculogenesis. In both forms of ovarian failure, the prognosis for fertility is poor; those with continuing follicular activity are likely to have a slightly better chance of pregnancy, but there are few data to support this assumption.

The majority of women presenting with secondary amenorrhoea will have disorders of gonadotrophin regulation. These may be related to specific diseases of the pituitary or hypothalamus (the most common of which is a prolactin-secreting pituitary adenoma), or may be the functional consequence of an underlying disorder, such as weight loss. Within this category of functional disorders of gonadotrophin regulation is a substantial subgroup (9% of cases of secondary amenorrhoea in one study), in whom the aetiology is unknown [1]. Primary deficiency of pituitary gonadotrophins is uncomon, occurring principally in women who have received previous ablative treatment for a pituitary or hypothalamic tumour [3].

The largest group of women with anovulation is the most difficult to classify in functional terms. These are the women with polycystic ovary syndrome, who may present with amenorrhoea or, more commonly, with anovulatory menses and oligomenorrhoea. The nature of the disorder of ovulation remains unclear, although recent data suggest that there may be impairment of FSH action at ovarian level by either endocrine or paracrine (locally produced) factors [4].

Investigation of Anovulation

Patients Presenting with Amenorrhoea

Differential diagnosis and appropriate selection of treatment can usually be accomplished by the use of a small number of investigations. The essential tests are measurements of serum concentrations of FSH and prolactin, and an assessment of oestrogen production [1,6]. The value of the serum FSH measurement is to exclude those subjects with primary ovarian failure, in which FSH concentrations are invariably elevated. Luteinising hormone (LH) levels will also be raised in primary ovarian failure, but elevation of LH concentrations is not specific to this group of patients. It can also be raised in women with polycystic ovary syndrome (see below). Measurement of serum prolactin will identify the 10%–15% of subjects with secondary amenorrhoea who have hyperprolactinaemia. Interpretation of a marginally elevated serum prolactin concentration (i.e. in the 500–1000 mu l^{-1} range) may be difficult, but persistent, moderate hyperprolactinaemia in a patient with oestrogen-deficient amenorrhoea is likely to be significant [5]. Dynamic tests are not required in the diagnosis of hyperprolactinaemia, but the finding of the raised level should be confirmed by a second and, if necessary, a third, sample. Underlying causes of hyperprolactinaemia, such as primary hypothyroidism and the use of dopamine receptor antagonist drugs (such as phenothiazines) can readily be excluded before considering imaging of the pituitary area. We consider this to be important in any patient with otherwise unexplained persistent hyperprolactinaemia; a moderately elevated level of prolactin may be the first endocrine marker of a large non-functioning tumour of the pituitary.

Assessment of oestrogen production may be made by measurement of serum oestradiol, but there is considerable overlap between the values obtained in hypogonadotrophic women and those observed in the early follicular phase of a normal menstrual cycle. The progestogen withdrawal test is a simple and reliable alternative to oestradiol measurements [6].

An ultrasound scan of the pelvis is not an essential investigation, but may be helpful, particularly in the management of clomiphene-resistant patients (see below).

Patients Presenting with Oligomenorrhoea or Irregular Cycles

In such cases it is unnecessary to perform a progestogen withdrawal test, but the investigations mentioned in the context of women with amenorrhoea also apply to those with anovulatory menses. Since polycystic ovary syndrome is highly prevalent in this group of women, many subjects with anovulatory menses also have hirsutism or persistent acne. In these patients (and this also applies to those who present with amenorrhoea), a serum testosterone measurement should be performed as a screening test to exclude more serious disorders, such as androgen secreting tumours of the ovary or adrenal. As a guideline, serum testosterone concentrations >5 nmol l^{-1} (approximately twice the upper limit of the normal range) require further investigation. An ultrasound scan is useful in confirming the diagnosis of polycystic ovaries.

Selection of Treatment

In women with primary ovarian failure there is no evidence that any form of induction of ovulation (whether or not this is preceded by a period of ovarian suppression) improves the chances of pregnancy [2]. On the contrary, it could be argued that in those women with continuing follicular activity, treatment which disturbs the ovarian/pituitary feedback loop may further reduce the chance of spontaneous ovulation.

In patients with normal or low levels of FSH but a raised serum prolactin, the treatment of choice is the dopamine agonist bromocriptine. This selectively reduces serum prolactin concentrations and leads to normalisation of the hypothalamic control of gonadotrophins. Bromocriptine may also be effective in reducing the size of a prolactin-secreting tumour and it is rarely necessary to perform pituitary ablative therapy in patients with prolactinomas [5].

Patients with normal concentrations of FSH and prolactin, and who have a positive response to progestogen withdrawal (well oestrogenised), either have a mild hypothalamic disorder of gonadotrophin regulation (e.g. those with a recent history of weight loss who have regained weight but not menstrual cyclicity) or, in most cases, polycystic ovary syndrome. In either case, the treatment of first choice is the antioestrogen clomiphene. If this proves unsuccessful, reassessment of the history and a pelvic ultrasound examination will aid the decision regarding further management (see below).

Induction of Ovulation

Attenuation should first be directed to treatment of an underlying cause which may then result in resumption of spontaneous ovulatory cycles. The most obvious example is management of anovulation which accompanies anorexia nervosa. Regaining weight in women with weight loss-related amenorrhoea usually requires the support of a counsellor or psychotherapist with experience in management of eating disorders. Weight reduction in obese women is also an important aspect of management of ovulatory disorders, particularly in the context of PCOS (see below).

Disorders of Gonadotrophin Regulation

Patients Without Oestrogen Deficiency

Most women who have normal serum concentrations of FSH and present with oligomenorrhoea or progestogen-positive amenorrhoea have PCOS [1,7]. There is no clear evidence of a primary hypothalamic disorder in patients with PCOS [8] but their management, in the first instance, is similar to that of well-oestrogenised women who have disordered gonadotrophin regulation (for example those with a history of anorexia who have regained weight and some degree of ovarian activity but who remain anovulatory). The treatment of first choice in such cases is antioestrogens of which clomiphene citrate is the most widely used. Induction of ovulation with clomiphene citrate is, in principle, very simple but, in practice, requires careful (but not necessarily frequent) monitoring to achieve optimal results. Our policy is to assess the response to the first course of treatment (50 mg per day for 5 days from day 2 of the cycle) by both ultrasound scanning and mid-luteal progesterone measurements [9]. It is in this way possible to identify those patients who either do not develop a dominant follicle in response to clomiphene or those who have multiple follicles. The dose can be adjusted up or down, accordingly, in a subsequent cycle. However, an important point is that there is no evidence that a dose greater than 100 mg per day improves the chance of ovulation in previously unresponsive patients. More than 80% of anovulatory cycles after clomiphene are characterised by lack of development of a dominant follicle and there is, therefore, no rationale for the routine administration of a "mid-cycle" dose of chorionic gonadotrophin (hCG) in clomiphene non-responders [9].

The management of patients who are unresponsive to treatment with clomiphene should be influenced by the underlying functional abnormality. In patients with a hypothalamic disorder of gonadotrophin regulation the most appropriate treatment is pulsatile gonadotrophin-releasing hormone (GnRH) as described in the following section. In women with PCOS, gonadotrophin therapy is more likely to be effective (see below). Although the diagnosis may be clear from history and clinical examination, the ovarian appearance on ultrasound is particularly helpful in deciding which form of treatment to choose in the well-oestrogenised but clomiphene-resistant patient. The ovaries may have a typical polycystic morphology but the presence of normal or multifollicular, normal-sized ovaries without increased stroma suggests a hypothalamic disorder [10].

Patients With Oestrogen Deficiency

Women with normal or low serum FSH concentrations and oestrogen deficiency have hypogonadotrophin hypogonadism due either to a hypothalamic or (rarely) a primary pituitary disorder of gonadotrophin regulation. Appropriate treatment of such patients should lead to normal fertility.

The specific treatment of hyperprolactinaemic amenorrhoea is dopamine agonist therapy, usually bromocriptine. With very few exceptions, bromocriptine is a effective and safe treatment for patients with or without a pituitary tumour and surgery is rarely required. (For a recent review see reference [5].)

In hypogonadotrophic subjects with normal serum prolactin concentrations, the mainstay of treatment for more than 30 years has been the administration of gonadotrophins, derived principally from menopausal and pregnancy urine [11,12]. Gonadotrophin treatment results in normal conception rates but a high rate (more than 20%) of multiple pregnancy. In the last 10 years, a more physiological method of treatment has become available for women with disordered gonadotrophin regulation – pulsatile GnRH. The principle of pulsed GnRH therapy is to mimic the hypothalamic signal to the gonadotroph and restore the normal pulsatile pattern of gonadotrophin secretion. The main advantage over gonadotrophin therapy is that, by allowing feedback of ovarian steroids at the level of the pituitary, it minimises the chances of multiple follicle development and, therefore, multiple pregnancy. Multiple pregnancies (almost all twins) are slightly more common than normal [13,14] but the risk of this can be much reduced by careful monitoring, particularly in the first cycle of treatment in which development of multiple follicles is most likely to occur. Pulsatile GnRH treatment results in normal fertility in women with hypothalamic disorders of gonadotrophin regulation and is surprisingly effective in patients with pituitary disease, some 50% of whom appear have sufficient gonadotrophin reserve to respond to exogenous GnRH [3,14]. The cumulative conception rates in these two groups of hypogonadotrophic subjects, after six cycles, are 93% and 100% respectively [3,14].

A small proportion (<10%) of women with hypogonadotrophic hypogonadism do not respond to pulsed GnRH and require very large doses of gonadotrophins to achieve follicle development. Studies in animals have highlighted the potential gonadotrophic role of insulin-like growth factor-1 (IGF-1) in folliculogenesis [15]. In an ingenious clinical application of these studies, Homburg et al. [16, 17] demonstrated that treatment of "resistant" hypogonadotrophic subjects with recombinant human growth hormone (hGH) increased circulating levels of IGF-1 and reduce the amount of gonadotrophin required to induce a dominant follicle. Intriguingly, recent in vitro studies of human granulosa cells suggest that it is also possible that growth hormone has a direct gonadotrophic action on the ovary which does not require mediation by IGF-1 [18,19]. But whatever the mechanism of action of hGH it is important to note that most of the subjects reported in Homburg's studies had evidence of more widespread hypopituitarism (and may well have been growth hormone deficient). It remains to be determined whether hGH will have a place in the routine management of women with hypogonadotrophic amenorrhoea or other causes of anovulation.

Patients With Polycystic Ovary Syndrome

The use of antioestrogen treatment in women with PCOS has been discussed above but, in terms of induction of ovulation, perhaps the most challenging group of patients with anovulatory infertility are those with clomiphene-resistant PCOS. Although the mechanism of anovulation is unknown, it is evident that treatment which increases serum FSH concentrations results in follicular maturation and, in most cases, ovulation [4,6]. Pulsatile GnRH stimulates FSH secretion but at the expense of a further rise in LH concentrations and the overall results of this form of treatment are poor [20].

Human Gonadotrophin Treatment

The Problem of Multiple Follicles. Human gonadotrophins (principally urinary-derived, human menopausal gonadotrophin; hMG) have been used for many years as the treatment of choice in clomiphene-resistant patients with PCOS. The results have been variable with pregnancy rates ranging from 16% to 78% in published series (about 30% overall) (for review see reference [21]). The particular problems of induction of ovulation with gonadotrophins in women with PCOS, compared with hypogonadotrophic subjects, is the increased tendency to develop multiple follicles, resulting in a higher prevalence of multiple pregnancies and of hyperstimulation syndrome. In a review of series of gonadotrophin-treated patients with PCOS, an average multiple pregnancy rate of 29% was found [22].

Other well-recognised complications of gonadotrophin treatment in women with PCOS are so-called premature luteinisation [23] and an alarmingly high prevalence (30%) of miscarriage [21]. There is no evidence that the use of purified FSH preparations, which have a low LH content, either increases the rate of fertility or has a significant impact on the prevalence of these complications.

Premature Luteinisation. Premature luteinisation, the inappropriate, preovulatory rise of progesterone during induced cycles, is, in most cases, a direct consequence of multiple follicles contributing to supraphysiological serum oestradiol concentrations and thereby triggering an early surge of LH. This phenomenon is associated with failure of ovulation or fertilisation. It can effectively be prevented by pretreatment with long-acting agonist analogues of GnRH which lower endogenous gonadotrophin concentrations [23]. This strategy is, of course, in widespread use for superovulation in in vitro fertilisation programmes but has also been employed, successfully, for ovulation induction in women with PCOS [23]. There is, however, a paucity of controlled trials in evaluation of the potential advantages of combined therapy over gonadotrophin alone and the place of GnRH analogue/hMG treatment remains uncertain. We would suggest that many of these problems associated with development of multiple follicles following the use of conventional doses of gonadotrophins can be avoided by using a treatment regimen in which the aim is unifollicular ovulation.

Miscarriage in Women with PCOS. Miscarriage, particularly early pregnancy loss, occurs in about 30% of pregnancies, regardless of the method of induction of ovulation. This frequency is twice that observed in hypogonadotrophic women [14,24]. The most important prognostic feature is a raised serum LH concentration in the mid-follicular phase of an ovulatory cycle [24,25]. Although hypersecretion of LH is characteristic of PCOS, there is a tendency for LH levels to fall as a dominant follicle emerges In some 25% of ovulatory cycles, however, LH concentrations remain elevated throughout and these cycles are either infertile or are associated with early pregnancy loss [25]. Obesity also has a negative influence on outcome of pregnancy. The higher frequency of miscarriage in obese compared with lean women appears to be independent of LH concentrations [26] (see below).

Low-dose Gonadotrophin Treatment in PCOS

Because of the unacceptably high frequency of multiple follicle development and the attendant problems of multiple pregnancies and hyperstimulation syndrome, a regimen for gonadotrophin therapy was devised with the specific aim of inducing a single ovulatory follicle. This schedule was based on the principle of "titrating" the dose of gonadotrophin against the ovarian response, using small, stepwise increments in dosage [27]. This method was adapted specifically for use in women with PCOS by using a low starting dose (52.5–75 units per day) and maintaining this for up to 14 days, increasing thereafter by 25–37.5 units at seven-day intervals until either a dominant follicle appeared (ultrasound was used as the primary method of assessment) or a designated maximum dose of 225 units/day was reached [25,28].

Initial experiments were performed using purified FSH (Metrodin, Serono) delivered by pulsatile subcutaneous infusion [28], but subsequent studies showed that a single, daily, intramuscular injection of gonadotrophin was equally effective and that there was no difference between the effects of FSH and hMG (Pergonal, Serono) [29]. In a series of 100 patients treated by this low-dose regimen (83 received hMG and 17 FSH, all by daily intramuscular injection) a 72% ovulation rate was found with a 73% prevalence of uniovulatory cycles [25]. The mean maximum daily dose of gonadotrophin was 95 units per day and the mean duration of the follicular phase (from start of treatment cycle to administration of hCG) was 14.2 days. There were 45 pregnancies in 42 women and, importantly, only two (4%) multiple pregnancies. There were no cases of severe hyperstimulation and only two patients developed moderate hyperstimulation syndrome. The cumulative conception rate was 55% at six months and rose to 63% if patients who never ovulated (five women) or were found to have persistently negative post-coital tests were excluded (Fig. 15.1). However, the early miscarriage rate was a disappointing 32% and was linked, as discussed above, to persistently elevated serum LH and to obesity.

The Effect of Obesity on Outcome of Treatment in PCOS

Obesity is a characteristic finding in women with PCOS. In a series of 263 consecutive subjects with PCO, presenting with symptoms of either hyperandrogenism or anovulation (or both), 35% had a body mass index (BMI) greater than 25 kg m^{-2} [30]. Obesity is associated with resistance to treatment with clomiphene [9] and (in clomiphene non-responders) gonadotrophins. Using the low-dose schedule described above, it was found that women with moderate

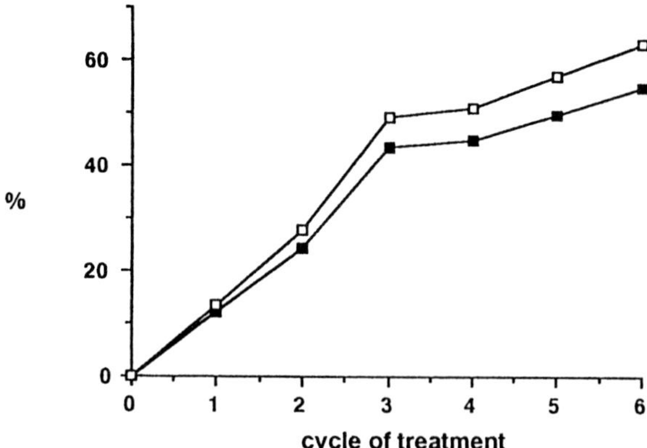

Fig. 15.1. Cumulative conception rate, by life table analysis, in patients with PCOS treated with low-dose gonadotrophin. ■ – ■ overall data in all patients, □ – □ after exclusion of subjects who never ovulated or with a male infertility factor. (From ref. [25], with permission.)

obesity (BMI 25.1–28) required significantly higher doses of gonadotrophin to induce ovulation and had a lower ovulation rate than their lean counterparts (Fig. 15.2) [26]. The pregnancy rate was similar in the two subgroups but obese women suffered a 60% rate of miscarriage compared with 27% in non-obese subjects. These results have prompted us to encourage obese patients with PCOS to undergo dietary treatment before starting induction of ovulation. The success of long-term calorie restriction alone, in terms of improvement of hyperandrogenaemia and ovulatory function [31], provides further support for the use of diet before considering induction of ovulation in obese women with PCOS.

Surgical Methods of Ovulation Induction in PCOS

Before the availability of gonadotrophins for induction of ovulation, the only method of restoring ovulation in women with PCOS was ovarian wedge resection. Until recently, this was still widely used in patients who had proved resistant to medical therapy and, in some centres, as an alternative to gonadotrophin treatment [21]. Despite reports of ovulation and pregnancy rates which approach those of medical treatment, the obvious disadvantages of major abdominal surgery, in particular the risk of postoperative pelvic adhesions, has

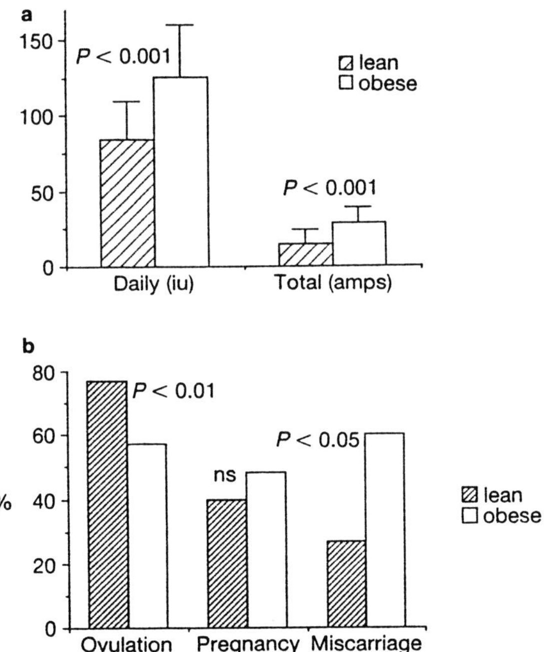

Fig. 15.2. **a** Daily and total dose of gonadotrophin required to induce ovulation in women with PCOS grouped according to BMI; lean: 19–25, obese: 25–28 kg m^{-1}. **b** Outcome of induction of ovulation with low-dose gonadotrophin in lean and obese women with PCOS (From reference [26], with permission.)

placed this procedure in the category of treatment of last resort. However, a less-invasive adaptation of this technique, laparoscopic ovarian diathermy (or electrocautery), has been advocated for the treatment of clomiphene-resistant women with PCOS [32,33]. Initial results have proved encouraging, particularly because (as with wedge resection) multiple ovulation is unusual and the miscarriage rate appears low. These results should be regarded with cautious optimism pending the outcome of further, more extensive studies with appropriate attention to criteria for selection of patients and with adequate, long-term follow-up data.

References

1. Franks S. Diagnosis and treatment of anovulation. In: Hillier SG, ed. Ovarian endocrinology. Oxford: Blackwell Scientific Publications, 1991; 226–59.
2. Hague WM, Tan SL, Adams J et al. Hypergonadotrophic amenorrhoea: aetiology and outcome in 93 young women. Int J Obstet Gynaecol 1987; 25:121–5.
3. Morris DV, Abdulwahid N, Armar NA et al. Induction of fertility in patients with organic hypothalamic–pituitary–gonadal dysfunction. Fertil Steril 1987; 47:54–9.
4. Franks S, Mason HD, Polson DW, Winston RM, Margara R, Reed MJ. Mechanism and management of ovulatory failure in women with polycystic ovary syndrome. Hum Reprod 1988; 3:531–4.
5. Franks S. Modern management of pituitary prolactinomas. Curr Obstet Gynaecol 1991; 1:84–92.
6. Hull MGR, Savage PE, Jacobs HS. Investigations and treatment of amenorrhoea resulting in normal fertility. Br Med J 1979; i:1257–61.
7. Fox R, Corrigan E, Thomas PG, Hull MG. Oestrogen and androgen states in oligo-amenorrhoeic women with polycystic ovaries. Br J Obstet Gynaecol 1991; 98:294–9.
8. Franks S. Polycystic ovary syndrome: a changing perspective. Clin Endocrinol (Oxf) 1989; 31:87–120.
9. Polson DW, Kiddy DS, Mason HD, Franks S. Induction of ovulation with clomiphene citrate in women with polycystic ovary syndrome: the difference between responders and nonresponders. Fertil Steril 1989; 51:30–4.
10. Adams J, Franks S, Polson DW et al. Multifollicular ovaries: clinical and endocrine features and response to pulsatile gonadotrophin releasing hormone. Lancet 1985; ii:1375–9.
11. Gemzell CA. Induction of ovulation with human gonadotrophins. J Reprod Med 1977; 18:155–62.
12. Lunenfeld B, Eshkol A, Titotzky D et al. Induction of ovulation: human gonadotrophins. In: Flamigini C, Givens JR, eds. The gonadotrophins: basic science and clinical aspects in females. London: Academic Press, 1982; 395–403.
13. Braat DDM, Ayalon D, Blunt SM et al. Pregnancy outcome in luteinising hormone-releasing hormone induced cycles: a multicentric study. Gynecol Endocrinol 1989; 3:35–44.
14. Homburg R, Eshel A, Armar NA et al. One hundred pregnancies after treatment with pulsatile luteinising hormone releasing hormone to induce ovulation. Br Med J 1989; 298:809–12.
15. Adashi EY, Resnick CE, D'Ercole AJ, Svoboda ME, Van Wyk JJ. Insulin-like growth factors as intra-ovarian regulators of granulosa cell growth and function. Endocrinol Rev 1985; 6:400–20.
16. Homburg R, Eshel A, Abdalla HI, Jacobs HS. Growth hormone facilitates induction of ovulation by gonadotrophin. Clin Endocrinol 1988; 29:113–7.
17. Homburg R, West C, Torresani T, Jacobs HS. Cotreatment with human growth hormone and gonadotrophins for induction of ovulation: a controlled trial. Fertil Steril 1990; 53:2540–60.
18. Mason HD, Martikainen H, Beard RW, Franks S. Direct gonadotrophic effect of growth hormone on oestradiol production by human granulosa cells. J Endocrinol 1990; 126:R1–R4.
19. Carlsson B, Bergh C, Bentham J et al. Expression of functional growth hormone receptors in human granulosa cells. J Clin Endocrinol Metab 1992 (in press).
20. Eshel A, Abdulwahid NA, Armar NA, Adams JM, Jacobs HS. Pulsatile luteinizing hormone-

releasing hormone therapy in women with polycystic ovary syndrome. Fertil Steril 1988; 49:956–60.
21. Hamilton-Fairley D, Franks S. Common problems in induction of ovulation. Bailliere's Clin Obstet Gynaecol 1990; 4:609–25.
22. Wang CF, Gemzell C. The use of human gonadotrophins for the induction of ovulation in women with polycystic ovarian disease. Fertil Steril 1980; 33:479–86.
23. Fleming R, Coutts JR. LHRH analogues for ovulation induction, with particular reference to polycystic ovary syndrome. Baillieres Clin Obstet Gynaecol 1988; 2:677–87.
24. Homburg R, Armar NA, Eshel A, Adams J, Jacobs HS. Influence of serum luteinising hormone concentrations on ovulation, conception, and early pregnancy loss in polycystic ovary syndrome. Br Med J 1988; 297:1024–6.
25. Hamilton-Fairley D, Kiddy D, Watson H, Sagle M, Franks S. Low-dose gonadotrophin therapy for induction of ovulation in 100 women with polycystic ovary syndrome. Hum Reprod 1991; 6:1095–9.
26. Hamilton-Fairley D, Kiddy DS, Watson H, Paterson C, Franks S. Association of moderate obesity with poor pregnancy outcome in women with polycystic ovary syndrome treated with low dose gonadotrophin. Br J Obstet Gynaecol 1992; 99:128–31.
27. Brown JB, Evans JH, Adey FD, Taft HP, Townsend L. Factors involved in the induction of fertile ovulation with human gonadotrophins. J Obstet Gynaecol Br Commonw 1969; 76:289–306.
28. Polson DW, Mason HD, Saldahna MB, Franks S. Ovulation of a single dominant follicle during treatment with low-dose pulsatile follicle stimulating hormone in women with polycystic ovary syndrome. Clin Endocrinol (Oxf) 1987; 26:205–12.
29. Sagle MA, Hamilton-Fairley D, Kiddy DS, Franks S. A comparative, randomized study of low-dose human menopausal gonadotrophin and follicle-stimulating hormone in women with polycystic ovarian syndrome. Fertil Steril 1991; 55:56–60.
30. Kiddy DS, Sharp PS, White DM et al. Differences in clinical and endocrine features between obese and non-obese subjects with polycystic ovary syndrome: an analysis of 263 consecutive cases. Clin Endocrinol (Oxf) 1990; 32:213–20.
31. Kiddy DS, Hamilton-Fairley D, Bush A, Short F, Anyaoku V, Reed MJ, Franks S. Improvement in endocrine and ovarian function during dietary treatment of obese women with polycystic ovary syndrome. Clin Endocrinol 1992; 36:105–11.
32. Gjonnaess H. The course and outcome of pregnancy after ovarian electrocautery in women with polycystic ovarian syndrome: the influence of body-weight. Br J Obstet Gynaecol 1989; 96:714–19.
33. Armar NA, McGarrigle HH, Honour J, Holownia P, Jacobs HS, Lachelin GC. Laparoscopic ovarian diathermy in the management of anovulatory infertility in women with polycystic ovaries: endocrine changes and clinical outcome. Fertil Steril 1990; 53:45–9.

Discussion

Winston: With regard to the high mid-follicular LH levels, is there much variation cycle to cycle or patient to patient?

Franks: Patient to patient yes, but those women who have persistently high LH levels during the follicular phase in the induced cycle seem to have it from cycle to cycle. At the moment we are selecting those women for combined analogue and gonadotrophin treatment and we do not have sufficient data to know whether that is successful. It is really quite important because all we are pointing out at the moment is an association and we do not know whether the high LH levels cause the increased risk of miscarriage or whether it is telling us something about the function of the ovary, or indeed the pituitary.

Baird: Is there a difference in the patern of follicular development and oestradiol levels in those patients as compared to the women in whom the LH level falls?

Franks: Not obviously. They are peak oestradiol levels. We do not have sufficient data on serial levels of oestradiol. Their peak oestradiol levels and their mid-luteal progesterone levels seem to be similar.

Baird: If the fall in LH in the women with single ovulation is a response to the feedback from the dominant follicle, and that is not occurring, then clearly the patient has no chance of getting pregnant.

Franks: That is quite right. But that is why we were keen to look at the oestradiol levels at the time the dominant follicle developed and in the preovulatory phase, and there were no differences. That is very interesting because it suggests there may be some other feedback factor from the ovary from the dominant follicle that influences LH levels, or it may be a differential, a sensitivity to oestradiol by the pituitary, although I doubt that.

Braude: There are people who would doubt that a raised LH can exist with a normal luteal phase progesterone.

Franks: There is a study on a group of more than 20 women who had a history of miscarriage and who had polycystic ovaries, most of them had polycystic ovaries on ultrasound, but were ovulating regularly [1,2]. They had abnormalities of LH secretion, but their progesterone profiles were normal in the luteal phase of the cycle. What is quite interesting is that their oestradiol excretion was different from the controls. They made more oestradiol in the luteal phase of the cycle than did the control subjects.

Fishel: In a preliminary study we have found that we should not ignore the androgens in the interplay of the regulation of ovulation after follicular stimulation for assisted conception.

Franks: I do not believe that androgen itself is an important feedback modulator of gonadotrophins in the female. But there may well be other factors. For example, Professor Templeton's gonadotrophin surge attenuating factor may be important. It would fit the bill. If there was an abnormality of that then that could account for raised LH levels or LH levels which fail to come down.

Templeton: Would the speakers comment on the potential use of recombinant FSH, particularly if it is possible to produce FSH with a shortened half-life. In multiple ovulation induction the problem is that to sustain the cohort of follicles one is interested in, the gonadotrophin level has to be such that there is continued recruitment of follicles that do not contribute to therapy but increase the risk of hyperstimulation. With shorter-acting FSH it might be feasible to achieve better manipulation.

Baird: Professor Templeton has said it. The other component, which we have ignored up to now, is the ratio of FSH to LH. There is increasing evidence that part of the selection process involves the action of LH on the non-dominant follicle and hastens their atresia. We have not begun clinically to explore the use of LH as an adjuvant to FSH to make the selection of a single follicle more efficient.

Hillier: Professor Franks spoke of the 75% of women with PCOS who are anovulatory.

Franks: That is not quite what I said; 75% of women presenting with anovulatory infertility have polycystic ovary syndrome.

Hillier: And we hear of the heterogeneity within that group in terms of LH levels and the outcome of treatment. A naive question from a non-gynaecologist, but how many of these patients have what I would understand as Stein–Leventhal syndrome and to what extent does the incidence of that syndrome correlate with any of these aspects of the outcome of therapy?

Franks: The women recruited for that study who were by definition anovulatory all had polycystic ovaries on ultrasound. All but one or two had either a raised LH level or a raised testosterone or both. Overall, 53% of the women had a raised basal LH. It really depends on what the definition is, but let us put it this way, they all had some biochemical marker that we would associate with the typical polycystic ovary syndrome. But there was nothing in particular about those patients, other than those with persistently raised LH levels, where we could pick out a predictive factor for response, and so this is a different group.

The baseline LH level was not as important as what happened to LH during the development of a dominant follicle, and that does seem to be a specific subgroup, those with persistently raised LH, who merge into the women with spontaneous cycles who have a history of recurrent miscarriage and have the same sort of profile. That is a very interesting group to look at.

Nieschlag: I have a question about the use of GnRH analogues. When we use agonists in our IVF programme, we first get a tremendous stimulation phase and then there is desensitisation. I always wondered what effect that may have on the later stimulation. When we use the antagonist to get the down-regulation equivalent, the fall in gonadotrophins is immediate. Could the two regimens have qualitative differences on the follicles that would show up later?

Baird: That is a difficult question to answer. The current hypothesis would say that the initial flare will destroy all antral follicles which are responsive to LH whose granulosa cells have sufficient LH receptors to respond to them. That is probably a good thing in terms of getting rid of those follicles that will not produce eggs that are ovulatory. But by the same token, I suspect it is just as easy to get rid of those follicles by totally withdrawing LH and FSH support by the administration of GnRH antagonists.

We do not have the information to answer the question and it is something that merits investigation.

Hull: Professor Asch asked whether there was a dose-related risk of multiple pregnancy and whether there was evidence for that. There is plenty of evidence. I recently did a metanalysis of all the papers I could find on gonadotrophin therapy in cases of PCO, conventional dosage and low dosage, and the results were quite clear. There was a halving of the multiple pregnancy rate with the low-dose therapy. But, the price to pay is a halving in pregnancy rates: the pregnancy rate per ovulatory cycle with standard dosage is 30% and with low dosage it was 15%. But, although we may be seeking solutions for the future, nonetheless I accept the pragmatic view that we must bring to current practice, which is that we have to accept there is a dilemma and there is a price to pay for low-dose therapy. Professor Franks's data fit in with those general results of a considerably reduced chance of conception.

Kerin: I was very attracted by that fine stepwise incremental regime that Professor Franks used to detect the fine window when the follicles take off. He mentioned that 50% of his patients had a spontaneous LH surge using that regime. Was that a good prognostic sign in terms of pregnancy outcome?

Franks: The short answer is that it did not make any difference to outcome whether they had spontaneous LH surge or were given hCG, with the exception of some patients who we gave hCG conventionally when the follicle diameter reached 18 mm, but we found in retrospect that their oestradiol levels were low and their endometrium had not thickened. Indeed we let them run for longer and they ovulated spontaneously with a follicle diameter of around 25–26 mm. So in individual subjects it did seem to make a difference, but overall it did not.

References

1. Watson H, Hamilton-Fairley D, Kiddy D et al. Abnormalities of follicular phase LH secretion in women with recurrent early miscarriage. J Endocrinol 1989; 123 (suppl):Abstract 25.
2. Watson H, Kiddy D, Hamilton-Fairley D et al. The effect of hypersecretion of LH on ovarian steroid hormone secretion in women with recurrent early miscarriage. J Endocrinol 1991; 129 (suppl):Abstract 206.

Section IV

Assisted Reproduction and Related Techniques

Chapter 16

Cryopreservation of Mammalian Oocytes

D. G. Whittingham and J. G. Carroll

Introduction

A mature viable oocyte is central to the success of in vitro fertilisation (IVF) programmes and other reproductive technologies. The ability to cryopreserve the oocyte would provide a means of banking oocytes for patients where loss of gonadal function is anticipated. The main clinical indications for oocyte storage are:

1. Donation to individuals lacking ovaries or with ovaries devoid of oocytes, e.g. premature menopause
2. Donation to carriers of severe genetic diseases
3. Prior to chemo- or radiotherapy treatment
4. Prior to oophorectomy
5. To avoid producing excess embryos in IVF treatment
6. Oocytes in excess of those transferred in a GIFT treatment.

In addition it would circumvent many of the ethical issues associated with preserving the human embryo. In contrast to the routine use of sperm banking techniques, and despite the clear need for a safe and reliable method for the storage of the human oocyte, as yet none is available.

The feasibility of preserving the mature mouse oocyte was first realised when live young were born after the transfer of embryos obtained from frozen–thawed mouse oocytes [1]. The procedure used was similar to that used in the first published reports of embryo cryopreservation [2,3], namely exposure to molar levels of DMSO at 0°C, seeding at −7°C, slow cooling to −80°C before transfer to liquid nitrogen (LN_2). Warming was slow (8–20°C min^{-1}) and DMSO diluted at 0°C. As techniques for the preservation of embryos were modified and refined many of them have become adapted for the preservation of oocytes including: slow cooling in DMSO to −40°C combined with rapid thawing [4] and also methods such as ultra-rapid freezing [5,6] and vitrification [7]. Oocytes from a number of different species have now been preserved but live young have only been produced from frozen–thawed oocytes of the mouse [1,8], rabbit [9,10] and human [11,12]. However, overall rates of survival remain low and the use of oocyte freezing programmes in place of relatively successful embryo freezing methods has not been justified.

One of the reasons for the limited success in oocyte cryopreservation is that the mature oocyte provides a number of potential problems that have not been encountered, or considered, during the freezing of the embryo. The mature oocyte is arrested at the metaphase II stage of meiosis when the chromosomes are arranged on a microtubular spindle which is extremely temperature-sensitive. It has long been suggested that disruption of the meiotic spindle during cryopreservation might lead to the loss of chromosomes from the meiotic spindle resulting in the production of an aneuploid embryo after fertilisation. Perhaps the most overt difference between the preservation of oocytes and embryos is that the frozen–thawed oocyte must be competent to undergo the specialised series of events that occur during the process of fertilisation. These include sperm penetration of the zona pellucida, membrane fusion, pronuclear formation and syngamy.

This chapter considers the present status of human oocyte preservation especially in relation to the work on animal models. We conclude that the transfer of embryos derived from frozen mature human oocytes after fertilisation in vitro should be allowed where the appropriate cryobiological expertise is available and all reasonable steps are taken to establish the normality of the resulting fetus in early pregnancy.

Cryopreservation of the Human Oocyte

Soon after the first reports that human embryos could be successfully frozen and thawed [13,14] interest rapidly turned to the mature oocyte. The first successful study was reported by Chen in 1986 [11]. High rates of survival (80%) and fertilisation (83%) were obtained after oocytes were exposed to 1.5 M DMSO at 0°C for 20 min, seeded at −7°C, before slow cooling (0.3°C min^{-1}) to −40°C and transfer to liquid nitrogen. Thawing was rapid, in a 45°C water bath, and DMSO was diluted at room temperature. After the transfer of embryos obtained from oocytes frozen using this protocol, one twin and one singleton pregnancy came to term. Subsequent studies have not reported similar levels of success. Low rates of survival (25%–36%) [15] and fertilisation (32%–58%) [16] as well as high levels of polyploidy [17] have been observed and only one other pregnancy has reached term [12]. Clearly cryopreservation of the human oocyte presents more problems than the freezing of embryos.

The explanation for such marked differences in the success of human oocyte preservation is not clear but there are a number of methodological differences between the initial studies by Chen and other methods. Most importantly, the techniques used by Chen were based, at least in part, on those used for the successful preservation of mouse oocytes [1], particularly with respect to the addition of cryoprotectant at low temperatures and the relatively short times used for equilibration. On the other hand, other studies had used procedures that were in use for human embryos. These involved addition of DMSO at room temperature or 37°C for longer periods (~50 min) before cooling began. The use of these protocols was presumably to avoid damaging the temperature-sensitive meiotic spindle but the adverse effect of cryoprotectants at ambient temperature is well established [18] and is generally known as "cryoprotectant toxicity". More recent experiments have highlighted the potentially harmful effects of cryoprotectants on mammalian oocytes (see below). Due to the problems associated with oocyte cryopreservation in the human there have been few studies published since 1987, and of these the report comparing the cryoprotection afforded by glycerol and DMSO, after taking into consideration the permeability of the oocyte to these agents, looks the most encouraging [19]. Most recent studies have examined the response of oocytes from other mammalian species to low temperatures.

The Effects of Cryopreservation on Mammalian Oocytes

Aneuploidy

Disruption of the meiotic spindle during freezing and thawing resulting in chromosome loss or non-disjunction in embryos obtained from frozen–thawed oocytes is a major concern. This has been addressed directly by determining the frequency of aneuploidy at the first cleavage division of mouse embryos obtained after the fertilisation of frozen–thawed oocytes. Two studies [8,20], have reported on 272 and 100 chromosome spreads, respectively, that there is no increase in the frequency of aneuploidy after cryopreservation. In contrast Kola et al. [21] found, in 47 chromosome spreads, that freezing and thawing resulted in a 3-fold increase in the frequency of aneuploidy. As a similar freezing method was used in all studies the reason for the difference between them is not clear but it might be explained by the relatively small sample size of the latter study. Further, additional support for normal chromosome segregation in frozen–thawed oocytes is provided by recent experiments where the number of chromosomes in frozen–thawed and control haploid parthenogenones were found to be similar [22].

The above studies suggest that in conditions used for oocyte freezing and thawing, normal chromosome segregation occurs. Recent studies have highlighted the potential for disruption to the spindle when cooling and exposure to cryoprotectants is inappropriate. For example, in mouse oocytes cooled to between 25°C and 4°C in the absence of DMSO, or on exposure of oocytes to DMSO at temperatures above 4°C the spindle is disrupted often leading to the dispersal of chromosomes. However, the combination of exposure to DMSO and low temperatures causes less disruption to the spindle and chromosome scattering is not seen. The reason why DMSO prevents spindle depolymerisation at low temperatures may be explained by its microtubule stabilising properties, favouring polymerisation of tubulin [10,23,24]. At ambient temperatures, DMSO promotes the formation of cytoplasmic microtubule asters for which tubulin is thought to be recruited from the spindle, eventually leading to its being dismantled. Low temperatures probably reduce this effect first, by limiting the amount of DMSO that enters the cell (in 1.5 M DMSO at 0°C return of the oocyte to isotonic volume takes about 50 min) and second, by slowing the rate of polymerisation in the cytoplasm. The observation that exposure to DMSO at low temperatures does not cause chromosome scattering endorses and supports the use of similar conditions, i.e. exposure to DMSO at 0°C for 10–20 min, that were used for the first [1], and subsequent [8], successful oocyte freezing programmes.

The weight of evidence supports the idea that frozen–thawed mouse oocytes undergo normal chromosome segregation. This observation provides a sound starting point for the human oocyte. There are, however, differences between the cytoskeletal architecture of mouse and human oocytes [25] which prevent direct extrapolation between the species. Ultimately the safety of oocyte preservation in the human depends on the frequency of aneuploidy in frozen–thawed human oocytes (see discussion below).

Finally, while chromosomal anomalies may result from the storage of mature oocytes, cleavage stage embryos going through mitosis are also at risk of becoming aneuploid, at least in some blastomeres. This possibility has not been considered seriously probably because mitosis is of relatively short duration during the embryonic cell cycle. Nevertheless, this clearly requires closer examination.

Polyploidy

Although cryopreservation has no effect on the frequency of aneuploidy, the level of polyploidy has been shown to be increased after the fertilisation of frozen–thawed mouse [8,26] and human [15] oocytes. In the mouse this is apparently due to an increase in digynic polyploidy caused by retention of the second polar body rather than an increase in polyspermic fertilisation [26]. The cause of polar body retention is not clear. The mouse oocyte has a cortical layer of actin microfilaments that is concentrated in the area of the meiotic spindle. Disruption of the distribution of actin inhibits polar body formation after fertilisation or activation [27]; cryopreservation may induce polyploidy by a similar mechanism. In certain conditions the cryoprotectants DMSO and propanediol disrupt microfilament organisation in mouse and rabbit oocytes [10,28,29] but this effect is less when DMSO is used at low temperatures [29]. The possible relationship between digynic polyploidy in frozen–thawed oocytes and disruption of microfilaments requires clarification.

The implications of an increased frequency of polyploidy in frozen–thawed human oocytes emphasises the need for careful determination of the number of pronuclei after fertilisation. This should not be considered a serious limitation since the assessment of pronuclei is a routine procedure in IVF programmes.

Fertilisation

A major source of oocyte loss after cryopreservation of the mature mouse oocyte is due to a decrease in the rate of fertilisation [8,26,30,31]. Thus recent studies have been aimed at determining the site in the oocyte at which fertilisation is inhibited, the component(s) of the freezing programme responsible, the mechanism of the inhibition, and how the decreased rate of fertilisation may be prevented. Some of these problems have been clarified.

Fertilisation may be inhibited at sperm binding and penetration of the zona pellucida, attachment and fusion of the sperm to the oocyte, or during pronucleus formation and the events of syngamy. Two independent studies have clearly demonstrated that frozen–thawed oocytes undergo normal rates of fertilisation if the zona pellucida is bypassed, either by zona drilling [30] or by its complete removal [31]. Thus cryopreservation induces changes in the zona pellucida that inhibit fertilisation. Recently, we have found that the type of macromolecule present during cryopreservation can influence fertilisation after thawing. Low rates of fertilisation occur after freezing in medium containing the non-proteinaceous molecule, polyvinyl alcohol (PVA), and also with some, but not all, batches of bovine serum albumin (BSA). In contrast, after freezing in the presence of fetal calf serum (FCS) the fertilisation rate is similar to unfrozen oocytes. This observation is supported by previous studies in which high rates of fertilisation were recorded after freezing in the presence of serum [32,33]. Thus the addition of serum to the freezing medium provides a simple solution to the decreased rates of fertilisation associated with cryopreservation.

The ability to preserve oocytes in medium containing PVA is a means of avoiding potentially contaminated serum. The low rates of fertilisation seen after freezing in medium PVA can be prevented provided serum (which may be obtained from the recipient) is used to supplement the medium during dilution of cryoprotectant. In addition to being a useful practical advance this study suggests that the changes in the zona pellucida that occur after freezing in a medium with PVA are manifested during the dilution of the cryoprotectant. This may not be the case when oocytes are frozen in the presence of BSA since the addition and removal of cryoprotectants without a freeze–thaw cycle does not affect the fertilisation rate [8,30]. Further studies are required to determine the effects of different dilution procedures on the fertilisation of frozen–thawed oocytes.

Precisely what the changes in the zona pellucida are that decrease fertilisation after freezing in the absence of serum and how addition of serum protects against them is unclear. This ability of FCS to protect the zona pellucida during freezing and thawing may be analogous to the inhibition of zona hardening in oocytes that are matured in vitro [34,35]. Zona hardening is characterised by a decrease in fertilisation and a decrease in the sensitivity of the zona pellucida to digestion with chymotrypsin [36]. This is correlated with the conversion of the secondary sperm receptor ZP2 into its inactive form, $ZP2_f$ [37]. The decrease in fertilisation and zona sensitivity to chymotrypsin together with the change in ZP2 are prevented when serum is added to the culture medium [38]. Fetuin, a glycoprotein that inhibits trypsin-like proteases has been proposed as a candidate for the protective molecule in serum [32]. Fetuin can be substituted for FCS in culture and prevent the conversion of ZP2 into $ZP2_f$ and the associated zona hardening during spontaneous maturation [32]. It is not known whether fetuin prevents the changes induced in the zona pellucida during freezing and/or thawing.

The conversion of ZP2 into $ZP2_f$ and the associated zona changes correlate with the release of cortical granules during meiotic maturation. The possibility that freezing-induced changes in the zona pellucida are caused by premature release of cortical granules has been investigated. Using a lectin that binds to cortical granule exudate there is no detectable staining on frozen–thawed oocytes [31], thus if cortical granule release does occur it must be below the sensitivity of this assay or consist of a subset of granules which are not stained by this lectin–UEA 1.

Exposure of oocytes to DMSO at 4°C also has no effect on the number of cortical granules in the oocyte cortex, fertilisation or zona hardening [39]. The adverse effects of DMSO at high temperatures are demonstrated by the observations that exposure to DMSO at 37°C, or cooling in the absence of DMSO, results in a decrease in the number of cortical granules which in turn is correlated with a decrease in fertilisation and sensitivity to chymotrypsin digestion as well as the conversion of ZP2 into $ZP2_f$. The numbers of cortical granules in frozen–thawed oocytes have not been determined.

Other studies have suggested that zona hardening is not due to cortical granule release and suggest that it may result from an oxidation reaction [36, 40]. Similar changes may be responsible for the changes associated with cryopreservation. Nevertheless, while further studies are required to understand the mechanism of these changes, they can be prevented by the use of serum in the freezing and/or dilution medium so that frozen–thawed oocytes are fertilised at rates simlar to controls.

Development

The ultimate measure of success of a cryopreservation programme is the ability of embryos from frozen–thawed oocytes to develop normally after transfer to recipients. Live young have been produced after the transfer of embryos from frozen–thawed mouse, rabbit and human oocytes. The frequency of development of embryos from frozen–thawed oocytes compared to unfrozen controls has been investigated mainly in the mouse. In a number of studies we have found similar proportions of implantation and fetal develoment in frozen–thawed and control groups [33]. In addition, development after transfer of embryos frozen in medium with different macromolecules is similar. On the other hand, other studies have reported that embryos from frozen–thawed oocytes implant and form fetuses at rates less than unfrozen controls [41]. The explanation for these differences is not clear, although in the latter studies it is unclear if the transfer of embryos from control and frozen groups was carried out contemporaneously (control levels of development may vary with time).

The effect of the macromolecules present in the freezing medium also had no effect on the ability of embryos to develop after transfer [32]. These studies suggest that, after cryopreservation in the presence of serum, oocytes can undergo fertilisation and development at rates similar to controls.

Vitrification

Vitrification is an attractive alternative to the other methods of low temperature cryopreservation since it avoids the potential damaging effects caused by the formation of intracellular ice. With mouse embryos, overall survival after implantation is similar to that obtained after conventional freezing when high concentrations of glycerol are used to achieve vitrification [42,43]. Limited success has been obtained with human oocytes [44] and further investigations have not progressed because of the reported high incidence of aneuploidy found in embryos derived from vitrified mouse oocytes [21] which might originate from alterations to the spindle and cytoskeleton brought about by prolonged exposure to high concentrations of cryoprotectants at suboptimal temperatures [25,28,29]. However, a recent study of the developmental potential of mouse oocytes previously vitrified in 6 M DMSO showed that the incidence of aneuploidy at first cleavage, the rate of implantation and development to late stage fetuses did not differ from control non-vitrified oocytes [45]. These results were obtained after taking precautions to reduce the risks of osmotic stress and prolonged exposure to high concentrations of cryoprotectants as well as preventing zona hardening. Further investigation of the effect of these parameters on the human oocyte is clearly warranted.

Prospects for the Storage of the Human Oocyte

Progress so far on the storage of mouse oocytes is encouraging. The problem of reduced rates of fertilisation after freezing or vitrification have for the most part been overcome. The incidence of aneuploidy is not increased when the appropriate methods of freezing or vitrification are used to minimise any adverse effects of cryoprotectants and prolonged exposure to suboptimal temperatures.

For the human oocyte it will be difficult to obtain clear-cut evidence of whether or not freezing or vitrification alters the incidence of chromosomal anomalies in the resulting embryos because there is a relatively high incidence of aneuploidy in embryos produced by fertilisation in vitro [45–48]. The incidence of aneuploidy recorded in the different reports varies between 11% and 40%. The data are confounded by the fact that some studies are made on oocytes that fail to fertilise, others on so-called spare oocytes and no account is taken of the type of infertility that affects the patients. In the cirumstance, large numbers of frozen/vitrified and control oocytes would have to be analysed cytogenetically at first cleavage in order to obtain definitive incidence of aneuploidy. Furthermore, there is no evidence that offspring derived from oocytes fertilised in vitro have more chromosomal abnormalities than babies born after natural conception. Thus, it would seem to be more acceptable to allow the transfer of embryos derived from frozen or vitrified oocytes provided all reasonable precautions have been taken to avoid polyspermy (similar to that taken in normal IVF) and to exclude abnormality as far as is feasible in the early stages of pregnancy.

References

1. Whittingham DG. Fertilization in vitro and development to term of unfertilized mouse oocytes previous stored at −196°C. J Reprod Fertil 1977; 49:89–94.
2. Whittingham DG, Leibo SP, Mazur P. Survival of mouse embryos frozen to −196°C and −269°C. Science 1972; 178:411–14.
3. Wilmut I. Effect of cooling rate, warming rate, cryoprotective agent and stage of development on survival of mouse embryos during freezing and thawing. Life Sci 1972; 11:1071–9.
4. Whittingham DG, Wood M, Farrant J, Lee H, Halsey JA. Survival of frozen mouse embryos after rapid thawing from −196°C. J Reprod Fertil 1979; 56:11–21.
5. Szell A, Shelton JN. Osmotic and cryoprotective effects of glycerol–sucrose solutions on day-3 mouse embryos. J Reprod Fertil 1987; 80:309–16.
6. Trounson A, Peura A, Kirby C. Ultrarapid freezing: a new low-cost and effective method of embryo cryopreservation. Fertil Steril 1987; 48:843–50.
7. Rall WF, Fahy GM. Ice-free cryopreservation of mouse embryos at −196°C by vitrification. Nature 1984; 313:573–5.
8. Glenister PH, Wood MJ, Kirby C, Whittingham DG. Incidence of chromosome anomalies in first-cleavage mouse embryos obtained from frozen–thawed oocytes fertilized in vitro. Gamete Res 1987; 16:205–16.
8. Al-Hasani S, Kirsch J, Diedrich K, Blanke S, van der Ven H, Krebs D. Successful embryo transfer of cryopreserved and in-vitro fertilized rabbit oocytes. Hum Reprod 1989; 4:77–9.
10. Vincent C, Garnier V, Heyman Y, Renard JP. Solvent effects on cytoskeletal organization and in-vivo survival after freezing of rabbit oocytes. J Reprod Fertil 1989; 87:809–20.
11. Chen C. Pregnancy after human oocyte cryopreservation. Lancet 1986; i:884–6.
12. Van Uem JFHM, Siebzehnrubl ER, Schuh B, Koch R, Trotnow S, Lang N. Birth after cryopreservation of unfertilized oocytes. Lancet 1987; i:752–3.
13. Trounson AO, Mohr L. Human pregnancy following cryopreservation thawing and transfer of an eight-cell embryo. Nature 1983; 305:707–9.
14. Zeilmaker GH, Alberda AT, Van Gent I, Rijkmans CMPM, Drogendijk AC. Two pregnancies following transfer of intact frozen-thawed embryos. Fertil Steril 1984; 42:293–6.
15. Al-Hasani S, Diedrich K, van der VEN H, Reinecke A, Hartje M, Krebs D. Cryopreservation of human oocytes. Hum Reprod 1987; 2:695–700.
16. Sathananthan AH, Trounson A, Freeman L. Morphology and fertilizability of frozen human oocytes. Gamete Res 1987; 16:343–54.
17. Mandelbaum J, Junca AM, Tib C et al. Cryopreservation of immature and mature hamster and human oocytes. Ann NY Acad Sci 1988; 541:550–61.
18. Mazur P, Schneider U. Osmotic responses of preimplantation mouse and bovine embryos and their cryobiological implications. Cell Biophys 1986; 8:259–86.
19. Hunter JE, Bernard B, Fuller B, Amso N, Shaw RW. Fertilisation and development of the human oocyte following exposure to cryoprotectants, low temperatures and cryopreservation: a comparison of two techniques. Hum Reprod 1991; 6:1460–5.
20. Bouquet M, Selva J, Auroux M. The incidence of chromosomal abnormalities in frozen–thawed mouse oocytes after in vitro fertilization. Hum Reprod 1992; 7:76–80.
21. Kola I, Kirby C, Shaw J, Davey A, Trounson A. Vitrification of mouse oocytes results in aneuploid zygotes and malformed fetuses. Teratology 1988; 38:467–74.
22. Bos-Mikich A, Whittingham DG. Parthenogenetic activation of frozen–thawed mouse oocytes. J Reprod Fertil Abstr Ser 1991; 7:18.
23. Algaier J, Himes RH. The effects of dimethylsulfoxide on the kinetics of tubulin assembly. Biochim Biophys Acta 1988; 954:235–43.
24. Himes RH, Burton P, Gaito JM. Dimethylsulfoxide-induced self-assembly of tubulin lacking associated proteins. J Biol Chem 1977; 252:6222–8.
25. Pickering SJ, Johnson MH, Braude PR, Houliston E. Cytoskeletal organization in fresh, aged and spontaneously activated human oocytes. Hum Reprod 1988; 3:978–89.
26. Carroll J, Warnes GM, Matthews CD. Increase in digyny explains polyploidy after in-vitro fertilisation of frozen-thawed mouse oocytes. J Reprod Fertil 1989; 85:489–94.
27. Maro B, Johnson MH, Pickering SJ, Flach G. Changes in actin distribution during fertilization of the mouse egg. J Embryol Exp Morphol 1984; 81:211–37.
28. Vincent D. Dimethylsulfoxide affects the organization of microfilaments in the mouse oocyte. Mol Reprod 1990; 26:227–35.

29. Vincent C, Pruliere G, Pajot-Augy E, Campion E, Garnier V, Renard JP. Effects of cryoprotectants on actin filaments during the cryopreservation of one-cell rabbit embryos. Cryobiology 1990; 27:9–23.
30. Carroll J, Depypere H, Matthews CD. Freeze–thaw induced changes of the zona pellucida explains decreased rates of fertilization in frozen–thawed mouse oocytes. J Reprod Fertil 1990; 90:547–53.
31. Wood MJ, Whittingham DG, Lee S-H. Fertilization failure of frozen mouse oocytes is not due to premature cortical granule release. Biol Reprod 1992; 46:
32. Schroeder AC, Champlin AK, Mobraaten LE, Eppig JJ. Developmental capacity of mouse oocytes cryopreserved before and after maturation in vitro. J Reprod Fertil 1990; 89:43–50.
33. Wood MJ, Carroll J, Whittingham DG. The addition of serum to the medium limits zona hardening in frozen oocytes. J Reprod Fertil Abstr Ser 1989; 3:33.
34. Downs SM, Schroeder AC, Epigg JJ. Serum maintains the fertilizability of mouse oocytes matured in vitro by preventing hardening of the zona pellucida. Gamete Res 1986; 15:115–22.
35. Choi TS, Mori M, Kohmoto K, Shoda Y. Beneficial effect of serum on the fertilizability of mouse oocytes matured in vitro. J Reprod Fertil 1987; 79:505–8.
36. Gulyas BJ, Yuan LC. Cortical reaction and zona hardening in mouse oocytes following exposure to ethanol. J Exp Zool 1985; 233:269–76.
37. Moller CC, Wassarman PM. Characterization of a proteinase that cleaves zona pellucida glycoprotein ZP2 following activation of mouse eggs. Dev Biol 1989; 132:103–112.
38. Ducibella T, Kurasawa S, Rangarajan S, Kopf GS, Schultz RM. Precocious loss of cortical granules during mouse oocyte meiotic maturation and correlation with an egg-induced modification of the zona pellucida. Dev Biol 1990; 137:46–55.
39. Vincent C, Pickering SJ, Johnson MH. The hardening effect of dimethylsulphoxide on the mouse zona pellucida requires the presence of an oocyte and is associated with a reduction in the number of cortical granules present. J Reprod Fertil 1990; 89:253–9.
40. DeFelici M, Siracusa G. "Spontaneous" hardening of the zona pellucida of mouse oocytes during in vitro culture. Gamete Res 1982; 6:107–13.
41. Freidler S, Giudice LC, Lamb EJ. Cryopreservation of embryos and ova. Fertil Steril 1988; 49:743–64.
42. Wood MJ, Rall WF. The preservation of embryos by vitrification. Hum Reprod 1987; 2:Suppl 1, 2–3.
43. Rall WF. Factors affecting the survival of mouse embryos cryopreserved by vitrification. Cryobiology 1987; 24:387–402.
44. Trounson A. Preservation of human eggs and embryos. Fertil Steril 1986; 46:1–12.
45. Bos-Mikich A, Candy C, Wood MJ, Whittingham DG. Cytogenetic analysis and developmental potential of vitrified mouse oocytes. J Reprod Fertil Abstr Ser 1992; 9:
46. Edirisinghe WR, Murch AR, Yovich JL. Cytogenetic analysis of human ooctyes and embryos in an in-vitro fertilization programme. Hum Reprod 1992; 7:230–6.
47. Tarin JJ, Gomez E, Sampaio M, Ruiz M, Remohi J, Pellicer A. Cytogenetic analysis of human oocytes from fertile women. Hum Reprod 1991; 6:1100–3.
48. Angell RR, Ledger W, Yong EL, Harkness L, Baird DT. Cytogenetic analysis of unfertilized human oocytes. Hum Reprod 1991; 6:568–73.

Chapter 17

Extracorporeal Development of Immature Ovarian Follicles

R. G. Gosden and N. I. Boland

The number of oocytes stored in the ovarian cortex at the beginning of reproductive life greatly exceeds that required for a lifetime of ovulatory cycles. It is, however, the number of mature cells rather than the much more abundant primordial and growing oocytes that is important for all practical purposes, such as treatment of infertility, animal production technology and investigative work. The oocytes available generally fall far short of needs.

The fate of most oocytes is atresia either at the primordial follicle stage or later in development. Atresia plays an essential role in follicle dynamics because the size of the cohort beginning to grow during each cycle is prodigious compared with the numbers of ovulations. Most oocytes that perish in atretic follicles are potentially normal judging by the fertility of those rescued after superstimulating doses of gonadotrophins. There are, therefore, no serious grounds for anxiety about oocyte quality should it become possible to rescue an even larger fraction from the pool of those undergoing atresia. Exogenous gonadotrophins and other superstimulating regimes can elevate the ovulation rate severalfold above the norm but only a fraction of a growing follicle cohort is recruited, the treatment should not be repeated too frequently and there is a small but definite risk of inducing ovarian hyperstimulation syndrome. Alternative methods for recovering even larger numbers of oocytes without jeopardising the well-being of the patient or her long-term ovarian function are needed.

All these aims could, in theory, be achieved by extracorporeal growth and maturation of immature oocytes. On average, each square millimetre of ovarian cortex in a young woman contains hundreds of follicles at primordial and preantral stages, representing more than the sum of a lifetime of ovulations. Yet the removal of just one such area will not significantly affect the course and extent of her menstrual lifespan. Should it become possible to produce from ovarian biopsies mature oocytes for use in IVF and GIFT operations and store the surplus, many additional cases of infertility could be helped and young cancer patients might be safeguarded against the ovarian effects of chemotherapy or radiotherapy.

This is admittedly an ambitious projection since the technology for growing follicular oocytes to maturity is still in its infancy and restricted to laboratory species. The span of oocyte development is lengthy and involves the maturation of a small cell to form a large, specialised one capable of resuming meiosis, undergoing fertilisation and endowing the zygote with materials for embryogenesis. The conditions required for these complex events are poorly understood, although establishment of effective culture conditions could interact with the progress being made in developmental biology of oocytes to their mutual benefit.

Rapid progress is now being made and it may not be long before complete development of murine follicular oocytes is possible in vitro. Because the patterns of growth and morphogenesis of follicles are similar in all mammals (Fig. 17.1) an emerging technology could one day have practical relevance for clinical medicine and agriculture. This chapter reviews the state of the art of growing follicular oocytes in vitro. The methods to be discussed in turn range from those involving the largest unit, the whole ovary, to the smallest one, the naked oocyte. Each has distinctive advantages as well as disadvantages, but we conclude that only two have real prospects of being applied to clinical problems.

Organ Culture

This type of culture carries the advantage of preserving normal tissue interactions but has serious drawbacks for long-term maintenance. A combination of a small surface area/volume ratio, a high rate of oxygen consumption in cortical

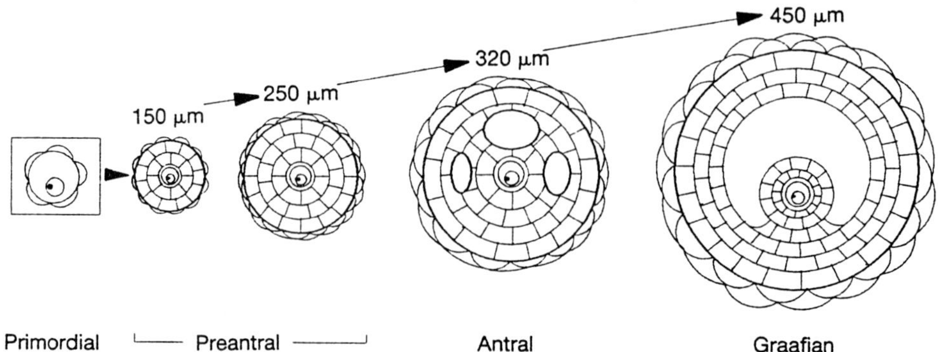

Fig. 17.1. Stages of follicle development in mouse ovaries (primordial follicles not drawn to scale).

tissue and dense stroma produces a steep Po_2 gradient and a reciprocal hydrogen ion gradient. The chemical microenvironment in all regions except the periphery is likely to be markedly unphysiological with cellular necrosis spreading from the centre (Fig. 17.2).

The severity of these problems can be diminished by a number of strategies. Immature ovaries perform better than adult organs in culture because they possess more oocytes and less connective tissue. The survival of delicate fetal ovaries is even greater and oogonia continue dividing and enter meiotic prophase, though follicles fail to form even when media are supplemented with follicle-stimulating hormone (FSH) and luteinising hormone (LH) [1–3]. If the larger mass of an adult organ is reduced to thin cortical slices, where most of the small follicles are found, necrotic changes can be minimised. Slices present a more favourable surface area/volume ratio for diffusion of dissolved gases and metabolites and a number of methods have been devised for suspending or floating organ cultures to optimise exchange at all surfaces [4], whereas Biopore membranes (Millipore Corporation) provide a commercially available alternative. Hyperbaric oxygen chambers have been used to increase the Po_2 at the centre of organ cultures but other unfavourable chemical gradients persist and oxygen toxicity is a potential problem.

The first attempts to culture fragments of human ovaries were performed in the pioneering days of tissue culture technology and produced outgrowths of fibroblasts with some epithelial sheets; no observations on oocytes were recorded [5]. With improved formulations and better techniques it became possible to maintain animal ovaries in vitro for weeks, allowing germ cells to

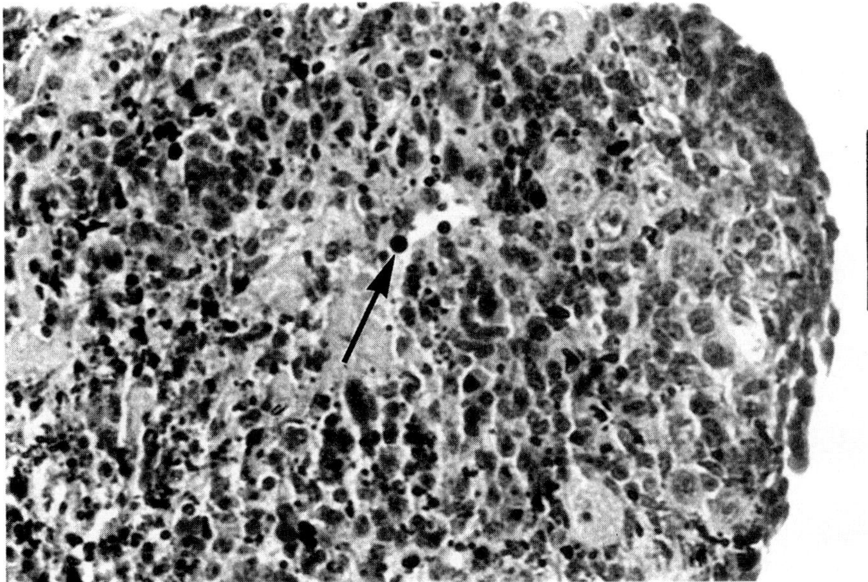

Fig. 17.2. Cortical mouse oocytes survive for at least a week in organ cultures, but many medullary cells die leaving traces as pyknotic bodies (arrow) (scale bar = 50 μm).

enter meiosis and grow [1,6]. Although a full span of follicle development lasting three weeks in mice is not readily obtained, preantral growth can be very successful in organ cultures because small follicles have lower metabolic demands than Graafian stages and are in favourable cortical locations. Small antra form and "ovulation" occurs in only a minority of follicles, but meiotic maturation is stimulated by gonadotrophins in short-term explants [7,8]. The achievement of ovulation is perhaps the most stringent indicator of good culture conditions since it is the culmination of follicle development and requires specific tissue responses to LH. Apart from the impressive results of ovarian vascular perfusion [9], conventional culture methods have not been very successful with the later stages of follicle development, and the difficulties become progressively more severe with organ size. Attention has therefore turned to the culture of isolated follicles.

Isolated Follicle Culture

Ovarian follicles are, in principle, almost ideal for culture. Like preimplantation embryos, they are developmental units possessing all the cell types required for morphogenesis and reproductive functions. Furthermore, granulosa layers enclosing oocytes are avascular and chemical gradients existing between the basement membrane and central oocyte, which are physiological, are not aggravated by culture conditions [10]. Despite such practical advantages, Graafian follicles of humans and farm animals do not resume full development to ovulation after explanting in vitro [11–13]. Perhaps this is because the theca of these species is a multilayered, vascular structure which, when no longer perfused by blood, imposes an additional barrier for diffusion into the follicle. This explanation cannot fully account for the failure of freshly explanted antral follicles from mouse ovaries to proceed to ovulation. They have a thin theca and can develop to Graafian sizes when cultured from preantral stages, as we shall shortly show. We conjecture that a period of adjustment to the culture environment contributes to their better performance.

Qvist et al. [14] were the first to show that primary mouse follicles can reach mature sizes in vitro, but the complex and unphysiological composition of their media and the possibility of follicle interactions complicated the interpretation of an otherwise important advance. Subsequent refinements by other investigators have shown that individual follicles with two or three granulosa cell layers (Fig. 17.1) can grow to full size during a culture period of 5–6 days.

Follicles are dissected with adhering stroma–theca tissue from 4-week-old mouse ovaries and incubated in Minimal Essential Medium (alpha-MEM) containing FSH, insulin and mouse serum. Stroma–theca cells are obligatory for development of antral follicles and the structural integrity of the granulosa epithelium must not be damaged during isolation or pipetting because follicles have little capacity for self-repair. Follicular oocytes remain arrested at the germinal vesicle stage in the absence of hypoxanthine or analogues of cyclic 3′, 5′-adenosine monophosphate (cAMP) and resume nuclear maturation after stimulation with LH. Follicles can be cultured either on porous, hydrophobic membranes [15] or in V-shaped wells of microtitre plates under a layer of mineral oil (Fig. 17.3) [16]. The second type of culture requires daily transfer to fresh wells to avoid cell adhesion and spreading over the plastic surfaces, but this minor drawback is offset by the advantages of concentrating follicle-secreted molecules in microdroplets of medium for studying metabolism. Approximately 60% of follicles cultured in this system reach Graafian sizes during a week of culture, far higher than the percentage that would have ripened in vivo and indicating that atresia is not programmed at an early stage of follicle development.

Both methods produce morphological and biochemical phenotypes that are virtually indistinguishable from follicles growing in vivo. Most follicles with 2–3 granulosa layers and measuring 150 μm on the first day of culture form an antrum on the third day and reach a Graafian size of > 400 μm within another 2–3 days (Table 17.1, Fig. 17.4a–c). Secondary follicular fluid, which normally accumulates during the final hours of ripening after LH stimulation [17], evidently does not occur under these conditions, but the healthiest follicles form stigmata and ovulate after an interval of about 12 h following hormonal stimulation (Fig. 17.4d). Oocytes from stimulated follicles undergo germinal vesicle breakdown and subsequently progress to metaphase II with the expulsion of a polar body.

Granulosa cells established as monolayer cultures differentiate to form lutein-like cells that switch to the production of progesterone rather than oestrogen [18] whereas those maintained in intact follicles evidently follow a more normal course of development because increasing amounts of oestradiol, but not progesterone, accumulate in the medium under the influence of FSH (Fig. 17.5). The ability to ovulate after LH stimulation is a further indication that physiology

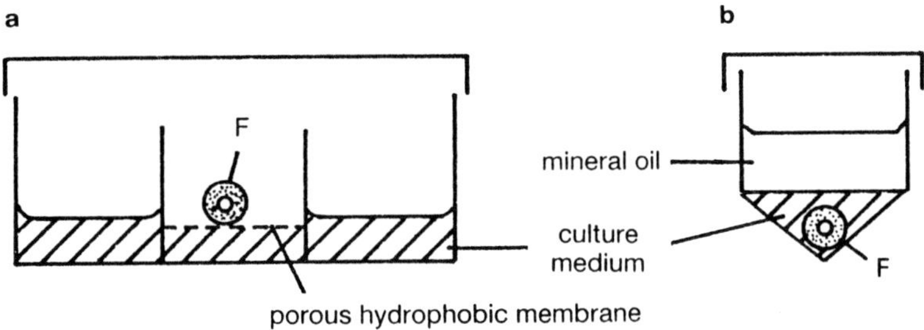

Fig. 17.3. Primary follicles isolated from mouse ovaries can be grown in vitro to Graafian stages either on (**a**) hydrophobic membranes or in (**b**) small V-wells under a layer of oil.

Table 17.1. Developmental capacity of preantral mouse follicles growing in V-well cultures[a]

Starting number	Graafian	GVBD[b]	Ovulating
30	19 (60%)	22 (73%)	10 (33%)

[a]Unpublished observations.
Germinal vesicle breakdown.

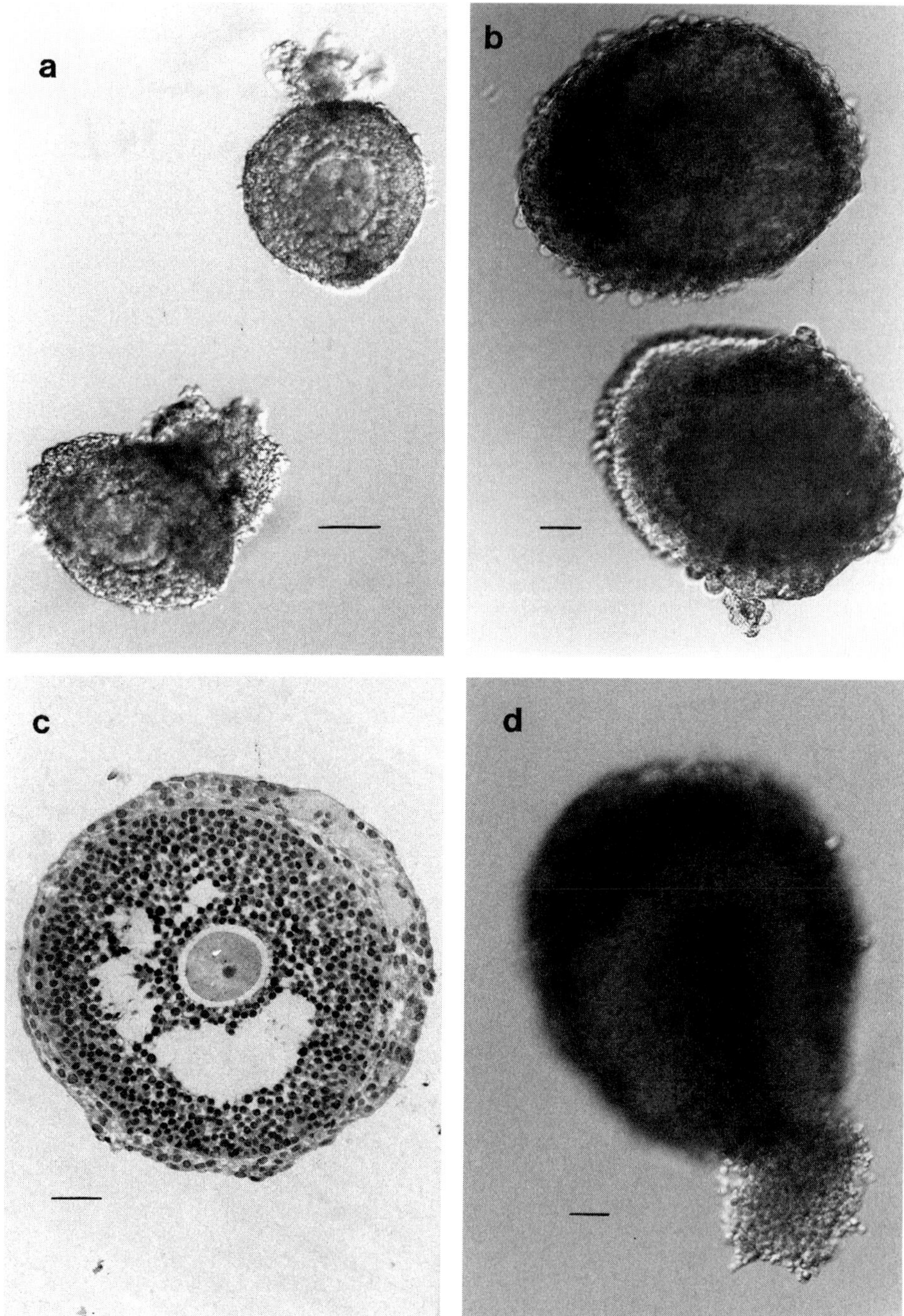

Fig. 17.4. Development of mouse follicles in vitro. **a** Freshly dissected primary follicles with 2–3 granulosa layers at the beginning of culture; **b** formation and expansion of antral cavities between days 4–6; **c** histological appearance of a comparable Graafian follicle showing an oocyte with germinal vesicle nucleus and normal granulosa and theca cell layers; **d** ovulation of a follicle grown for 6 days from the primary stage (bar = 50 μm).

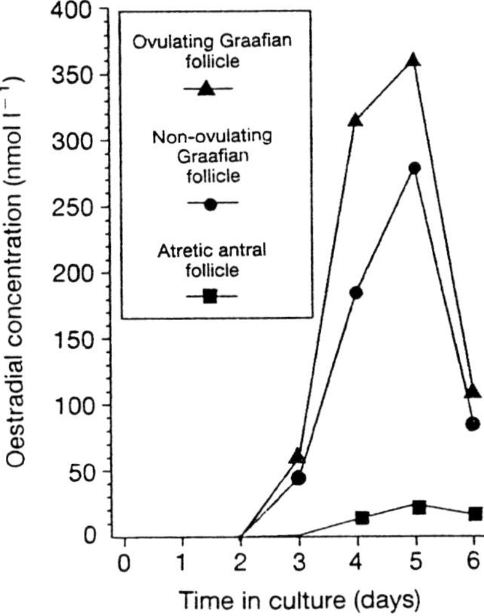

Fig. 17.5. Representative 24 h production of oestradiol for three mouse follicles grown in V-well cultures with different developmental outcomes (unpublished observations).

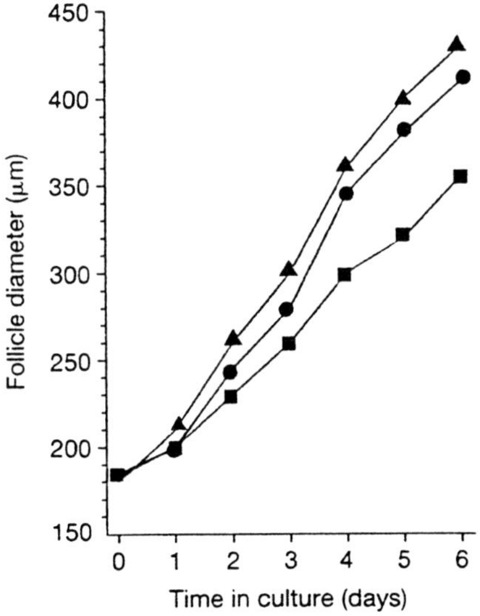

Fig. 17.6. Growth curves for the three follicles represented in Fig. 17.5.

is being mimicked, those that grow fast and produce more oestradiol during the antral phase being more likely to respond (Figs. 17.5 and 17.6).

There are grounds for believing that oocytes developing in these follicles are fertile, although this has not yet been formally proven. During the period of culture, they grow to full size, acquire the capacity to resume meiosis and form a zona pellucida capable of binding spermatozoa. Evidently, LH receptors are expressed during development because the hormone stimulates germinal vesicle breakdown and mucification of the cumulus mass.

The ultimate aim of these experimental studies is to grow primordial follicles to Graafian stages in continuous culture for which good prospects exist because small follicles initiate growth in vitro to form small primary follicles. The hiatus between this stage and that from which full development can be achieved is not great, but bridging it will probably require methods for colonising follicles with a theca layer [19].

Granulosa–Oocyte Complexes

Methods for culturing intact follicles can be effective and mimic physiological conditions but are labour-intensive because each unit requires individual attention. If a large-scale operation is required alternative methods may be needed.

Up to 50 preantral follicles can be recovered from a pair of immature mouse ovaries after incubating for 30 min in neutral protease ("Dispase") or collagenase [20]. Disaggregation yields granulosa–oocyte complexes that are denuded of theca cells and have little basement membrane material remaining. When transferred to conventional tissue culture vessels, spheroidal morphology is lost because granulosa cells attach to and spread over the plastic substrate; because intercellular adhesion is more delicate in primordial stages oocytes are often extruded. The three-dimensional integrity of complexes deriving from preantral follicles can be preserved in collagen gels. These histotypic cultures are prepared by adjusting the pH of ice-cold solubilised collagen to normal immediately before adding the cells and raising the temperature to set the gel. Primordial follicles begin to grow and granulosa cell layers multiply in the floating gels [21]. Stromal cells grow vigorously under these conditions and colonise vacant spaces in the gel until it assumes the appearance and disadvantages of an organ culture (Fig. 7). Follicles do not form antral spaces under these conditions despite the presence of stroma and FSH and/or LH in the medium, although their potential for becoming Graafian and producing fertile oocytes can be demonstrated by grafting gels in a host animal [22,23].

Granulosa–oocyte complexes cultured in more conventional environments, however abnormal in morphology, have important practical advantages over gel cultures. Oocytes at mid-growth stages resume meiosis after a 10-day culture period and, following fertilisation and embryonic development in vitro, can even produce viable conceptuses in host animals [24,25]. Although dispersion of the complexes over the membrane promotes development, mutual inhibition does not appear to pose a problem for "farming" large numbers of oocytes in a single culture. The original formulations of culture media required cyclic AMP analogues or pharmacological concentrations of hypoxanthine to prevent spontaneous germinal vesicle breakdown, but the introduction of collagen-impregnated membranes and Waymouth's medium has produced better results under more physiological conditions [26,27]. The granulosa cells extend to form a stalk of cumulus-like cells bearing the oocyte while the base resembles mural

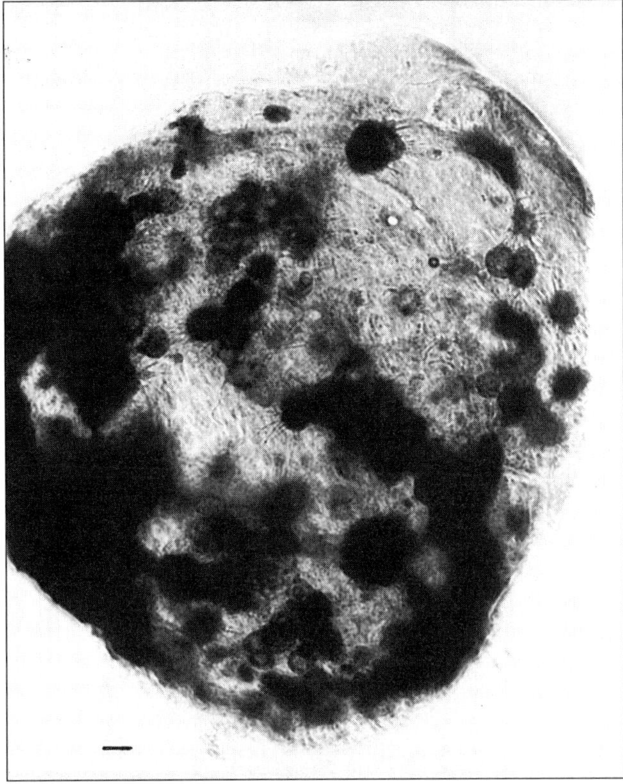

Fig. 17.7. Primary follicles isolated from mouse ovaries by collagenase digestion and cultured in floating collagen gel. After several days, granulosa cells from degenerating follicles and stromal cells merge into solid masses (bar = 50 μm).

granulosa cells (Fig. 17.8). Those in collagen gel cultures, on the other hand, behave more strictly like mural cells, which may explain a lower rate of oocyte maturation in this model (Evelyn Telfer, unpublished observations).

Culture of Naked Oocytes

The acme of oocyte culture technology would be the growth of naked primordial oocytes to full size and fertility either in complete isolation from other cell types or on a "lawn" of somatic "feeder" cells. Unfortunately, it is far from clear whether this can be achieved even in principle. Oocytes can be separated from their granulosa cells in calcium-free medium containing trypsin and isolated either by pipetting or on Percoll density gradients. But small oocytes cultured on their own degenerate after a few days and those co-cultured with

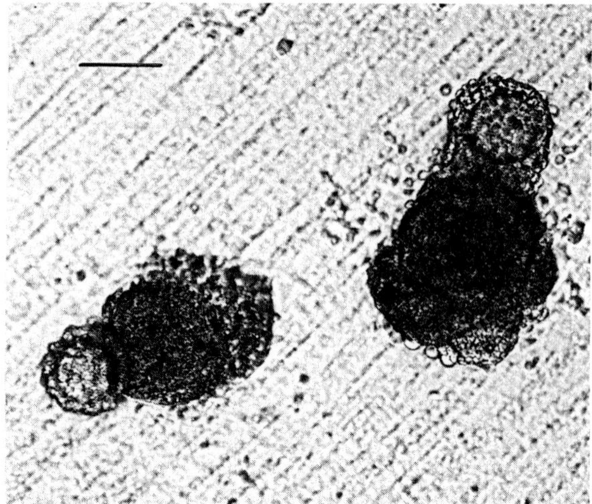

Fig. 17.8. Granulosa–oocyte complexes grown on collagen-impregnated membranes lose their spherical morphology to form stalks of cumulus-like cells carrying the oocyte (bar = 50 μm). (Reproduced from Eppig [39], with permission.)

fibroblast or granulosa cell "feeder" layers or their conditioned media, although surviving longer, do not complete development [28,29].

Such findings confirm the belief that intimate physiological interactions between granulosa cells and oocytes are obligatory for the normal development of both partners. Somatic cells not only provide physical support for the oocyte but are metabolically coupled to permit the uptake of nutrients and transfer of informational molecules such as cyclic nucleotides [30–33]. Evidently, the oocyte is partially emancipated from its somatic cell environment because it can grow without concomitant multiplication of granulosa cells and is responsible for producing its own zona pellucida [33]. The synthesis of macromolecules by oocytes is energetically expensive and energy substrates for oxidation are probably derived from granulosa cells [34]. These substrates can be provided in the culture medium but the extent to which the oocyte is dependent on its somatic cell environment for other molecules is largely unknown. Whether vital steps in development can be mimicked in vitro in the absence of union with granulosa cells will need to be addressed experimentally before the present limitations of isolated oocytes can be understood and perhaps overcome.

Storage of Follicular Oocytes at Low Temperatures

Storage of surplus mature oocytes at liquid nitrogen temperatures is immensely important for preserving scarce biological resources and avoiding the controversial issue of frozen embryo storage (Whittingham, Chapter 16, this volume).

Unfortunately, oocytes are not hardy cells and do not tolerate cooling and/or thawing well, perhaps because their ooplasm is highly specialised. Cooling promotes depolymerisation of microtubules in the spindle apparatus and cryprotectants are potentially cytotoxic, either of which could contribute to poor fertilisation and cleavage rates and chromosomal anomalies [35]. Storage of immature oocytes promises greater success because the cells are smaller, have decondensed chromatin in the germinal vesicle nucleus and time to repair damaged organelles between thawing and fertilisation. These predictions are borne out by the first results of frozen storage of isolated preantral follicles [23]. The majority of follicles excluded trypan blue dye, indicating that cell membrane damage was minimal, and, following grafting and culture, fertilised oocytes produced viable young. Subsequently, it was found that primordial follicles tolerate storage at low temperatures and microdissected primary follicles resume normal growth to Graafian size in vitro after similar treatment (unpublished observations). These encouraging results suggest that if it becomes possible to grow human follicular oocytes to maturity in vitro cryopreservation technology need not delay applications to infertility treatment.

Practical and Clinical Perspectives

The mouse has proved to be the ideal species for developing models for growing follicular oocytes in vitro just as it was for pioneering the culture of preimplantation embryos. We are now at a similar stage to embryo culture technology of 30 years ago, but it is premature to predict rapid progress towards the final aim of growing small oocytes to maturity entirely in vitro. Additional problems are anticipated with humans and other large species because their ovaries are much more fibrous, have a lower follicle density and the full course of oocyte development is far longer than in mice.

Which of the four strategies reviewed carries the best chance of success with human follicular oocytes? Organ culture is not the method of choice since its limitations soon become evident even with mouse ovaries. It is between the extremes of whole organs and isolated oocytes that the most promising strategies are to be found, namely, intact follicles and granulosa–oocyte complexes. Since healthy tissue from premenopausal ovaries is not routinely available for experimental work sheep ovaries have been used as general models of fibrous organs to adapt methods that have proved successful with mice. Follicles < 0.5 mm diameter were harvested by microdissection from cortical slices because the bulk of connective tissue would require a prohibitively long period for enzymatic digestion. The foilicles grew in culture but became arrested and necrotic after about 4 days, when the theca was overgrown (unpublished observations). These pilot results confirm that technical hurdles lie ahead for growing human immature follicles, and these will undoubtedly be more acute at later stages of development when the diameters of Graafian follicles exceed those of mice by more than an order of magnitude. The simplicity of culturing granulosa–oocyte complexes on coated membranes will then be attractive. It is too early to be sure whether these techniques will eventually prove successful, but the experimental effort is worthwhile because the rewards for basic and applied science are assured if we improve on the harvest of oocytes presently obtained by ovulation or aspiration from antral follicles [36–38].

Acknowledgements. We gratefully acknowledge the support of The Wellcome Trust and the Leukaemia Research Fund and assistance from Alison Murray, Vivian Bryce and Kay Grant.

References

1. Blandau RJ, Warrick E, Rumery RE. In vitro cultivation of fetal mouse ovaries. Fertil Steril 1965; 16:705–15.
2. Blandau RJ, Odor DL. Observations on the behaviour of oogonia and oocytes in tissue and organ culture. In: Biggers JD, Schuetz AW, eds. Oogenesis. Baltimore: University Park Press, 1972; 301–20.
3. Baker TG, Neal P. Oogenesis in human fetal ovaries maintained in organ culture. J Anat 1974; 117:591–604.
4. Freshney RI. Culture of animal cells – a manual of basic technique. New York: AR Liss, 1983; 225–8.
5. Wolff EK, Zondek B. Die Kultur menschlichen Ovarial- und Amniongewebes. Virchows Arch [A] 1925; 254:1–16.
6. Martinovitch PN. The development in vitro of the mammalian gonad – ovary and ovogenesis Proc R Soc Lond B 1938; 125:232–49.
7. Baker TG, Neal P. Gonadotrophin-induced maturation of mouse Graafian follicles in organ culture. In: Biggers JD, Schuetz AW, eds. Oogenesis. Baltimore: University Park Press, 1972; 377–96.
8. Ryle M. Morphological responses to pituitary gonadotrophins by mouse ovaries in vitro. J Reprod Fertil 1969; 20:307–12.
9. Lambertsen CJ, Greenbaum DF, Wright KH, Wallach EE. In vitro studies of ovulation in the perfused rabbit ovary. Fertil Steril 1976; 27:178–87.
10. Gosden RG, Byatt-Smith JG. Oxygen concentration gradient across the ovarian follicular epithelium: models, predictions and implications. Hum Reprod 1986; 1:65–8.
11. Moor RM, Hay MF, McIntosh JEA, Caldwell BV. Effect of gonadotrophins on the production of steroids by sheep ovarian follicles cultured in vitro. J Endocrinol 1973; 58:599–611.
12. Baker TG, Neal P. Organ culture of cortical fragments and Graafian follicles from human ovaries. J Anat 1974; 117:361–71.
13. Baker TG, Hunter RHF, Neal P. Studies on the maintenance of porcine Graafian follicles in organ culture. Experientia 1975; 31:133–5.
14. Qvist R, Blackwell LF, Bourne H, Brown JB. Development of mouse ovarian follicles from primary to preovulatory stages in vitro. J Reprod Fertil 1990; 89:169–80.
15. Nayudu PL, Osborn SM. In vitro studies of factors influencing the rate of preantral and antral growth of isolated mouse ovarian follicles. J Reprod Fertil 1992 (in press).
16. Boland NI, Humpherson PG, Leese HJ, Gosden RG. Carbohydrate metabolism by mouse ovarian follicles. J Reprod Fertil (Abstr Ser) 1991; 8:8.
17. Gosden RG, Hunter RHF, Telfer E, Torrance C, Brown JB. Physiological factors underlying the formation of ovarian follicular fluid. J Reprod Fertil 1988; 82:813–25.
18. Hillier SG. Sex steroid metabolism and follicular development in the ovary. Oxford Rev Reprod Biol 1985; 7:168–222.
19. Spears N, Boland N, Murray A, Gosden R. Thecal colonization of mouse ovarian follicles. J Reprod Fertil (Abstr Ser) 1991; 7:55.
20. Eppig JJ. Mouse oocyte development in vitro with various culture systems. Dev Biol 1977; 60:371–88.
21. Torrance C, Telfer E, Gosden RG. Quantitative study of the development of isolated mouse pre-antral follicles in collagen gel culture. J Reprod Fertil 1989; 87:367–74.
22. Telfer E, Torrance C, Gosden RG. Morphological study of cultured preantral ovarian follicles of mice after transplantation under the kidney capsule. J Reprod Fertil 1990; 89:565–71.
23. Carroll J, Whittingham DG, Wood MJ, Telfer E, Gosden RG. Extraovarian production of

mature viable mouse oocytes from frozen primary follicles. J Reprod Fertil 1990; 90:321–7.
24. Eppig JJ, Schroeder AC. Capacity of mouse oocytes from preantral follicles to undergo embryogenesis and development to live young after growth, maturation and fertilization in vitro. Biol Reprod 1989; 41:268–76.
25. Daniel SAJ, Armstrong DT, Gore-Langton RE. Growth and development of rat oocytes in vitro. Gamete Res 1989; 24:109–21.
26. Eppig JJ, Schroeder AC, van de Sandt JJM, Ziomek CA, Bavister BD. Developmental capacity of mouse oocytes that grow and mature in culture: the effect of modification of the protocol. Theriogenology 1990; 33:89–100.
27. Eppig JJ. Maintenance of meiotic arrest and the induction of oocyte maturation in mouse oocyte–granulosa cell complexes developed in vitro from preantral follicles. Biol Reprod 1991; 45:824–30.
28. Bachvarova R, Baran MM, Tejblum A. Development of naked growing mouse oocytes in vitro. J Exp Zool 1980; 211:159–69.
29. Canipari R, Palombi F, Riminucci M, Mangia F. Early programming of maturation competence in mouse oogenesis. Dev Biol 1984; 102:519–24.
30. Moor RM, Smith MW, Dawson RMC. Measurement of intercellular coupling between oocytes and cumulus cells using intracellular markers. Exp Cell Res 1980; 126:15–29.
31. Helller DT, Schultz RM. Ribonucleoside metabolism by mouse oocytes: metabolic cooperativity between the fully grown oocyte and cumulus cells. J Exp Zool 1980; 214:355–64.
32. Herlands RL, Schultz RM. Regulation of mouse oocyte growth: probable nutritional role for intercellular communication between follicle cells and oocytes in oocyte growth. J Exp Zool 1984; 229:317–25.
33. Wassarman P. The mammalian ovum. In: Knobil E, Neill J, eds. The physiology of reproduction. New York: Raven Press, 1988; 69–102.
34. Biggers JD, Whittingham DG, Donahue RP. The pattern of energy metabolism in the mouse oocyte and zygote. Proc Natl Acad Sci USA 1967; 58:560–7.
35. Bouquet M, Selva J, Auroux M. The incidence of chromosomal abnormalities in frozen–thawed mouse oocytes after in vitro fertilization. Hum Reprod 1992; 7:76–80.
36. Sirard MA, Parrish JJ, Ware CB, Leibfried-Rutledge ML, First NL. The culture of bovine oocytes to obtain developmentally competent embryos. Biol Reprod 1988; 39:546–52.
37. Cha KY, Koo JJ, Ko JJ, Choi DH, Han SY, Yoon TK. Pregnancy after in vitro fertilization of human follicular oocytes collected from nonstimulated cycles, their culture in vitro and their transfer in a donor oocyte program. Fertil Steril 1991; 55:109–13.
38. Johnston LA, Donoghue AM, O'Brien SJ, Wildt DE. Rescue and maturation in vitro of follicular oocytes collected from nondomestic felid species. Biol Reprod 1991; 45:898–906.
39. Eppig JJ. Intercommunication between mammalian oocytes and companion somatic cells. BioEssays 1991; 13:572.

Discussion

Shaw: There have been a number of publications, particularly from Germany, reporting frozen eggs, or perhaps "eggs", since virtually all of those publications really refer to pronucleate embryos. The only data in the human have been those few cases from a group in Australia, now moved to Singapore, and I am not sure that that has been repeated since.

Whittingham: There is nothing in the literature. It stops at 1987.

Shaw: One of the problems in a mature human oocyte is permeability of the oocyte for cryoprotectants – getting the cryoprotectants in. Nothing was said about the differences between human and mouse in terms of permeability coefficients.

Whittingham: Taking those into consideration, as Professor Shaw and his colleagues have done in their laboratory, the method is very similar to the one currently being used for the mouse and hamster work and survival is 70% or 80%.

Fishel: Are Dr Gosden's cultures dynamic or static? Were they changing the concentration of the constituents throughout the day prior to ovulation, or was it spontaneous?

Gosden: Taking up the point that Dr Leese made in an earlier discussion about stirring, these cultures are static but we do change the medium daily. The reason for that is that even though the membranes are not tissue culture treated, they do not have the adherent surface. Nevertheless the granulosa cells are treated with a lot of fibronectin and after a day or two the cells start to plate down and spread out and we lose the normal morphogenesis. So we keep them moving every day.

The protocol to ovulation is that we add the normal constituents to the medium plus 1 mU ml^{-1} on day 6. We do not expose the follicles at an earlier stage to LH apart from the contamination we have in the FSH preparation because we are not using recombinant FSH. It is a matter of considerable interest to us; the contamination may be important for quality and this is obviously an area we want to investigate, especially because of the clinical importance of premature LH elevation.

Fishel: I wanted to look further into the future. If we take patients whom we know may have functioning or almost functioning ovaries for a very short time in their lives, would we see a possible scenario that we would take out part of the ovarian tissue, freeze it and at a later stage look to thawing and then culture with either granulosa cells or intact follicles, rather than getting out the oocytes?

Whittingham: Slices of ovarian tissue can be frozen quite effectively. That was done back in the 1950s where they transferred the frozen graft, or grafted them back, and got live young – mainly in the rat.

Gosden: We have got an experiment of this kind at the moment, taking just a cortical slice to see whether we can get good survival of tissues.

Hillier: What may be the underlying mechanism for the effect of insulin on, as I understood it, the proliferation of the granulosa cells in intact follicles? Is the presence of insulin or a similar factor required in cultures of just oocytes and granulosa cells in order to achieve maturation?

Gosden: Those are important questions. The insulin levels do seem to be fairly critical: it is an absolute requirement for growing these follicular tissues, at least at the early stages. Something we do not know about is whether, when the granulosa cells start to express IGF_1, they have autonomous maintenance by self-stimulation.

It seems that insulin is a potent growth factor in these tissues. We have not yet investigated thoroughly, but in some of the cultures it seems that if we double the dose, the granulosa cells cause a breach of the follicle wall: it breaks open and the cells stream out. Certainly we do not get that kind of picture other than by physically damaging the follicle. There may be quite a narrow normal range for insulin stimulation at these early stages, but I threw this out as an interesting observation. We have not investigated it in any depth.

We do not know whether the granulosa/oocyte cultures have such a critical dependence. It is a non-physiological system. The whole follicle unit is a developmental unit and has presumably all the information it needs there.

The complexes have got insulin in them in the medium but we do not know whether that is critical. That will be looked at.

Whittingham: With regard to the follicles which ovulate in vitro, are those eggs viable?

Gosden: We have had a couple of two-cell embryos so far and we have not carried any through to live births by transfer.

Whittingham: Then in relation to John Eppig's work, what ages are they? The age at which he takes them from the prepubertal animal influences the outcome of their actual viability.

Gosden: The follicle stage is very similar in that the primary follicles have about two, maybe at the most three, cell layers – maybe just one cell layer in some cases. He obtains his material, as we did, from prepubertal animals because we can harvest so many more follicles. But for the whole-follicle culture work we have started with animals that are about pubertal age, because there may be physiological differences in the follicles starting growth very early in life versus those which are destined for adulthood.

Braude: A propos Professor Whittingham's reservations about the moratorium on freezing of oocytes. Given the fact that mouse oocytes are much more robust in terms of their spindle, the way they can be handled and the lower likelihood, probably, that they would develop aneuploidy, what should we be doing now in terms of freezing oocytes?

Whittingham: We should be looking to a method where we have good survival using the modifications of a technique where we shall get high rates of fertilisation, that is taking into consideration the protection of the zona pellucida as a barrier to the penetration of sperm. Using methods such as vitrification which will take us through the barrier of depolymerisation, with high rates of cryoprotectance which will stop depolymerisation, with high rates of cryoprotectance which will stop depolymerisation, we should get no changes in the spindle, and taking that into consideration, I do not think we should get any more aneuploidy than we get under normal handling of the egg for IVF.

Braude: It was suggested that we should not be adhering to a moratorium, but perhaps be getting on with freezing and doing it as treatment in the light of what has been done in the mouse.

Whittingham: Sure.

Franks: What happens if primordial follicles are taken out and put in culture medium, and either FSH or insulin is omitted?

Gosden: We have not reached the stage of examining in that sort of detail. When we take out primordial follicles and put them in culture, a rather remarkable transformation takes place. We find that the cells plate down very readily and the oocyte pops out and floats away. It seems that the oocyte has a very much more delicate association with the granulosa cells at this stage. Certainly if they are put in collagen gel, or on hydrophobic membrane, the unit does remain intact and one can get growth of the very early stages. They will initiate their development in vitro. We have not explored their requirements: we have used a standard medium.

The problem of taking them through does not seem to be one of the medium composition so much as that these very early ones do not seem to have association with the stroma/theca cells. The basement membrane seems to change its properties around the primary stage, which allows colonisation by thecal stroma cells, and if we are to be able to culture them through from the primordial stage it will mean that we shall have to introduce them to some stroma theca cells at the appropriate stage. We found that this was possible, but we still have not managed to carry them right through. I think this will be the trick, rather than the medium.

Whittingham: At what stage can one destroy the oocyte and still get continued differentiation of the follicle?

Gosden: We are trying this experiment at the moment. All the indications would be that the oocyte is sending information to the follicle cells and it would be very interesting to know what the molecules involved are. The critical thing is to make an ovectomised follicle which does not sustain too much damage because we know that basement membrane damage has a very bad prognosis for these follicles.

I cannot answer that question but it is a very interesting problem.

Shaw: This is an area of great potential development, but at the moment still in its very embryonic developmental stages!

Chapter 18

Egg and Embryo Donation: Implantation Aspects

R. H. Asch

Oocyte donation is the only method by which women with primary or secondary ovarian failure can achieve a pregnancy. Recently, this procedure has also been utilised in some women with functional ovaries, e.g. carriers of undesirable genetic traits (X-linked or autosomal dominant), poor responders to controlled ovarian hyperstimulation (COH), inaccessible ovaries for oocyte retrieval and following repeated failure of fertilisation in vitro (Table 18.1). In 1984, Lutjen et al. [1] reported the first pregnancy and delivery following in vitro fertilisation and embryo transfer to the uterus (IVF-ET) in a patient with primary ovarian failure treated with steroid replacement. In 1986, our group reported the first pregnancy obtained using gamete intrafallopian transfer (GIFT) with donated oocytes and husband's sperm in subjects with premature ovarian failure [2]. We have also recently published a series of 11 cycles with pregnancies and births of normal children resulting from oocyte donation and tubal embryo transfer (TET) in agonadal patients [3].

Table 18.1. Indications for oocyte donation

Premature ovarian failure
Gonadal dysgenesis
Iatrogenic: surgery, radiation, chemotherapy
Poor responders to controlled ovarian hyperstimulation
Inaccessible ovaries for oocyte retrieval
Carriers of undesirable genetic traits: X-linked or autosomal dominant
Repeated failure of fertilisation in vitro
Unexplained repeated pregnancy wastage

The purpose of this chapter is to compare three different methods of assisted reproductive technologies, IVF-ET, GIFT and TET in the oocyte donation programme at the University of California, Irvine (UCI) and to discuss the physiological and practical implications of the results. Special emphasis will be given in the discussion section to recent developments on the hormonal control of implantation, the effect of genetic versus gestational age and the characteristics required for an oocyte donor.

Oocyte Donation and Different Assisted Reproductive Technology (AT) Techniques

Patient Population

Recipients

Our study population consisted of 106 transfer cycles in 89 recipients performed from January 1986 to July 1990. Recipient's ages ranged between 23 and 49 years. All the agonadal patients (n = 74) had amenorrhoea, elevated levels of serum follicle-stimulating hormone (FSH) and luteinising hormone (LH), and oestrogen deficiency. Fifteen recipients were normo-ovulatory with normal levels of circulating gonadotropins. Table 18.2 shows the aetiological factors in gonadal and agonadal recipients.

During the transfer cycle all patients received a similar regimen of hormone replacement [4]. Basically, they received the following steroid replacement protocol: micronised 17β-oestradiol (E2) orally (Estrace; Mead Johnson, Evansville, IN), 2 mg on days 3–8; 4 mg on days 9–11; and 6 mg on days 12–28, patients received oestrogen for an average of 12.7 days before onset of progesterone administration. Progesterone injections (50–100 mg) were started on the day of aspiration of the donated oocytes. Patients underwent GIFT on the same day as oocyte collection or embryo transfer on the third day of progesterone replacement. In the gonadal recipients basically we used the same protocol in addition to leuprolide acetate 1 mg day^{-1} subcutaneously (Lupron,

Table 18.2. Aetiology for oocyte donation

	Number of patients
Premature ovarian failure	74
Idiopathic	59
Gonadal dysgenesis	10
Iatrogenic	5
Genetic disease	1
Poor responders to COH	14
Total	89

TAP Pharmaceuticals, IL) in the cycle before the steroid replacement, to achieve a medical oophorectomy.

The selection criteria used to decide which ART technique (IVF-ET, GIFT or TET) should be applied in each recipient were applied in a non-randomised fashion. Historically, at the outset of the programme all subjects underwent GIFT as the method of oocyte donation. Later on, IVF-ET and TET were incorporated into the programme and offered to the patients in some cases exclusively (i.e., patients with tubal factor associated: IVF-ET or with an obvious male factor: TET) and in others as part of a research programme to study the efficacy of different ART in the oocyte donation model. In those cases in which a tubal transfer was chosen, tubal patency was confirmed 3–6 months before the GIFT or TET procedure by either laparoscopy or a hysterosalpingogram.

In the GIFT cycle, as previously described [5], 1–2 preovulatory oocytes plus 100000 motile sperm per tube were transferred. In the fertilisation in vitro procedures (IVF-ET, TET), oocytes were inseminated 3–8 h after follicular aspiration,; the embryos produced in vitro were transferred 45–50 h later to the uterus (IVF-ET) or to the midampullary portion of the fallopian tubes through a minilaparotomy (TET) [6]. If pregnancy was confirmed two weeks after transfer by a positive β human chorionic gonadotropin measurement (β-HCG, RIA) and later a gestational sac was seen by transvaginal ultrasound, micronised E2 and progesterone were maintained at 8 mg day^{-1} and 100 mg day^{-1}, respectively until day 100 of gestation. Levels of β-HCG, E2 and progesterone were determined every two weeks until day 100.

Donors

The mean age of the donors was 30.4 ± 4.5 years (range 19–39 years). In most cases oocytes were obtained from patients who underwent ART procedures. They agreed to donate their extra oocytes not being used for transfer. In these cycles, the donation was anonymous, a consent form was signed prior to the procedure, separately by both the recipient and the donor. The donation was not financially reimbursed. The donors were matched to selected recipients for oocyte donation who were within 12–14 days of oestrogen replacement from our recipient waiting list. The selection of recipients for oocyte donation was based primarily upon the timing of their cycles and their ethnic similarities to the potential donor. In other cases donors were provided by the recipients themselves (sister to sister). In these non-anonymous donations, recipients started the hormone replacement on the second day of the donor's menstrual cycle. Donors received the following protocol for induction of follicular development: Lupron 1 mg day^{-1} subcutaneously initiated between day 22 and 24 of a previous cycle; follicle stimulating hormone (FSH, Illetrodin; Serono Laboratories Inc. Randolph, MA) 150 IU daily days 2 and 3 combined with human menopausal gonadotropin (hMG, Pergonal, Serono Laboratories Inc.) 150 IU daily from day 2 until two follicles were greater than 20 mm by transvaginal ultrasound and E2 levels were at least 250 pg ml^{-1} per large follicle. Human chorionic gonadotropin (hCG, Profasi, Serono Laboratories, Inc.) 10000 IU i.m. was given and 36 h later follicular aspiration under transvaginal ultrasound guidance was performed. Statistical analysis of the data was performed using chi square test.

Results

The recipients' mean age was 36.8 (± 7.0) years (range 26–48 years) (IVF-ET group), 34 (± 5.4) years (range 26–42 years) (GIFT group) and 35.5 (± 5.4) years (range 23–45 years), (TET group). Overall the mean age in the pregnant patients was not significantly different than in the non-pregnant recipients; 35.3 vs. 35.0 years, respectively.

Table 18.3 presents a comparison of the three different techniques. In the IVF-ET group 51 transfers resulted in a clinical pregnancy rate of 37%, the average chance of implantation per embryo transferred was 14%, and the abortion rate was 10%. In the GIFT group we performed 29 transfers, with a clinical pregnancy rate of 48% and an implantation rate of 16% per oocyte transferred and the abortion rate was 28%. In the TET group we performed 26 transfers with a clinical pregnancy rate of 57%, an implantation rate of 23% and an abortion rate of 13%. No ectopic pregnancies occurred in this study.

Table 18.4 shows the pregnancy rate in relation to the indication for oocyte donation and divided by procedure. Fig. 18.1 analyses the pregnancy rate in

Table 18.3. Comparative results with assisted reproduction techniques (ART)

	IVF-ET	GIFT	TET
Number of transfers	51	29	26
Clinical pregnancy	19 (37%)[a]	14 (48%)[a]	15 (57%)[a]
Oocytes/embryos transferred	184	109	102
Number of gestational sacs	26	18	24
Implantation rate[c]	14%[b]	16%	23%[b]
Abortion (%)	2 (10%)	4 (28%)	2 (13%)
Ectopic pregnancies	0	0	0

[a]N.S.
[b]$P < 0.02$.
[c]Implantation rate: no. of gestational sacs seen by ultrasound/no. of oocytes or embryos transferred.

Table 18.4. Pregnancy distribution according to reason for oocyte donation

	Transfers	IVF-ET	GIFT	TET	Rate (%)
Premature ovarian failure					
Idiopathic	73	11	10	11	45
Iatrogenic	7	1	1	0	28
Gonadal dysgenesis	10	3	2	0	50
Genetic	1	1	0	0	100
Poor responders	15	3	1	4	53

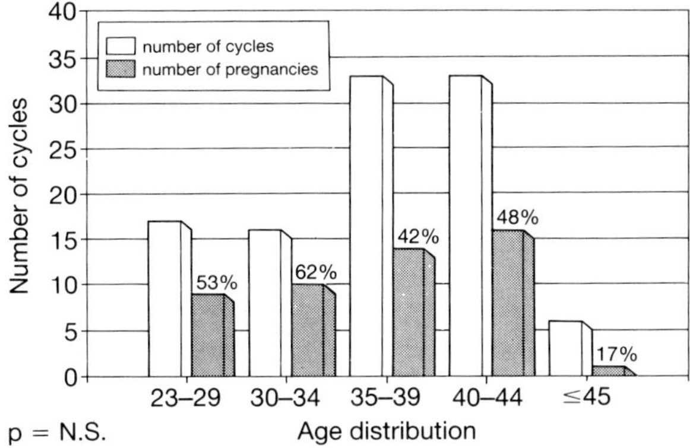

Fig. 18.1. Pregnancy rate distribution by age.

the entire population by age groups. This analysis shows no significant difference in the clinical pregnancy rate according to the recipient's age.

Comments

In this study we present an update of the University of California, Irvine, experience in ART and oocyte donation. The overall pregnancy rate of 45% is significantly higher than the 30% obtained during cycles in which patients undergo COH and transfers performed with their own gametes [7]. The difference in the results could be due to detrimental effects of ovarian stimulation on endometrial receptivity or to better quality of the oocytes selected for donation cycles. Paulson et al. [8] found a significantly higher implantation rate with IVF-ET in hormonal replacement cycles (HR) when compared with age-matched controls during stimulated cycles (SC). The authors concluded that the higher implantation rate in their oocyte donation programme is due to better endometrial receptivity. Age is an important factor determining pregnancy outcome after ART in COH patients [9]. There is a lower pregnancy rate and increased abortion rate in patients over 40 years of age. In our experience with oocyte donation, the chance of pregnancy and embryo implantation in recipients over 40 years of age does not show a significant difference compared with the population under 40 years of age. Considering that the source of oocytes for both age groups is the same (donors), these data imply that endometrial receptivity does not change importantly with age, when hormone replacement is adequate, and suggest that embryo quality is an important factor in determining the possibility of implantation.

Yovich et al. [10] showed a higher pregnancy and implantation rate in hyperstimulated subjects, when oocytes or embryos were transferred to the fallopian tubes (GIFT, TET) when compared with IVF-ET. Even though in the present study, the implantation rate was significantly higher for TET than for IVF-ET, in the oocyte donation model there does not appear to be a definite advantage of one technique over the other in terms of pregnancy rate.

If no advantages are proven, as IVF-ET is a non-surgical procedure it may become the preferred method in oocyte recipients without cervical difficulties for transfer. The only way scientifically to compare pregnancy and implantation rates between tubal and uterine embryo transfers in an oocyte donation programme, would be to perform a randomised prospective study where similar matched subjects would be alternately allocated to either group. There have been no studies of such a nature, most probably due to the practical difficulties to carry them out in traditional ART settings. However, recently in our unit Balmaceda et al. [11] carried out a small prospective randomised study between 22 patients that underwent uterine transfer versus 20 that underwent tubal transfer, in an embryo donation and hormone replacement therapy (HRT) programme. A total of 23 pregnancies were achieved, 12 (54.5%) after uterine transfers and 11 (57.9%) after tubal transfers. Implantation rates were 17.4% and 21.5%, respectively. Pregnancy and implantation rates were statistically not different among patients undergoing uterine or embryo transfers. The conclusions of this study show that in uniform endometrial stimulation with HRT, there is no apparent advantage in transferring embryos to the fallopian tubes versus to the uterine cavity.

Non-surgical intratubal gamete or embryo transfer has been used as an alternative that obviates the need for an invasive surgical procedure [12,13] and could become an interesting option. However, since the GIFT procedure has excellent and reproducible pregnancy rates, it may be considered in many centres the treatment of choice because it avoids the technical and economical aspects of in vitro fertilisation.

To date we have treated a total of nine patients with gonadal dysgenesis (four with normal chromosomal complement and five with demonstrable abnormalities). These nine patients underwent 13 cycles of oocyte donation. We achieved nine pregnancies in five patients (three had pure gonadal dysgenesis and two were mosaic Turner's syndrome). In the IVF-ET group all six transfers resulted in pregnancies, with an implantation rate per embryo transferred of 22%. Two of the six uterine transfers were performed with cryopreserved embryos. In this group there was one abortion resulting from a false negative β-hCG (laboratory error) at the time of first testing, and the subsequent cessation of HRT. The pathology report confirmed trophoblastic tissue in the material expelled by the patient. Of the seven tubal transfers, two pregnancies were achieved with GIFT, for a pregnancy rate of 33% and an implantation rate of 18%.

The conclusions of this study were that, (a) oocyte donation and HRT is an excellent option for patients with gonadal dysgenesis and (b) in contrast to other conditions, uterine embryo transfers result in better pregnancy and implantation rates than tubal embryo transfers. A possible explanation for this may be defective gamete or embryo transport through a dysgenetic dysfunctional fallopian tube.

The Effect of Age on the Results of Different ART Procedures

We compared the efficacy of three different ART procedures (IVF-ET, GIFT and TET) in women above the age of 40 years [14], dividing all groups into two: (a) subjects undergoing COH and ART; and (b) subjects undergoing oocyte donation and HRT.

The results of the study are shown in Table 18.5. It is clear from this study that in women above the age of 40 years undergoing ART procedures using their own oocytes during a COH cycle, the transfer of gametes or embryos to the fallopian tubes when possible, presents great advantages over transferring them to the uterus. It is tempting to speculate that the factor(s) in the fallopian tubes that facilitates early embryo development, are accentuated with age, or that hostile uterine factors that create an inhospitable environment for the embryo as a result of ovarian hyperstimulation [8] are increased in older women.

In contrast, and similarly to the results of Balmaceda et al. [11] the efficacy of tubal versus uterine gamete or embryo transfers in women above the age of 40 years undergoing oocyte donation and HRT did not reveal any statistically significant difference. In addition, this study demonstrated similar pregnancy rates and live birth rates in oocyte donation and HRT between recipients less than 40 and over 40 years old, indicating an unchanged endometrial receptivity with advanced age. This apparent lack of ageing of the human uterus towards implantation prompted us to perform a collaborative study between Bourn Hall in Cambridge, UK and our Center for Reproductive Health at UCI in Orange, California, USA [15]. In this report the results of IVF-ET were analysed in terms of implantation and pregnancy rates in women of different ages, and the differences seen in oocyte/embryo recipients were compared according to their history of menstrual regularity. The results of the study show that in both the UK and US programmes, rates were higher in previously amenorrhoeic (acyclic) women than in agonadal women, irrespective of age and number of embryos replaced. Oocyte quality as determined by age of oocyte donor, in vitro growth of embryos and proportion of defective embryos in culture, did not contribute to the difference, nor did type of cycle during therapy. The study indicated that among recipients in embryo transfer programmes, acyclic women are more fecund than cyclic women, more of them became pregnant and to older eggs.

Three different possibilities are listed as reasons for our findings: (a) acyclic women might have a higher inherent fecundity; (b) donated embryos might be of better quality than those derived from an infertile woman's ova; (c) there may be other factors such as antigenic disparity between acyclic women and the donated oocytes.

If indeed, future studies confirm that a uterus that has not been subjected to regular menstrual cycles is a favourable factor for increasing the chances of implantation in humans, then a period of induced amenorrhoea (induced by LH-RH analogues or steroids) might be beneficial for infertile women undergoing ART procedures.

Table 18.5. Outcome of ART procedures in women over the age of 40 years

	Tubal transfers			Uterine transfers		
	Pregnancy rate (%)	Implantation rate (%)	Live birth rate (%)	Pregnancy rate (%)	Implantation rate (%)	Live birth rate (%)
	1	2	3	1	2	3
Group A COH + ART	38	12	19	9	5	34
Group B Oocyte donation + HRT	60	18	47	38	10	29

1 and 2 = $P < 0.05$; 3 = $P < 0.001$.
COH, controlled ovarian hyperstimulation.
ART, assisted reproduction techniques.
HRT, hormone replacement therapy.

The Importance of Luteal Oestrogens on the Process of Implantation in Humans: Effects on Endometrial Histology and Pregnancy Outcome

The role of post-ovulatory oestrogens in the process of implantation has never been identified in primates. Even though oestrogens are extremely important in inducing endometrial changes necessary for progesterone action, their function after the corpus luteum begins actively secreting progesterone has not been studied before. An interesting clinical model to investigate the role of oestrogens in implantation and occurrence of pregnancy, is the one of agonadal patients undergoing oocyte/embryo donation and HRT. We have recently carried out a study that investigated the effects of luteal oestradiol in the endometrial histological changes of agonadal women during the evaluation cycle prior to oocyte/embryo donation [16].

To this end, 22 agonadal patients that were potential recipients in our oocyte donation–HRT programme, were included and divided into two groups. In Group A, 11 patients received micronised E_2 orally, 2 mg on days 3–7, 4 mg on days 9–11, 6 mg on days 12–28. In Group B, 11 patients received the same hormone replacement until day 15 after which day the oestrogens were discontinued. All patients in both groups received progesterone in oil intramuscularly 100 mg day^{-1} on days 16–28. The cycle of hormone replacement was monitored by several measurements of E_2 and progesterone using radioimmunoassay four times during the cycle of evaluation (days 9,12,15 and 21). An endometrial biopsy was performed by aspiration curettage on either day 6 or day 9 after initiation of progesterone administration and the biopsy was performed according to the classic criteria of Noyes et al. [17]. The mean ages of patients in groups A and B were not different, 42 ± 4 and 39 ± 5 (years ± SD), respectively.

Adequate (similar to the follicular phase in normal cycling women) serum levels of E_2 were observed prior to day 15 in all patients. However, on day 20 of the evaluation cycle subjects in group B showed significantly lower levels of E_2 than those in group A (38 ± 26 pg ml^{-1} versus 367 ± 124 pg m^{-1}, respectively) ($P < 0.001$). Serum progesterone levels on day 2 of the cycle showed "normal postovulatory" levels in subjects in both groups (19 ± 3 ng ml^{-1} and 22 ± 5 ng ml^{-1}, respectively).

The endometrial tissue showed a discordance between the timing noted by biopsy when compared to the expected cycle assuming that the first day of progesterone administration corresponds to the first day after ovulation. Delayed maturation, of the endometrial glands was noted in nine of 11 of the biopsies from patients in group A. These changes were observed on the tissue samples from day 6 as well as day 9. Similar results were observed in patients in group B, ten of eleven biopsies showed a delayed appearance for the appropriate day of the menstrual cycle. These changes were also observed on samples from both day 6 and day 9. In two biopsies, one from day 6 and one from day 9, we observed a premenstrual endometrium (i.e. cycle days 26 and 27). In three biopsies, the glands were delayed while the stroma was advanced resulting in stromal-glandular disparity. There was no striking difference in the endometrial response between patients in group A or group B.

In addition, we have documented in three agonadal subjects of our oocyte donation–HRT programme, that suspension of oestrogens in the luteal phase of the transfer cycle does not affect the outcome of the ART cycle. Three agonadal subjects, ages 26, 31 and 36 that underwent GIFT (two cases) and TET (one case) conceived on their first attempt of oocyte donation and HRT. These three patients discontinued the oestrogen administration in error 1 or 2 days after progesterone initiation. In all of them documentation of lack of oestrogen intake was demonstrated by serum oestradiol determinations by radioimmunoassay. All patients continued with uneventful pregnancies; to term in two cases (one singleton and one twin) and to 35 weeks in the other (singleton).

Furthermore, three regularly cycling patients were down-regulated with LH-RH analogues in order to prepare them for HRT and frozen embryo transfer to the uterus (two cases) and to the tube (one case). These three patients also discontinued oestrogen admininstration by mistake, after beginning progesterone administration. Two of them had continued pregnancies (one quadruplet and one singleton) and one had a spontaneous miscarriage at 19 weeks of gestation.

All these data together, the histological and the preliminary clinical, cast doubt about the physiological role of luteal oestrogens in the implantation process in the human, as well as their need to maintain an early gestation. Randomised studies will certainly be necessary to conclude that oestrogens are not required at all to be administered to egg donation recipients, after progesterone administration has been initiated.

The Selection of Donors in a Programme of Oocyte/Embryo Donation

We have recently reported on the characteristics of the oocyte donors evaluated in our programme at UCI [18]. We analysed the different variables that affected the efficacy of egg donation in our ART unit. Potential donors (n = 108) were divided into two groups: (a) partial donor (PD), subjects that underwent ART for the treatment of their infertility, who agreed altruistically to donate extra eggs not used during their own cycles (n = 31); and (b) full donors (FD), subjects that donated all the eggs recovered during their COH cycle (sister to sister donation, volunteers, diagnostic or sterilisation laparoscopies) (n = 77). The entire donor population was divided into two subpopulations by age, under 30 years and 31–39 years old, and the pregnancy rates in their recipients were studied. A higher pregnancy rate was observed when the oocytes were provided by the younger group of donors (54% versus 36%). No significant difference was found in the recipients' pregnancy rates, according to the previous reproductive performance in their donors. Also even though FD and PD yield similar pregnancy rates during the fresh cycles, the availability of cryopreserved embryos in the former group enhances the cumulative pregnancy rate.

References

1. Lutjen P, Trouson A, Leeton J, Findlay J, Wood C. The establishment and maintenance of pregnancy using in vitro fertilization and embryo donation in patient with primary ovarian failure. Nature 1984; 307:174.
2. Asch RH. Pregnancy following GIFT (gamete intra Fallopian transfer) in premature ovarian failure. Hum Reprod 1986; 1:17.
3. Rotsztejn DA, Remohi J, Weckstein L, Ord T, Moyer D, Balmaceda J, Asch RH. Results of tubal embryo transfer in premature ovarian failure. Fertil Steril 1990; 54:348.
4. Borrero C, Remohi J, Ord T, Balmaceda JP, Rojas F, Asch RH. A program of oocyte donation and gamete intrafallopian transfer. Hum Reprod 1989; 4:275.
5. Asch RH, Balmaceda JP, Ellsworth LR, Wong PC. Preliminary experiences with gamete intrafallopian transfer (GIFT). Fertil Steril 1986; 45:366.
6. Balmaceda JP, Gastaldi C, Remohi J, Borrero C, Ord T, Asch RH. Tubal embryo transfer as a treatment for infertility due to male factor. Fertil Steril 1988; 50:476.
7. Rotsztejn DA, Asch RH. Effect of aging on assisted reproductive technologies (ART): experience from egg donation. Semin Reprod Endocrinol 1991;
8. Paulson RJ, Sauer MV, Lobo RA. Embryo implantation after human in vitro fertilization: importance of endometrial receptivity. Fertil Steril 1990; 53:870.
9. Craft I, Al-Shawaf T, Lewis P et al. Analysis of 1071 GIFT procedures – the case for a flexible approach to treatment. Lancet 1988; ii:1094.
10. Yovich JL, Blackledge DG, Richarson PA, Matson PL, Turner S, Draper R. Pregnancies following pronuclear stage tubal transfer. Fertil Steril 1987; 48:851.
11. Balmaceda JP, Alam V, Rotsztejn D, Ord T, Snell K, Asch RH. Embryo implantation rates in oocyte donation: a prospective comparison of tubal versus uterine transfers. Fertil Steril 1991; 57:362–5.
12. Jansen RPS, Anderson JC, Sutherland PD. Nonoperative embryo transfer to the fallopian tube. 1988; 319:288.
13. Bustillo M, Munabi AK, Shulman JD. Pregnancy after nonsurgical ultrasound guided gamete intraFallopian transfer. N Engl J Med 1988; 319:313.
14. Asch RH, Ord T, Stone S, Balmaceda JP, Rotsztejn DA. Assisted reproductive techniques in women over 40 years of age: IVF, GIFT, ZIFT, egg donation: is there a best alternative? 39th Annual Meeting of the Pacific Coast Fertility Society Supplement; O-016:pg. A15.
15. Edwards RG, Morcos S, Macnamee M, Balmaceda JP, Walters DE, Asch RH. High fecundity of amenorrhoeic women after oocyte donation. Lancet 1991; 338:292–6.
16. Asch RH, Rotsztejn D, Balmaceda JP, Pauerstein CJ. Role of estrogens in implantation and early pregnancy maintenance. (in press).
17. Noyes R, Hertig W, Rock J. Dating the endometrial biopsy. Fertil Steril 1950; 1:3.
18. Rotsztejn D, Ord T, Balmaceda JP, Asch RH. Variables which influence the selection of an egg donor. Hum Reprod 1992; 7:59–62.

Chapter 19

The Social Aspects of Ovum Donation

H. I. Abdalla

Pregnancy following sperm donation was first reported in 1884 [1] but it was not until 100 years later that successful pregnancies following ovum donation were achieved [2,3]. Currently the success rate from ovum donation exceeds other methods of assisted conception. This appears to be due to the use of young donors of proven fertility; the age of the recipient having little or no influence on the success rate [4,5]. In theory this means that women of any age could potentially have children and when the technical difficulties of cryopreservation of ova are resolved, young women could have their eggs stored indefinitely. This would give them freedom to choose the age at which they may conceive. In addition, with the ready availability of donated sperm it could even remove the need for a stable relationship.

Doctors and scientists together are evolving new methods of achieving parenthood in women who previously would have been childless. These new developments undoubtedly initiate wide controversy and debate in society and among different religious and ethnic groups. This chapter addresses some of the issues surrounding ovum donation, but cannot possibly hope to provide all the answers. It is our wish that the social and ethical dilemmas resulting from oocyte donation stimulate and provoke discussion and debate not only in medical circles but also in society in general.

The Definition of Motherhood

With the possible exception of the Virgin Birth a mother has previously been regarded as the woman who both produced the ova and carried the child. Adoption led to the concept of a mother whose contribution was solely in rearing the child. With the advent of oocyte donation and surrogacy the terms; "genetic mother", "carrying mother", "surrogate mother", have been added. To those undergoing infertility treatment the definition of motherhood depends on their predicament. To the couple undergoing surrogacy, the genetic source of the oocytes is of paramount importance. On the other hand couples on the receiving end of oocyte donation would have no hesitation in considering that the mother is the one who gives birth to the child. Are they both right?

In the UK, the Human Fertilisation and Embryology Act 1990 has defined the mother as the woman who gives birth to the child, although in cases of true surrogacy there is an amendment that allows the woman who donated the ova (together with her husband) to adopt the child. The adoption is, however, dependent on the consent of the women who gives birth to the child. This law means that a woman who conceives from ovum donation is considered to be the legal mother of the child, even though it carries none of her genetic material.

Indications for Ovum Donation

This technique offers the only hope of pregnancy for young women with ovarian failure from any cause. It may also be indicated in women who carry genetic disorders, particularly if prenatal diagnosis is not available or they are averse to termination of pregnancy. Because of the success of ovum donation, it has been suggested that it should routinely be offered as part of the counselling procedure for all women with genetic disorders. These indications are the least controversial.

Techniques of assisted conception have low success rates in women over 40 years and largely because of financial constraints in the NHS, such patients are often denied treatment. As the success of ovum donation is not dependent on the age of the recipient (Table 19.1), many of these women with long-standing infertility may now have the chance of having a family. However, there are also increasing enquiries and demand for ovum donation from women who are postmenopausal. Although at first sight this may seem unreasonable, it should be remembered that there are well-publicised cases of men fathering their children well beyond the age of 70 years. Many women will spend a third or more of their life after the menopause and we do not know whether it is appropriate for them to be denied ovum donation. In many developing countries the grandmothers have prime responsibility for rearing the children, whilst the mother works. In developed societies the children of working mothers are often raised by non-family members.

At the Lister Fertility Clinic we have currently decided not to offer ovum donation to women who are over the age of 50 years (the age of natural menopause), or in whom the combined age of the couple exceeds 105 years. All women referred for ovum donation are extensively counselled by a clinician, the egg donation co-ordinator and an independent counsellor before being accepted into the programme. Despite all these restrictions, a wide public debate was stimulated, when two women achieved pregnancy at the age of 50. Recently an Italian woman gave birth to a child at the age of 56 following

Table 19.1. Outcome of oocyte donation (fresh/frozen), compared to IVF–GIFT related to age of patients (Lister Hospital 1988–1991)

	Age of patients (years)						
	20–24	25–29	30–34	35–39	40–44	45–50	Total
Fresh–OD[a] PREG/CYCLE	–	8/25 (32%)	16/43 (37%)	13/33 (39%)	30/58 (52%)	9/23 (39%)	76/182 (42%)
Frozen–OD PREG/CYCLE	–	3/14 (21%)	6/37 (16%)	9/35 (26%)	13/64 (20%)	9/39 (23%)	40/189 (21%)
Total–OD PREG/CYCLE	–	11/39 (28%)	22/80 (28%)	22/68 (32%)	43/122 (35%)	18/62 (29%)	116/371 (31%)
IVF–GIFT[b] PREG/CYCLE	26/63 (41%)	140/379 (37%)	230/711 (32%)	183/608 (30%)	44/274 (16%)	4/34[c] (12%)	627/2069 (30%)

[a]OD = Ovum Donation.
[b]All fresh cases.
[c]All four cases miscarried.

oocyte donation and a famous 54-year-old actress is already applying to the same programme. This has caused much alarm in society and promoted discussion about the merits of treating older women.

Navot et al. [6] has argued that the greatest impact of ovum donation will be on the ageing infertile population rather than on the unusual group of women with premature menopause. Although a woman's life expectancy has increased by a mean of 30 years in the last century, the age of the menopause has only progressed by 3–4 years [7]. The longevity is due in part to changes in the socioeconomic climate together with improvements in medical care. Should advances in fertility treatment be denied to older women when other improvements are not? Egg donation, for this group of women, should be viewed as an attempt to lessen the disparity between women's life expectancy and their static reproductive potential. Against this argument, from the child's point of view, older parents are more likely to be old, infirm or dead while the children are still in their late teens.

In previous generations it was not unusual for sibling births to occur throughout the reproductive life span of the mother. This frequently led to grandchildren being the same age as their uncles and aunts. In 1844 almost 20% of legitimate births in Scotland occurred to women of 35 years or over with some 7% of mothers being over 40 years of age [8]. In England the birth rate was first recorded according to the age of the mother from the 1930s. At that time 15% of babies were born to women over the age of 35, with some 4% being born to women over 40; this is twice the incidence seen in the 1970s [9]. There is currently a revival of the trend towards later parenting both in the UK [10] and in the USA [11].

The Older Mother

Berryman [12] noted that the medical view of later motherhood tends to be problem oriented. Psychosocial studies are scarce, but suggest that older mothers have qualities that make them as good, though different from their younger counterparts. She argues that the recent trend in later parenthood produces women who are more committed to their children and are better informed about their pregnancy; they actively decide to become pregnant [12]. There is a positive correlation between maternal age at birth and the subsequent intellectual development of the child that is independent of confounding variables such as social class or maternal education [13,14], possibly because such mothers encourage independence and verbalisation [15].

Whether planned, unplanned or following infertility treatment, first pregnancies in women over 40 are a source of much happiness to both partners [16]. The same survey however, showed that the usual response from family and friends was of shock, horror or disgust.

Older women have the same incidence of breast feeding as their younger counterparts but tend to continue for longer. Berryman and Windridge [17] showed that one-third of women over 40 ceased full-time employment to look after their child and that only a third accepted the arrangement of both parents in full time work, with the child being looked after by paid help. Their survey cannot be considered representative of all women over 40 years of age, but in their sample only 2% of women reported the pregnancy as a bad experience. This survey also showed that the majority of women with unplanned pregnancies were shocked to be offered the option of termination based on their age alone [16,17].

Should Infertility Treatment for Women Over 45 Be Banned?

Taken to its logical conclusion, if infertility treatment is banned for women over 45 years of age on the basis that they are less than ideal mothers, then society should insist on permanent methods of contraception for all women over 45 years, or otherwise prohibit sexual intercourse. If you start the process of prohibition, it will not be long before an eleventh commandment is adopted stating *"Thou shalt not have a child beyond society's decreed age"*.

As fertility in men has no natural boundaries, should society measure the age of man by his productivity and the age of woman by her reproductivity? Criticisms of men in their 70s and 80s who father children are rare. Although a large age gap between the mother and child has disadvantages, such as being taken for a granny at the school gates [18] there is clear evidence that older parents have more time for their children and far more experience with which to bring up their children. Finally, older women receiving ovum donation do not have the same risk of chromosome abnormalities or of miscarriage as their peers who achieve spontaneous pregnancies.

At the Lister Hospital, assisted conception and ovum donation occur alongside care and advice for postmenopausal women. Changes in medical science and life expectancy are slowly changing our concept of age. We believe that the decision to use the natural time of the normal menopause as an age limit to ovum donation is workable at the current state of knowledge. If projects suggest that pregnancy in later life is beneficial to the mother, the child and society, then this arbitrary age limit will need to be reviewed.

Donors

In the UK human tissue donation has usually been a "gift" from the donor or the relatives. This has applied to blood transfusion and organ donation, a sperm donor, however, receives a small fee. Despite the fact that ovum donation involves a great deal more commitment the Human Fertilisation and Embryology Authority (HFEA) has deemed it undesirable to offer a financial incentive. The demand for ovum donation far outstrips the supply of donors. If society accepts ovum donation, how is it to attract suitable donors?

Donated ova come from women undergoing assisted conception, who donate extra oocytes, those undergoing intercurrent surgery, and philanthropists. Cryopreservation of supernumerary embryos limits the number of oocytes available from women undergoing assisted conception treatment. Robertson [19] argues that unless eggs are randomly selected from assisted conception patients, those chosen for donation may be morphologically less desirable. However, random selection, while increasing the recipient's chance of pregnancy may lower the donor's own chances, making her less likely to participate in the ovum donation programme. However, random selection of eggs appears to be unnecessary because pregnancy rates from frozen embryos created from supernumerary eggs from assisted conception donors are similar to pregnancy rates from frozen embryos produced by fertile donors [20]. This is despite the fact that the morphologically best oocytes are given to the assisted conception donor. There is evidence that women who donate eggs from assisted conception cycles have a high pregnancy rate, and those who do not fall pregnant do not regret their actions. Well over 90% of all such donors are willing to donate on another occasion [21]. A further strategy to increase the supply of eggs from assisted conception donors would be to allow donors and recipients to share the cost of treatment. This has the additional advantage of reducing the cost of treatment to the donor. This cannot really be seen as inducement as the donor already wishes to undergo assisted conception. Anonymity could be maintained by crossing over donors.

Attempts to increase ovum donation from women undergoing tubal ligation have met with a peculiarly British hurdle. Although these women were not offered a financial inducement, it was felt that agreeing to be a donor in the private sector meant that they jumped the NHS queue and had their operations in luxurious surroundings by highly experienced operators. The inconvenience to the donor of ovarian stimulation and strict operative timing has been overlooked. Furthermore, this source has the advantage of imposing little surgical risk on the donor. However, ovum donation in this situation is perceived to be against the altruistic motives of tissue donation in the UK. This source of donors will only be increased if this issue can be overcome.

Women who volunteer to donate altruistically are by far the greatest potential source of donors but pose the greatest ethical problems. While not insignificant, the risks of ovarian hyperstimulation syndrome (OHSS) and surgical oocyte retrieval appear to be small. At the Lister Fertility Clinic a total of 191 women donated oocytes from 243 cycles between the years 1988 and 1991 (161 donated on one occasion, 23 donated twice, four donated on three occasions and six donated on four occasions). No woman had any operative or postoperative complications, and only two women developed moderate OHSS, requiring hospital admission for 2 days in both cases. All those who wanted to become pregnant again following donation succeeded. Increasing the numbers in this group could be done by asking the recipient to find her own donor, by advertising or by offering payment. Known donation raises a problem of anonymity but this could be preserved by crossing over pairs of donors and recipients. As long as potential donors are adequately counselled it is probable that cases in which undue pressure has been applied would be vetted out.

Advertising for donors by private clinics could be seen to be exploitational if the clinics are making money out of this benevolent donation when the donors are not. At the Lister Fertility Clinic the increased cost of ovum donation charged to the recipients is accounted for by the additional costs of providing the donors with the necessary drugs and no extra profit is being made on egg donation compared to standard assisted conception treatment. One possible way to increase this source of donors would be for the HFEA to advertise and the formation of a central agency such as is recommended by the Polkinhorne Report on the use of fetal tissue and research. Central advertising may, however, cause offence to people who are opposed to assisted conception. Likewise, direct payment for ovum donation in the UK is considered unethical as it lowers human dignity. The most serious ethical concerns arising out of payment are that such payment may allow the donor (or others) to consider that parts of the human body can be bought and sold, and that women in financial difficulties may be forced to undergo donation. There is also the concept that the genetic material may result in a new individual who has been purchased.

The alternative arguments are that it is obligatory to pay donors [22] as they contribute significant time and effort, and the payment recognises the donor's part in the enterprise. This argument is strengthened by the fact that sperm donors are paid, even if it is only to cover expenses [23]. The ban on payment would be acceptable if it did not interfere with an infertile woman's access to donated eggs [19]. In fact, such a ban on paying donors would face constitutional difficulties in the United States [24]. However, surveys among donors in both the UK and Australia [21,25] demonstrated that most did not feel that payment was an incentive and this view was also held by recipients [26].

The HFEA has in 1990 decreed that donors should be under the age of 35 years in order to reduce the likelihood of chromosomal abnormalities. This is a misinterpretation of statistics because however low the risk of chromosome abnormalities, if your baby has Downs syndrome, it is 100%. Setting an arbitrary upper age limit is therefore illogical, especially in times of shortage. Patients should be counselled as to the likelihood as for any other patient. Our experience of ovum donation pregnancies is that there is no decrease in the fecundity rate from donors aged 35–38 years as compared with donors aged 20–24 years (Table 19.2). The supply of oocytes could be increased by relaxing this restriction. The potential recipient would be told of the age of the donor, and she could then make an informed decision. If the recipient is older than the donor then the genetic risks are reduced compared with the recipient's natural chances.

The use of a known donor will avoid the possible risk of consanguinity, allow the recipients to choose who provides the genetic material for the child and may also allow the donors the choice of whether they wish to get to know the child. In the UK the known donor situation was felt to put undue pressure on all who are involved, especially the child. When adoption registers were opened it was of interest that a large proportion of the children wished to know who their genetic parents were. We await the psychological studies of the effects of this knowledge on these adopted children.

Women who donate oocytes have often thought deeply about their actions and presently receive little or no reward other than knowing that they may have benefited someone else. Should they be allowed to know if a pregnancy has resulted from their efforts?, Should they be allowed to have a say about the age or type of recipient to whom the eggs are donated?.

All the problems of finding sufficient eggs are magnified in ethnic minority groups. If these are not to be further discriminated against, better ways must be found of attracting donors. If a recipient from such a group requests cross-cultural donation, should she be denied?, Currently there are no guidelines as to whether it is right to perform such a procedure.

What to Tell the Child?

Whether or not the parents plan to tell the child of his or her genetic origin is an important issue and it is recognised that secrets can have a negative effect on the family unit by placing a lie at the centre of the most basic of human

Table 19.2. Outcome of oocyte donation in recipients related to the age of the donor (Lister Hospital 1988–1991)

	Age of donors (years)			
	20–24	25–29	30–34	35–38
No. recipients[a] cycles	33	106	175	57
No. pregnant	7	31	60	18
No. miscarried	1	7	24	8
No. ectopics	0	2	2	0
Preg/Cycle	21%	29%	34%	32%
Take home/Cycle	18%	21%	19%	18%

[a]There was no significant difference in the mean age of recipients between any of the groups.

relationships. Kirkland et al. [26] found that patients who were receiving oocyte donation were equally divided with regard to the issue of telling the child. Two-thirds of the donors thought that they would not want to know of their genetic origin had they been born from a donated egg [21]. It is tempting to liken oocyte donation to adoption [27]. On the other hand Robertson [19] argues that egg donation is preferable to adoption because each rearing parent will have either a genetic or a gestational relationship to the offspring and he believes that the separation of female genetic material and parenting that occurs in egg donation appears to pose the least risk of family conflict or psychological confusion for the offspring.

Whichever decision is made, couples should acknowledge the profound responsibility of following that decision. If they tell their child they should decide, from the start, how and when the child should be told and perhaps seek advice from adoption agencies about the best way to approach such an issue. If the couple decide not to tell the child about his/her genetic origin, they should be helped to acknowledge the burden which secrecy may place on them, and on their marriage. They should also be encouraged to discuss strategies which they can use to cope with these burdens. For example, what would they do if one party changes his or her mind, or if one of them accidentally tells a different story.

Anonymity

The need for donor anonymity is complex. Donors may fear contact with their offspring, the couple may feel vulnerable and confused about how the child may feel about his/her identity. In the medical community we fear that non-anonymity might reduce the supply of donors thus endangering the survival of donor programmes [28]. In the UK the Interim Licensing Authority advised that egg donation should be anonymous except in special circumstances and, in fact, one clinic was threatened with withdrawal of its licence because it performed sister-to-sister egg donation. This anomaly has been corrected with the HFEA which allows donation from known or anonymous origins. The attitudes of donors and recipients in relationship to anonymity was studied by Kirkland et al. who found that most patients are not secretive, since 86% of donors and 71% of recipients have told at least one person other than their partner about their efforts. Donors and recipients differed when the relationship between them was examined. A total of 63% of donors would still donate if the recipient was told their name, but only 23% of recipients would agree to receive oocytes if their name was made available to the donor. Likewise, 70% of donors would prefer to donate to a known person, whereas only 44% of recipients were prepared to take eggs from a known donor. Although 55% of donors had no objection to the children contacting them as adults, 90% of recipients were against the donor contacting the child later in life. Donors and recipients expressed a preference for anonymity (71% of donors and 86% of recipients). Finally, although 86% of volunteer donors would like to know the outcome of their donations, more than 80% of them denied any connection with that child. These figures indicate that both donors and recipients consider gestational parentage more important than genetic parentage [26].

A matter of concern, however, arising from these data is that only 12% of recipients were in favour of the child making contact with the donor [26]. This large discrepancy between donors and recipients could lead to conflict, should the child want to pursue its genetic mother. This could be especially acute since half the recipient mothers were proposing to tell their children. Given the strong desire of people to know their heritage, this may cause problems later for the recipients. Couples who choose to tell their children about the means of conception therefore need to have social, personal and medical information about the donors which will facilitate a greater understanding for the child. Donors should be told that records are kept and, in this country, centrally registered, as a large proportion of recipients will tell their children about their mode of conception. As the law stands in the UK, children born from egg donation are entitled, at the age of 18, to seek non-identifying information about their genetic mother and for medical and social reasons such records are being centrally held. There is, however, a potential for conflict if the law were to change and allow children to have identifying information about the donor. Problems might arise in later years for the women willingly agreeing to donate eggs on the understanding that their donation was anonymous, given the fact that 50% of recipients would tell their children their origin. We believe that the donors must be counselled about such an eventuality and, as 50% of them have no objection to the child contacting them in the future, perhaps records should be kept as to whether the donor would accept contact from the child. If donors have no desire for such contact their wishes should be paramount. Their views of course may change with time.

Counselling

Although infertile couples are not sick, they have been patients as long as they have been on treatment. Their lives often revolve around their fertility problem, and sex becomes mechanical, subservient to conception. They have intrusions on the most basic human relationship, and slowly they lose the closeness they once had. Because of the pressure they are under, they become vulnerable, and a gradual loss of confidence develops. Feelings of guilt and jealousy become most upsetting, and they have a profound feeling of grief. Although many of them may be critical of their medical care, when offered any opportunity to have a child they will be highly susceptible to any suggestions. Gamete donation, although it may initially sound appealing, evokes a wide range of emotional, social and religious thoughts. When a couple feel desperate, they may accept such a suggestion without thinking carefully about the implications on both their relationship and on future children. In some situations apparent acceptance may result not from a positive attitude, but from a sense of weakness and failure [27]. As Mahlsted and Greenfeld [27] discussed; the most important component of preparation is time. It is therefore extremely important that clinics performing such treatment do provide extensive counselling which will help the couple to discuss, identify with, and accept the issues involved in donor conception. Counselling should also encourage the couple to resolve the way infertility has affected them individually and as a couple. Physicians should be aware that couples may begin treatment with donor gametes before they are emotionally prepared. They should therefore encourage couples to examine the implication of their choices and to discuss the issues of secrecy and anonymity. Since, however, the thoughts and beliefs of people differ it is of the utmost importance that our advice and counselling reflects this diversity of human thoughts.

References

1. Finegold WJ. Artificial insemination with husband's sperm. Springfield: Thomas, 1976; 6.
2. Trounson A, Leeton J, Besanko M, Wood C, Conti A. Pregnancy established in an infertile patient after transfer of a donated embryo fertilised in vitro. Br Med J 1983; 286:835–8.
3. Lutjen P, Trounson A, Leeton J, Findlay J, Wood C, Renou P. The establishment and maintenance donation in a patient with primary ovarian failure. Nature 1984; 307:104–5.
4. Serhal PF, Craft I. Simplified treatment for ovum donation. Lancet 1987; i:687.
5. Abdalla HI. Ovum donation. Curr Opin Obstet Gynecol 1991; 4:4.
6. Navot D, Bergh P, Williams M, Garrisi G, Guzman I, Sandler B, Grunfeld L. Poor oocyte quality rather than implantation failure as a cause of age-related decline in female fertility. Lancet 1991; 337:1375–7.
7. Soules MR, Brenner WJ. The menopause and climacteric: endocrine basis and associated symptomatology. J Am Geriatr Soc 1982; 30:547–56.
8. Flinn M (ed.) Scottish population history: from the 17th century to the 1930s. Cambridge: Cambridge University Press, 1977.
9. Macfarlane A, Mugford M. Birth counts: statistics of pregnancy and childbirth. London: HMSO, 1984.
10. Birth Statistics. Birth Statistics 1987. Series FMI no. 16. London: HMSO, 1989.
11. Berkowitz GS, Skovron ML, Lapinski RH, Berkowitz RL. Delayed childbearing and the outcome of pregnancy. N Engl J Med 1990; 322:659–63.
12. Berryman JC. Perspective on later motherhood. In: Phoenix A, Woolett A, Lloyed E, eds Motherhood: meanings, practice and ideologies. London: Sage, 1991.
13. Ragozin AS, Basham RB, Crnic KA *et al.* Effects of maternal age on parenting role. Dev Psychol 1982; 18:627–34.
14. Zybert P, Stein Z, Belmont L. Maternal age and children's ability. Percept Mot Skills 1978; 47:815–18.
15. Seth M, Khanna M. Child rearing attitudes of the mothers as a function of age. Child Psychiatry Q 1978; 11:6–9.
16. Berryman JC, Windridge K. Having a baby after 40:I. A preliminary investigation of women's experience of pregnancy. J Reprod Infant Psychol 1991; 9:3–18.
17. Berryman JC, Windridge K. Having a baby after 40:II. A preliminary investigation of women's experience of motherhood. J Reprod Infant Psychol 1991; 9:19–33.
18. Jeffries M, ed. You and your baby: pregnancy to infancy. London: British Medical Association, Family Doctor Publication, 1985.
19. Robertson J. Ethical and legal issues in human egg donation. Fertil Steril 1989; 52:353–63.
20. Abdalla H, Baber R, Kirkland A, Leonard T, Power M, Studd J. A report on 100 cycles of oocyte donation: factors affecting the outcome. Hum Reprod 1990; 5:1018.
21. Power M, Baber R, Abdalla H, Kirkland A, Leonard T, Studd J. A comparison of the attitudes of volunteer donors and infertile patient donors on an ovum donation programme. Hum Reprod 1990; 5:352–5.
22. Andrews LE. My body, my property. Hastings Cent Rep 1986; 16:28.
23. Burton G, Abdalla H, Studd J. Ethical problems of recruiting oocyte donors. Editorial. Br J Hosp Med 1990; 44:239.
24. Robertson JA. Embryos, families and procreative liberty: the legal structure of the new reproduction. So Cal L Rev 1986; 59:939.
25. Leeton J, Harman J. Attitudes towards egg donation of 34 infertile women who donated during their in vitro fertilisation treatment. J In Vitro Fert Embryo Transf 1986; 3:374–8.
26. Kirkland A, Power M, Burton G, Baber R, Studd J, Abdalla H. Comparison of attitudes of donors and recipients to oocyte donation. Hum Reprod 1992; 7:355–7.
27. Mahlsted PP, Greenfeld DA. Assisted reproductive technology with donor gametes: the need for patient preparation. Fertil Steril 1989; 52:908–14.
28. Robinson JN, Forman RG, Clark AM, Egan DM, Chapman MG, Barlow DH. Attitude of donors and recipients to gamete donation. Hum Reprod 1991; 6:307–9.

Chapter 20

Egg and Embryo Donation: Legal Aspects

D. J. Cusine

What is "Donation"

The title of this chapter contains the word "donation". In dealing with gamete and embryo donation, the Human Fertilisation and Embryology Act 1900 ("the 1990 Act") provides in Section 12 that no money or other benefit shall be given or received in respect of any supply of gametes or embryos unless authorised by directions, and by virtue of Section 41, anyone who contravenes that provision commits an offence. The Code of Practice (paragraph 3.29) provides that the Human Fertilisation and Embryology Authority (HFEA) will give directions on this matter. The words "or other benefit" should not be construed too literally. If a baby results and is regarded as a benefit, then you might, if the Act is to be construed literally, find yourself in prison!

Directions were given in July 1991. No donor may be paid more than £15 per donation plus reasonable expenses. Although it is clear that the object of the prohibition in the Act is to prevent a trade in gametes and embryos, the Act does not impose any parameters on the directions. An obvious example of a benefit is to reimburse travelling expenses which seems to be envisaged and is currently the practice in some centres for semen donors. However, the directions seem to exclude loss of earnings or other payment to recognise the time involved in being a donor and that there may also be some degree of inconvenience. In the case of ovum donation, there might be some discomfort because the procedure is more invasive than semen donation. Invasive treatments carry greater risks, and another issue is whether these greater risks are to be reflected in some form of compensation. However, having regard to the practice in drug trials, no such compensation will be paid. In any event, no establishment may give a donor more than was paid to donors in the year ending 31 July 1991 and that would rule out compensation.

Supply of Goods or Services

The next issue is what is being provided in egg or embryo donation. Is it a supply of goods or the provision of services or perhaps both? In any court action against a doctor or a clinic, it might be argued that the product, i.e. what was donated, was defective and that liability arises under the Consumer Protection Act 1987 which imposes strict liability for defective products or under the Sale of Goods Act 1979 which requires goods to be of merchantable quality and reasonably fit for the purpose for which they were intended. In some contexts, body products such as blood and semen, once separated from the body and in someone's control, are capable of being owned. That is clearly sensible, because otherwise, no offence would be committed by the removal of a blood or semen sample from the possession of the prosecuting authorities. It is unfortunate that the 1990 Act does not make matters clearer by declaring that gametes and embryos are not within the ambit of legislation dealing with the sale of goods or the supply of services. In the USA a blood bank has been held liable for the supply of defective blood [1]. Also in the USA in *Del Zio* v *Presbyterian Hospital* [2] the District Court of the Southern District of California appeared to treat an embryo as a chattel or piece of property, where the obstetrician in charge who was opposed to embryo research had destroyed an embryo. On the other hand, in the more recent *Davis* case in Tennessee [3] the Court at first instance clearly regarded the embryo as a person. The Directions issued on 23 July 1991 provide that suppliers of gametes or embryos, who are not donors, may be given their reasonable expenses for the supply. That possibly brings such supply within the ambit of the Consumer Protection Act (1987).

Other issues arise also. If an embryo is removed from one EC State to another, is that free movement of goods or the free movement of persons to which different considerations apply? Warnock recommended that there should be no right of ownership in a human embryo [4]. That is reflected in the Act by the fourteen-day period [5] which implies that an embryo, if it is to be regarded as a person at all, will not be so regarded until the expiry of the fourteen days and implantation. That still leaves the "ownership" issue unsettled. For example, if a storage facility was destroyed or materially damaged by fire, does the section of the insurance policy dealing with damage to property cover the embryos destroyed?

Selection of Donors

The Code of Practice (paragraphs 3-41–3-49) has detailed provision about suitability of donors, testing, and persons who are unsuitable. While it is outwith the remit of this chapter to say anything about suitability which is a matter of clinical judgement, and, while the same could be said of scientific testing, the Code contains two statements which require comment from a legal point of view. One is "Centres should adopt whatever is current best practice in the scientific testing of semen samples and of donors of gametes and embryos" (Paragraph 3.42). Is this current best practice set out anywhere, because the Code might be read as suggesting that it is? If it is not, the law regards the standard as that of a reasonably competent practitioner in that particular field and, of course, there may be a variety of different practices and it is important to bear in mind that the mere fact that there are several schools of thought does not mean that one school of thought is the best and that the others are unacceptable.

What is more important, however, are the provisions about unsuitable donors. The Code provides that the Centre should record the reason for rejecting a potential donor, the reasons should be explained to the donor and he or she should be given assistance in appropriate cases about treatment and counselling (Paragraphs. 3.47–3.49). My question is "What is the point of the Centre recording this information but not being obliged to pass it on to HFEA so that other Centres might have access to it"? If the information is to be supplied to HFEA, there would have to be an obligation posed on all Centres to check with HFEA on all potential donors and it may be that the benefits to be gained from that far outweigh any inconvenience. A donor who is rejected by one Centre may conceal information from another Centre and be accepted as a donor. If there were a central register of unsuitable donors, such a scenario might be avoided. It is perhaps worth noting that both the Act and the Code of Practice are silent, for example, on the use of children (i.e. those under 16) as donors, or the use of prisoners. There should be no objection at common law to using either provided they are capable of understanding the risks involved and, particularly in the case of prisoners that their consent is freely given.

The obligation to disclose the information to HFEA may have been omitted from the Code and the legislation because of the general principle that Centres should maintain confidentiality. The provisions in the Code are, however, somewhat odd and they are odd because the Act is likewise peculiar. The Act deals with confidentiality in Section 33 and the most important part is Sub Section 5 which states "No person who is or has been a person to whom a licence applies and no person to whom directions have been given shall disclose any information falling within Section 31(2) of this Act which he holds or has held as such a person". The information under Section 32 is (a) the provision of treatment services for any identifiable individual, or (b) the keeping or use of the gametes of any identifiable individual or of an embryo taken from any identifiable women, or if its shows that any identifiable individual was, or may have been, born in consequence of treatment services. It follows, therefore, that the Centre is not permitted to tell the patient's general practitioner (GP) that she has been treated, nor that she is pregnant. Had the Act not contained this bizarre provision, the matter would have been dealt with perfectly adequately at common law which sensibly would not regard the disclosure of information by the GP to the Centre or by the Centre to the GP as a breach of the patient's confidentiality. That would be either because the law would regard such disclosure as being between individuals who are each bound by a duty of confidentiality towards the patient, or that the disclosure was essential for the proper treatment of the patient, or that the patient had consented to the disclosure by being referred by the GP to the Centre. This point is mentioned for two reasons. One is to indicate that the common law is sufficient and the second is because it is quite clear that at common law a patient may waive confidentiality. The Act does not seem to exclude waiver by the patient and so it would seem appropriate that the consent form used by Centres should authorise disclosure to the patient's GP and the outcome of any treatment. It is difficult to agree with the provisions in the Code of Practice (paragraph 3.6) which suggest that the Centre should write to the patient and ask the patient to reveal the information by passing on a copy of the letter. If the above suggestion was adopted and the patient for some reason did not wish the information to be disclosed, that part of the consent form could be deleted.

The Act is rightly concerned about breaches of confidentiality, but one vexed issued which it does not address is whether an obligation to a third party may permit, or even require, confidentiality to be breached. If, for example, a child born following donation suffers from some inheritable or transmissible disorder, is the doctor under an obligation to report that to the donor and anyone else who may be at risk? This could be left to the common law, although it would have to be made clear that the common law seas here are largely uncharted.

The Status of the Child and What To Tell It

The status provisions in the 1990 Act are clearly of great significance because the status of children conceived as a result of artificial procreation has for too long remained uncertain. The 1990 Act deals with mothers, fathers and posthumous children. Children conceived by donor in semination and by more recent artificial reproductive techniques raise the issue of what to tell the child about its origins.

1. Mothers: Section 27 provides that the carrying mother of a donated egg or embryo is to be regarded for all legal purposes (except succession to titles, honours or dignities) as the mother of the child, irrespective of the child's genetic origins.
2. Fathers: Section 28 provides that where a married woman gives birth to a child following egg or embryo donation, her husband is to be treated as the father of that child unless he can show that he did not consent to the treatment. The Section also deals with the case where a married woman receives treatment along with someone other than her husband and her husband does not consent or where an unmarried woman seeks treatment with a man. In each case that man is to be regarded for all legal purposes (except those already mentioned) as the father of the child.

Because of the implications arising from consent, the Code of Practice emphasises the need to obtain written consent from the woman and her husband if she has one (Paragraphs 5.6–5.8).

One can readily see that if a husband does not consent to his wife's treatment, the subsequent child may be legally fatherless. The Act provides that if the donor's egg or embryo is used in accordance with the consent which has been given then he or she is not to be regarded as either the mother or the father of the child and that the parent of the child is the carrying mother and, if she has one, her husband will be regarded as the father. If, therefore, the gametes or embryos are used in accordance with the consent the donor is not the father, but if the husband of the wife does not consent he is not the father and so the child has no legal father.

Posthumous Children

The policy behind the Act (Section 28) is to discourage posthumous procreation and it does that by making the child illegitimate and legally without a father if procreation takes place after the death of the husband. That would not be the position, however, if, in the interval, the woman had remarried. In that case, there would be a common law presumption that her husband was the father of the child and, if he consented, he would be the father in terms of the Act. It is perhaps unfortunate that the Act has taken this absolute stance in relation to posthumous children and given that the law has, on a number of fronts, attempted to do away with the distinction between legitimate and illegitimate children, it is perhaps strange that the Act continues to insist upon making the distinction in certain instances.

What To Tell the Child?

This is not a matter of law; there is no legal obligation to disclose the child's origins to it and likewise, no legal obligation not to disclose. This issue is dealt with in another chapter.

Legal Action Against the Doctor/Clinic

It will be obvious that any doctor/clinic which disregards the prohibitions in the Act may be liable criminally. There is also the possibility of civil action against the doctor which may be based on negligence, for example arising from an alleged failure to screen a donor or otherwise to reject an unsuitable donor. Even if the donor is adequately screened, there is a possibility that the legislation mentioned earlier may be used against a clinic if the gametes or embryos supplied were alleged to be defective. It may be that the supply of gametes or embryos is not covered by the Consumer Protection Act and even if it is, the defences in the Act may be adequate to protect the doctor or clinic but it is unfortunate that the matter has not been put beyond doubt by the 1990 Act.

One form of civil action which may become more common in the future is an attempt to review a decision by a clinic to reject a donor, or more likely to review a decision which has resulted in the rejection of a recipient or a couple. This is perhaps more likely to arise in connection with the application of Section 13(5) of the Act which states that "A woman shall not be provided with treatment services unless account has been taken of the welfare of any child who may be born as a result of the treatment (including the need of that child for a father), and of any other child who may be affected by the birth". If, therefore, a recipient or couple was rejected on the basis that there would be some risk to the welfare of the subsequent child, the potential recipient may attempt to have that decision reviewed and reversed. In a case involving St Mary's Hospital, Manchester [6] a woman who had been a prostitute believed that she had been excluded from continuing treatment services when the clinic discovered her history. She challenged the decision to exclude her and also the advice of the local Ethical Review Committee which had concurred in the decision of the clinic in withdrawing the services. Her claim was dismissed by the judge, but he held that, while the Local Ethical Review Committee was not under a duty to investigate her case before giving its advice and that it was no more than an informal body whose function was to provide a forum for discussion, the Court could nevertheless review the policy of an Ethics Committee. He was in no doubt that the Committee would be required to give lawful advice and further that a decision would be reviewed if it was one which no reasonable Committee could have reached in the circumstances.

There was a similar case in France involving Alain Parpalaix [7] who was warned that treatment for testicular cancer might cause infertility. In advance of the treatment, he deposited his sperm with the National Sperm Bank (CECOS) and later married the woman with whom he had been living. After his death, she sought the Court's assistance in forcing CECOS to deliver his sperm to her and in the action against CECOS, the Court ordered that the sperm should be handed over to the widow because it held that that was her dead husband's intention. Furthermore, the Centre had not made it clear that it would refuse to hand over the sperm and the Court went on to say that there was no legislation or other regulation which prevented the request being made and, since one of the aims of marriage was procreation, there was nothing in natural law prohibiting the sperm being given to her.

It is clear that in terms of Schedule 3, and in particular paragraphs 5, 6 and 7, the consent of the donor or donors is paramount. It may be that a donor would wish to donate on condition that certain use was not made of his or her gametes or embryos, for example, that they should not be implanted to anyone of a particular race. It is perhaps unlikely that that direction would be successfully reviewed by a Court, but clinics would have to look very carefully at their policies and be able to justify the exclusion of particular groups within society. It would not be possible, for example, to exclude persons on the grounds of race or sex without there being other reasons and it would be advisable and indeed proper for the clinic's policy on recipients to be set out quite clearly. While, therefore, a clinic's decision never to treat single persons or lesbians might be subject to successful judicial review, a decision in a particular case not to treat a lesbian might not be the subject of a successful judicial review, if the clinic's policy was to treat, say, in the first instance married couples, secondly, those in a stable family relationship where the partners were of different sexes, thirdly single persons and fourthly lesbian couples. If the clinic could establish that there were a sufficient number of persons within the first two categories who were potential recipients, a decision not to treat a lesbian or a single person would probably be upheld.

Consent to the Storage and Use of Gametes

The provisions of the Code of Practice (paragraphs 5.9–5.16) provide, reflecting the Act, that anyone who consents to the storage of gametes or embryos must state what is to be done with them in the event of the donor or donors dying or becoming incapable of varying or revoking the consent. It goes on to say that if the intention is to donate the gametes for the treatment of others, the donor must consent in writing to that use. That is clearly a sensible provision, but one is left wondering whether the storage authority must comply with the directions which are given. If, for example, a couple were to store their embryo and provide that in the event of their failing to use it, it should be passed on for the use by any of their children who might turn out to be infertile, both the Act and the Code of Practice are silent on the issue of whether the Storage Authority must comply with that request. That also raises the possibility that directions given by donors might conflict with the Storage Authority's own policy in relation to suitable recipients.

Conclusion

The 1990 Act is a commendable attempt to achieve a balance between, on the one hand, the interests of those involved in infertility services to improve their practices and develop their research, against, on other, the concerns in some sectors of the public that these activities be carried on in a way which the public would regard as acceptable.

The Act, however, does pose a number of problems, the main ones being (a) that there is no requirement to maintain a central register of unsuitable donors; (b) that some of the provisions on confidentiality are unduly restrictive, and (c) that the status of the embryo has been left in doubt, a doubt which ought to have been removed.

That said, practitioners should not have much difficulty in operating within the Act and the code of practice. However, even if they do comply, that does not exclude the possibility of their being involved in legal action. Negligence may give rise to litigation and it may be that someone who is rejected for treatment may seek to have that decision reviewed by the courts.

References

1. Cunningham v MacNeal Memorial Hospital 266 NE 2d 897 (1970); Belle Bonfils Memorial Blood Bank v Hansen 579 P 2d 1158 (1978).
2. 74 Civ. 3588 (SD, NY) (1976).
3. 15 FLR 2097 (1989).
4. Report of the Committee of Inquiry into Human Fertilisation and Embryology Cmnd 9314 (1984) para. 10.11.
5. Section 3.
6. R v Ethical Committee of St. Mary's Hospital (Manchester) ex parte Harriot [1988] 1 FLR 512.
7. Parpalaix v CECOS (unreported).

Discussion

Templeton: I have some difficulty with the study looking at replacement techniques in women aged over 40 years because I understand that the replacement techniques were dictated by the underlying condition causing the infertility. There are two variables, yet Professor Asch has described the assessment of only one of those variables.

Asch: That is a very important variable. However, if anything it would work in favour of patients that receive intrauterine transfer.

At the very beginning of the history of ART, IVF, GIFT, etc., it was thought, and documented at the time, that patients with severe tubal factors got pregnant more easily and would have better pregnancy rates. Later that proved not to be the case, like many other things in our specialty. The last two registries in the United States, and the one in Australia, and now I believe in the UK, show that despite the aetiology of the infertility, whenever we place embryos the results are the same.

To answer the question, I do not think that that variable affects the results that I showed. If anything it should be in favour of IVF.

Templeton: But we do not really know?

Asch: I do not know.

Hull: One of the intriguing issues is if the tubal environment confers any benefit, does it confer the benefit at the initial stage of fertilisation or the later cleaving stage. We have a well-controlled study from Bristol about to be published showing that there is a significantly better implantation rate with GIFT than with embryo transfer to the uterus.

One of my difficulties with the data from California was that in the two studies Professor Asch reported, one well controlled and the other the comparative study in the over-40s, they were transferring at a different stage to the tubes. In the well-controlled study in women with primary ovarian failure, they were transferring at the cleaving stage, whereas in the over-40s it was ZIFT.

Fishel: I have a difficulty when I only see mean numbers of embryos, or eggs if it is being compared with GIFT. Are there any data that compare equal numbers of embryos being transferred per treatment?

Second, Professor Asch gave virtually all the reasons why the donation recipients are much better than the IVF recipients but I wondered whether transient hyperprolactinaemia has been considered and ruled out as a problem.

Asch: We have not studied transient hyperprolactinaemias and I cannot answer. We shall need to look at it in the future.

I do not think I understood the first question. An implantation rate is not a mean.

Fishel: Although one may have an overall mean in a group of patients, it does not necessarily give us a handle on the implantation rate per embryo, or per patient given an equal number of embryos with particular technique.

Asch: The way we calculate implantation rates is by the total numbers of embryos transferred among all the patients and the numbers of sacs seen by ultrasound.

Fishel: That is literally implantation. If we talk about pregnancy rates, the mean may be fairly similar but more patients may be getting four embryos in one technique and with GIFT there may be six oocytes going back. I wondered if the comparisons for pregnancy rates were equivalent.

Asch: I do not know if I compare as Dr Fishel describes but I think they are well compared. The numbers of embryos or eggs transferred are very similar.

The problem arises when we talk about implantation rates and add GIFT into the equation. It is not possible to determine the implantation rate per egg. We can be sure that not every egg we put in will turn into an embryo.

Fishel: That is exactly the point.

Asch: Sure: one should correct for that. I am sure that the fertilisation in vivo is not better than the fertilisation in vitro, in the dish. In fact I think it is worse.

Shaw: We still have problems with implantation rates per embryo when we cannot assess embryo quality in terms of replacement. The data are still fuzzy.

Sheldon: Because there are possibly other confounding factors – which might be underlying disease but also several other things – without randomisation, or without every close matching for every conceivable confounding factor, it is really very difficult to interpret a number of those differences.

Asch: That is absolutely right. But let me make one point. Most of the progress we have seen these past few years in reproductive medicine has come from non-prospective studies.

Sheldon: But even with randomisation, with such small numbers no one can guarantee they have got equivalence. The point about means illustrates that when the numbers are so small one might have to think about matching and trying to enforce more comparability. Unless one does that, the *P*-values do not mean much, because the statistical analysis is based on the assumption that some form of randomisation matching was done.

Lastly, the numbers issue. In some of the trials there are quite large differences, but we are told that they are not significant and no one has the power to detect them as significant. We need to be careful. It is something that has to be improved on in the future. It is really difficult to interpret what is happening.

Asch: We should. But the statisticians have to understand that we have limitations.

Shaw: We are all well aware of that. We know that to get a 5% improvement in pregnancy rates probably needs 1000 patients in each arm of the studies for IVF. No single centre can achieve that.

Winston: If, when someone is using a particular technique, they are much less than averagely successful, and they then perceive an improvement in their success using a difference to that technique, can any conclusions be drawn from that?

Sheldon: Not necessarily.

Winston: That is another problem that I have. The implantation rates with Professor Asch's IVF programme are half what we would observe in ours, and I would have thought therefore that we cannot easily interpret the results from that point of view either.

Sheldon: One could be getting regression to the mean as well. There may be variation between centres. It is really difficult.

Shaw: We all accept the problems of statistics in this area.

Baird: In that last study there were no abortions at all in the surrogates. That must be a statistical fluke. It would be most unlikely to have a situation where there were no spontaneous miscarriages.

Asch: In the general population, yes, but this is not the general population. These are selected women.

Baird: Where are those eggs coming from? The majority of spontaneous miscarriages, perhaps 75%, are due to chromosome abnormalities, and the vast bulk of those abnormalities reside in the oocyte. What Professor Asch is saying is that somehow or other they have selected out oocytes with no chromosome abnormalities. That would seem *a priori* quite unlikely.

Asch: The data on the percentage of early miscarriages that are due to chromosomal abnormalities are very debatable.

Several: No!

Winston: Absolutely not.

Braude: It is an average figure. The incidence of miscarriage in primagravidae or in women who have not miscarried before is extremely low; it is about 5%.

Asch: But one is not putting in one embryo!

Braude: I accept that it is a statistical fluke and not the norm. It could have been 1 in 10, but it could be lower than that.

Shaw: Let us leave statistics and move on to social aspects of oocyte and embryo donation.

Dickson: Mr Abdalla said very clearly that eggs are in short supply. So why treat the menopausal woman, the 40–50-year-old?

Abdalla: Eggs are in short supply but several of these patients provide the donors themselves.

It is a question of policy. Do we say that anyone beyond a certain age does not have the same rights to get eggs? It is true that eggs are in short supply and it could be argued that the younger woman should get eggs more quickly. But the other side of the argument is that the older woman – someone of, say, 46 years – needs to become pregnant more quickly and the younger woman can stay longer on the list. We see both arguments and the only way we can deal with it is to allow them on the list.

A good number of the donors are coming in because the patients are campaigning for them.

Shaw: But there is age restriction for other donation, e.g. kidney transplants. Any patient who is over a certain age will not get into a kidney transplant programme. So it is being applied in other areas of medicine.

In most of these oocyte donation programmmes there is a financial barrier in the first instance. Only those who can afford it are getting into it. So there is another preselected group.

Inglis: Are there figures on the uptake of counselling by these couples and the number of sessions that they have before they go into treatment?

Abdalla: About 80% of all patients. They are always advised at the consultation.

Inglis: Advice is not counselling.

Abdalla: No. They are normally seen by the doctor and by the egg donation coordinator, who herself is a trained counsellor. After that they have an appointment made for them to see our independent counsellors. Two independent counsellors come to the clinic twice a week and appointments are made with them.

About 70% of our recipients opt to see the counsellors. They normally have one session with the counsellors but some of them opt for more sessions.

Inglis: It is my experience that it takes at least three sessions before couples can engage with the psychological implications. It is quite difficult to reflect on some of them. My concern is that they may be going into treatment without giving informed consent.

Abdalla: I could not agree more. I wanted to stress that these patients are extremely vulnerable and the only way one can counsel them is with time, seeing them again and again and again.

Inglis: But most of them have only one session.

Abdalla: The problem is we cannot force them to come for counselling. We cannot say they must come again, and again, and again.

Inglis: But it does depend how it is presented, whether it is taken up and used.

Abdalla: We cannot do more.

Shaw: We can come back to this when we discuss the role of counselling later. It is important for both recipient and donor to be adequately counselled.

Sheldon: There is a natural variation in the age at which women have babies. It is suggested that within the righthand tail of that distribution there are the women aged 45–50 years who would have a baby naturally and therefore the technique should be able to reproduce the whole spectrum of that distribution. That is what was implied. Can we not go later and later?

Abdalla: No.

Sheldon: But one of the points raised was the justification that since it happens naturally, why can we not do it this way. I think we cannot apply that. I cannot think of many examples where we would apply this reasoning in other clinical areas. Because someone else is more beautiful would one do cosmetic surgery so that the patient is the most beautiful? Would we intervene in terms of any form of surgery so that someone lives the longest? Or do we somehow have some idea of what is normal and acceptable?

Guidelines could perhaps be taken from adoption. I see parallels between adoption and donation and I wonder whether some of the guidelines around adoption could be incorporated into this field.

Abdalla: I did not say we should extend beyond the age of the natural menopause. I did not try to put it in those terms. I was talking of the 45–50 year age group rather than beyond that. Of course we are providing a greater chance of pregnancy in that situation.

There are several reasons why we might refuse. In general we do not like to take on any woman aged over 45, not only because of the rights and wrongs of her becoming pregnant, but also because the chances are very low. In that situation we are providing a better chance. In my opinion if we deny this group of women the ability to become pregnant, we should campaign in society that others of a similar age should not. But if they have that potential to become pregnant, by whatever method, then should we provide the opportunity or not? I cannot see that we can say no, that she is 45 and she cannot become pregnant because it is wrong.

Shaw: This is where we overlap into the issue of the welfare of the child. Are there any good data on the outcome for children that are born into families whose mothers are perhaps 50 and the fathers 60 or 70, on how they progress into their teens and adulthood compared to other people?

Abdalla: I got what I could from the literature but the only studies I found were in women who had become pregnant after age 40. I incorporated some of the conclusions of these studies.

Shaw: They are often presumably with other children in the family.

Abdalla: Half of them were first-time mothers.

Sheldon: It is not about judging the mother. It is a more general issue of clinical intervention and when one decides it is acceptable to intervene. Does one intervene in order to get everybody up to the righthand end of that distribution?

Drife: Is this a big problem? We are not likely to be facing a stampede of 500 000 women deciding to have their babies at age 50 rather than 30.

Winston: It is a problem.

Drife: It is perceived as a problem in specialist clinics in the capital city of a country with 50 million people but I do not believe that it is a big problem.

Winston: We are slightly at risk of treating a biological condition with a medical technology, and I wonder if that is wise. There is no doubt that women aged over 40 years are biologically infertile and that this becomes increasingly severe.

The other point is the question that Professor Cusine raised, that we have to be careful to do things that are seen to be acceptable by society in general. There is some evidence, loose evidence, that a very large proportion of people feel threatened by the idea of treating women much over the age of 40 years for infertility. It is just an observation and I do not think we can make hard rules, but we have to be very cautious.

Shaw: Professor Cusine's main concern related to confidentiality has been well recognised as the major problem in terms of clinical functioning of the HFEA Act and is being addressed, or there are hopes that it will be addressed in some way, with amendments. But they are not yet with us. The thoughts on the Consumer Protection Act and other issues that perhaps have not been thought about in this way are intriguing. There may be concerns too in terms of the embryologist's role and whether people can be sued because of failure to fertilise or failure to get conceptions.

Drife: To reiterate the point I was trying to make. Today (9 April 1992) is a day when the running of the country is being left to the commonsense of the non-experts and in formulating laws we have to try to bring our commonsense into laws. I sometimes sense around a table like this an urge to make decisions on behalf of patients because we as experts know best. But there are two considerations. One is an ability to delegate to the commonsense of the people, which is what I was trying to get at by asking whether we should be setting an upper age limit. Patients themselves will set an upper limit by not wanting the treatment. The other is when formulating laws to apply commonsense. The law on confidentiality becoming ridiculous because it was logical but not commonsense is an example.

Fishel: I was interested in Professor Cusine's hypothetical surmise and I wondered how far it could be taken in terms of protection against litigation. If things go wrong there is clearly the possibility of litigation against the donor, against the practitioner, or maybe the clinic, and ethics committees might be in trouble. And how well protected is the HFEA?

Cusine: I suspect the HFEA is not in the firing line except in one sense, and this has been mentioned. If the HFEA went beyond the strict limits of the legislation, then some brave soul might seek to have their activities reviewed judicially. I would suggest that that should not be done concurrent with an application for a licence!

It is unlikely that local ethical committees would be subject to legal action. Ethical committees simply give advice on some research project, and unless they went completely off the rails, which is highly unlikely, they would not be sued in that respect. But there is every prospect that some individual doctor, or a clinic, could be sued and the way to avoid that is for centres to draw up guidelines which meet not just with the approval of the centre but with the approval of a larger group of individuals. It is not foolproof, but one of the ways to protect oneself against medical negligence is to point to the fact that what has been done is in accordance with procedures which are generally accepted.

Fishel: Which is my point. Presumably if one does work within these generally accepted procedures the chances of litigation are small, unless the whole body itself is at risk of litigation.

Cusine: They are small, but they are not entirely impossible. For example, in a survey in the States about donor insemination, something like 95% of those who were asked would reject a donor who had Tay–Sachs disease yet only 1% actually did the test. Someone being sued for omitting the test could say the other 99% were also doing this, but the obvious answer is that that was just not good enough. While there is safety in numbers, it is not absolute safety.

Barlow: In the days of the ILA meetings for centres where various questions on legal aspects were raised, it was always very unclear what the status was of those who practised fetal reduction. Is the situation any clearer in 1992 than it was those few years ago? Do those who remove one or two fetuses of a multiple pregnancy have any protection from the Abortion Act or from other legislation, or are they very exposed?

Cusine: I do not know the answer to that.

Shaw: It was not within Professor Cusine's remit.

Cusine: There is a very good argument that selective reduction, looking at the strict terms of the Abortion Act, may not be an abortion at all.

Adballa: There was talk of central registration of donors. When we reject them, do we have to give their details to the Authority? My understanding is that they are centrally registered and we do send in information about them.

Whittall: Information is sent in about those who are accepted but not the others.

Cusine: To me it seems bizarre that someone who is regarded as unsuitable is not centrally known.

Whittall: The intention within the Code of Practice was not so much to mark up that individual for anybody else to become aware of him, but rather to ensure that in making that kind of decision, the centre had got fairly rigorous procedures or had clarified its own idea of what it regarded as a suitable or an unsuitable donor.

Cusine: Which re-emphasises my point. If the centre uses a regular procedure and as a result of that has come to the conclusion that X is not suitable, then it is highly likely that another centre would come to exactly the same decision. But, they do not have access to the fact that a centre somewhere else has said X is not on our list.

Templeton: There is a problem. Donors are registered centrally with their permission and to ask permission of the donor who has been rejected might be difficult.

Cooke: Professor Drife puts the onus on society to decide what the standards might be and I think that is remarkably premature. The characteristic of society at large is that it is ignorant, and it is largely ignorant because we have not presented the issues as clearly as we might, particularly in the area that we are discussing. A democratic process brings out over-representation of pressure groups and the group that is defined as society is unrepresentative. There is usually a stack of individuals who are grossly biased against the whole procedure, not against the subtleties of individual points, so there is a huge inbuilt bias.

The argument about the righthand end of the distribution is irrelevant because in medical practice the people who present with symptoms nearly all lie at the righthand end of the distribution. That is true with patients in this area. If we had a larger proportion we would not cope. We only have the demand now because it is a relatively selected group at the extreme end of the distribution.

Much of the regulation, for example with respect to the number of gametes (and it is more relevant to sperm than it is to eggs, but the same issue would arise if we ever get to the egg in vitro development and maturation programme) is that philosophical issues are put forward for limiting the number of gametes without any real discussion of practicalities of providing the service. What happens is that "society" makes a decision about the philosophy without any understanding whatever about the practicalities, and there is a big divide there that needs to be bridged.

Kerin: In principle a central registry for suspect donors, whether they be sperm donors or egg donors, might have some virtue, but in practical terms it may be difficult to set up. We already have problems recruiting donors and it might be offputting, because they have to be aware of this problem. And how would it be made foolproof? The donor who is rejected and who is determined to get on to a programme could easily change his or her name, and there may need to be some sort of an Interpol, or a fingerprint register, or sperm DNA profiling, to make it workable.

Cusine: There is no way that we can legislate for ingenuity, I fully accept that. But there is some benefit to be had in recording centrally the fact that somebody has been declared unsuitable. If they do have the wit to change their name, and conceal features, and so on, I agree we may not be able to pick them up. But it would be very unfortunate – and I can see litigation arising out of this – if Centre X knew about a particular donor, had rejected that individual, and he or she walked down the road to another centre, some horrendous result came from the use of that donor, and then someone asked why that individual had been used when Centre X had rejected him for perfectly sound reasons. If it is not practical, then by all means dismiss it.

Shaw: Knowing the problems with credit ratings and suchlike, this is much more likely to generate litigation than anything else.

Chapter 21

Embryo Micromanipulation

A. Dokras, I. L. Sargent and D. H. Barlow

Steady advances in the in vitro fertilisation (IVF) technique since the birth of the first baby [1] have now established it as a form of treatment for a variety of causes for infertility. With improvements in pregnancy rates [2] the number of embryos transferred to the uterus at any given time has been reduced to three or even less. Despite this trend, induction of superovulation continues in order to allow selection of embryos for transfer. This has provided embryologists with research material to develop techniques for the micromanipulation of human gametes and embryos. The application of some of these techniques, which are widely used in animal husbandry, has further widened the scope of human IVF and its ability to result in successful pregnancies. The manipulation of human embryos for various applications has been reported: one or more cells can be removed for preimplantation genetic analysis [3,4]; an artificial slit made in the zona pellucida has been shown to increase the rate of blastocyst hatching [5] and implantation [6]; and removal of the extra pronucleus after polyspermic fertilisation could potentially return the embryo to a genetically normal state [7,8]. In addition, the use of micromanipulation for germ line gene therapy and to induce artificial twinning is a possibility. Of the above applications, single cell biopsy for preimplantation diagnosis [9] and slitting the zona pellucida to improve implantation rates [6] have already resulted in the birth of live, normal babies.

Despite this preliminary success, it is critical to evaluate the effects of the artificial conditions that micromanipulation imposes on the embryos. The safety of the entire procedure as assessed by in vitro and in vivo development needs to be unequivocally demonstrated. There have been studies to establish animal models for preimplantation diagnosis [10–12] and assisted hatching [13], but the direct extrapolation of these results to the human is not always reliable. After the efficacy of a technique has been established in an appropriate animal model, critical evaluation of the approach using donated human embryos also needs to be carried out. This chapter will discuss some of the established and potential applications of embryo manipulation and address the problems which may have consequences for the safety of the procedures.

Preimplantation Diagnosis

The Need for Preimplantation Diagnosis

In couples with autosomal or X-linked genetic disorders, there is a 25%–50% chance of producing an affected child. This risk is the same in each pregnancy and unrelated to the number of previously affected pregnancies. Using techniques in the prenatal period (chorion villus sampling and amniocentesis), diagnosis can be available only after the pregnancy is established (10–20 weeks gestation). As a result, couples with a high risk for genetic disorders may be repeatedly posed with the difficult decision – whether or not to terminate the pregnancy if the fetus is found to be affected. This could lead to significant psychological and physical trauma. Moreover, some may find termination of established pregnancies unacceptable on moral and religious grounds. The problem is further aggravated in 10% of these couples who are also infertile [14] or in those who already have an affected child. Therefore, many women decide not to continue to try to have a normal child rather than submit to the trauma of repeated abortions.

The possibility that preimplantation diagnosis could be performed was first suggested by Edwards and Gardner [15], who, in 1968, successfully sexed rabbit blastocysts by examining cells removed from the trophectoderm layer for the presence of sex chromatin [16]. If the diagnosis could be made in humans in the preimplantation period only the "normal" embryos could be transferred to the uterus, leading to the establishment of a "normal" pregnancy. This would overcome the drawbacks associated with repeated termination of pregnancies. Moreover, it would allow couples to be committed to a pregnancy from the very outset, as they would be reassured about the diagnosis in the preimplantation period. The advances in in vitro fertilisation and the ability to micromanipulate human embryos in the early cleavage stages, have now made it possible to obtain cells for diagnosis during this period. In addition, the development of techniques for analysing gene sequences in very small quantities of DNA [17] and microassays for biochemical analysis [18,19] will allow the diagnosis of inherited diseases from the biopsied cells.

Biopsy Methods for Preimplantation Diagnosis

A number of techniques have been described in animal models for obtaining cells at the early developmental stages of the embryo (Table 21.1). The selection of a technique would depend on its minimal effect on in vitro and in vivo development and the possibility of obtaining unequivocal results using the biopsied material. Broadly, the biopsy can be performed at one of two stages: either before differentiation has occurred, when all the blastomeres in an

Table 21.1. Different approaches described for obtaining material from preimplantation animal and human embryos for genetic analysis

Biopsy method	Animal model	Human model
1. Polar body removal	Gordon and Gang [20]	Monk and Holding [21] Verlinsky et al. [22]
2. Separation of blastomeres from a 2-cell embryo	Tarkowski and Wroblewska [23] Epstein et al. [24] Willadsen [25] Nijs et al. [26]	
3. Removal of 1–2 blastomeres from a 4–8 cell embryo	Monk et al. [28] Monk and Handyside [27] Wilton and Trounson [29] Krzyminska et al. [12]	Hardy et al. [30]
4. Bisecting the embryo at the morula or blastocyst stage	Ozil et al. [32] Ozil [33] Van Blerk et al. [35] Wang et al. [34]	
5. Biopsy from the trophectoderm layer at the blastocyst stage	Gardner and Edwards [16] Monk et al. [10] Summers et al. [11] Nijs and Steirteghem [36]	Dokras et al. [4,5] Muggleton-Harris and Findlay [37]
6. Aspiration of cells from the blastocoelic cavity	Hartshorne and Avery [38]	

embryo are totipotent and removal of one or more blastomeres will not affect further development, or after differentiation has occurred, when removal of cells from only the extra-embryonic region will not affect further development.

The use of the first polar body to genotype the oocyte before fertilisation without affecting its further development, has been described [20–22] (Table 21.1). In the absence of crossing over, the first polar body will be homozygous for the presence or absence of the allele to be detected. If crossing over occurs, the first polar body will be heterozygous. Therefore the oocyte will be selected for insemination only if the first polar body is homozygous for the abnormal allele. The second polar body has the same genetic composition as the secondary oocyte and could therefore be used to type the secondary oocyte. However, both these approaches will offer information of the maternal genome only, whereas samples obtained at the embryonic stage will be fully representative of the embryonic genome.

At the embryonic stage, one approach in animal models has been to separate blastomeres at the two cell stage, using one for analysis, the other being transferred to the uterus after the diagnosis [23–26] (Table 21.1). It is evident, however, that manipulation at such an early stage of development impairs further in vitro and in vivo growth in the mouse [26]. This approach has not yet been applied to human embryos.

A second approach involves the removal of one or two totipotent blastomeres from a four- or eight-cell embryo, respectively [12,27–29] (Table 21.1). The blastomeres removed can then be used for diagnosis and the remaining three- or six-cell embryo is transferred to the uterus. It has been demonstrated that this technique does not impair in vitro or in vivo development in mouse [12] and human embryos [9,30]. However, the main drawback in all these biopsy methods is the limited material available for diagnosis. The accuracy of any diagnostic technique using only one or two cells has been questioned [31].

In farm animals, embryos at the morula or early blastocyst stage are split to produce monozygotic twins with a high rate of successful transfers [32–35] (Table 21.1). Although bisection of the mouse embryo at this stage does not affect implantation there is a significant reduction in the ability to form a fetus [34]. This is due to the absence of egg cylinder development, which correlates with the reduced number of cells in the inner cell mass (ICM) of the bisected embryo. Therefore, even though this technique would offer more cells for diagnosis, the removal of embryonic tissue after differentiation has occurred in human embryos, could be detrimental.

Alternatively, it is possible to remove only extraembryonic cells from the trophectoderm layer of the blastocyst [4,5,10,11,16,37] (Table 21.1). The trophectoderm eventually forms the placenta and does not contribute directly to the formation of the fetus. It has been demonstrated that the removal of a few to several cells from this region does not impair further development and implantation in the marmoset [11]. Furthermore, the increase in the number of cells available for analysis will increase the accuracy of the diagnostic procedure.

The possibility of using cells obtained by aspirating the blastocoelic fluid has been reported [38]. However, it has been suggested that the blastocyst may shed "abnormal cells" into the cavity during development [39] and hence make interpretation difficult.

Single Cell Biopsy: a Feasible Approach?

The feasibility of removing one or two blastomeres from a six–ten-cell human embryo as an approach to preimplantation diagnosis has been reported [3]. The effect of the biopsy procedure on the further in vitro development of the embryos was assessed by the rate of blastocyst formation and hatching, measurements of pyruvate and glucose uptake and cell numbers in the trophectoderm and ICM at the blastocyst stage [30]. The nutrient uptake of biopsied embryos decreased in proportion to the number of cells removed. The total cell numbers were reduced especially on day six, but the ratio of ICM to trophectoderm cells was maintained. Subsequently, 22 embryos diagnosed as "female" were transferred to the uterine cavity of eight women on day three [9]. Five of these women got pregnant and a chorion villus biopsy was performed in all ongoing pregnancies at 10 weeks' gestation. One pregnancy was terminated following the diagnosis of a male fetus instead of a female. Further details will be presented in Chapter 22.

The main criticism of this technique is that the diagnosis is based on a single cell, which could be an anucleate cytoplasmic fragment or the second polar body instead of a blastomere. The lack of replicate samples limits further analysis if the results obtained are equivocal, if technical failure occurs or if validation of a diagnostic error is required as in the case mentioned above. On the other hand, since IVF embryos are currently transferred to the uterus on the second or third day after insemination, this approach allows the screening of embryos within this time period.

Blastocyst Biopsy

Trophectoderm biopsy at the blastocyst stage has several advantages. Most important is that at this stage the embryo undergoes differentiation and reaches a maximum cell number prior to implantation, and the majority of these are trophectoderm cells [40]. Therefore, removal of cells from the trophectoderm layer would increase the sample available for genetic analysis and hence provide for the loss of material during preparation and allow duplicate analysis. Second, trophectoderm biopsy can be likened to a chorion villus sampling at a very early stage of development since the trophectoderm cells removed would normally differentiate to form the placenta whereas the inner cell mass develops into the embryo proper. The biopsied material is therefore extraembryonic, unlike a single cell removed from a four–eight-cell embryo. Hence there should be fewer ethical objections to this technique. Third, if the biopsied trophectoderm cells can be induced to proliferate in vitro as has been demonstrated in the marmoset [41], preimplantation diagnosis for chromosomal disorders by karyotyping should be possible. The drawbacks of this approach are discussed later.

Trophectoderm Biopsy Technique

The technique described here involves the removal of cells from the mural trophectoderm of human blastocysts [4]. A flame-polished holding pipette (~50 μm inner diameter) was used to immobilise the blastocyst in a microdrop chamber for manipulation (Fig. 21.1a). A siliconised microneedle was then introduced through the zona pellucida opposite the inner cell mass so that its tip entered the perivitelline space. The needle tip was then moved over the trophectoderm without penetrating it and pierced back through the zona, so that an area of the zona was "trapped" between the two points pierced by the needle (Fig. 21.1b). The human zona offered more resistance as compared to the mouse zona. The size of the "trapped" area was standardised to less than one-quarter the circumference of the blastocyst. The blastocyst was then released and the lower end of the holding pipette was rubbed on the "trapped" area of the zona until the blastocyst fell off the needle. The manipulations were performed at 37°C and usually took 1–2 minutes after which the zona-slit blastocysts were immediately returned to the incubator.

The initial herniation of trophectoderm through the slit was usually observed within 6–18 h (Fig. 21.1c). The herniating trophectoderm cells were biopsied when the size of the herniation was equal to (large biopsy) or less than (small biopsy) the diameter of the blastocyst. The biopsy was performed free hand under a stereo dissecting microscope by gently rubbing the end of a microneedle across the waist of the herniation against the bottom of the dish. This procedure usually took less than a minute (Fig. 21.1d).

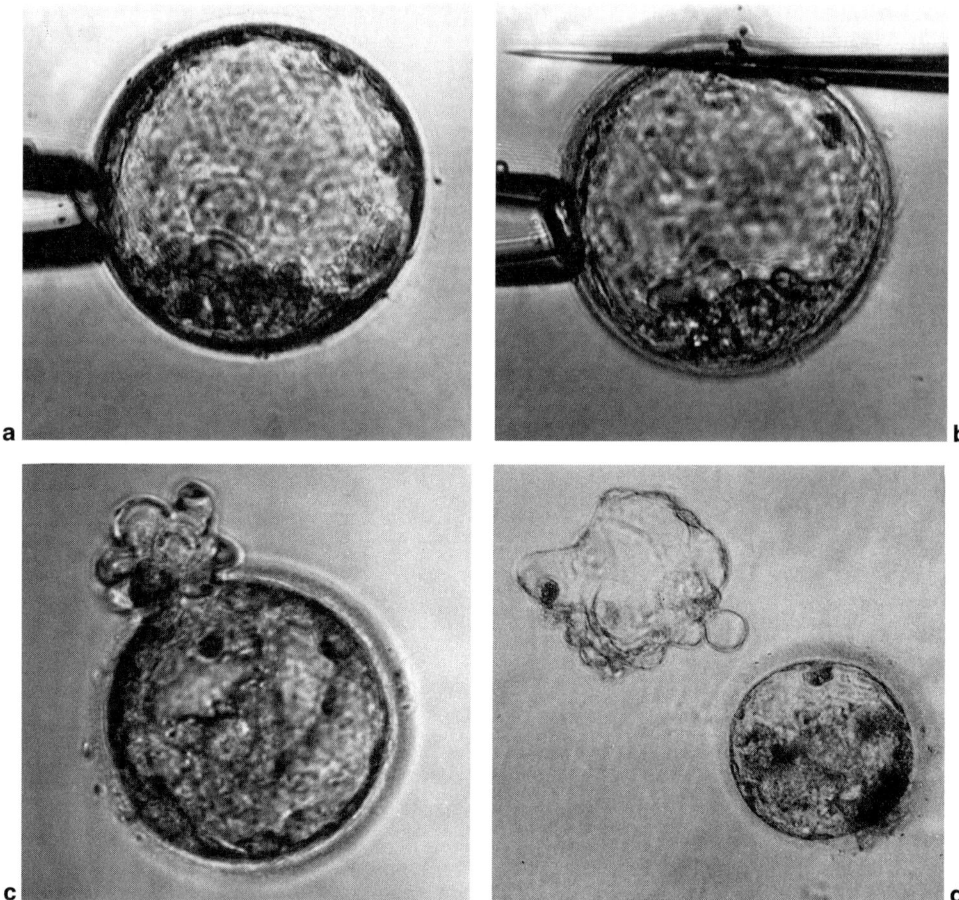

Fig. 21.1. Technique of trophoectoderm biopsy in human blastocysts. **a**, Day 5 expanded human blastocyst is held by a holding pipette. **b**, A microneedle is introduced into the perivitelline space opposite the inner cell mass, without piercing the trophoectoderm layer. **c**, A part of the trophoectoderm layer is seen to herniate out of the slit made in the zona pellucida (magnification × 200). **d**, The herniating cells are separated using a microneedle. The photograph shows the biopsied blastocyst and the biopsied trophoectoderm cells (magnification × 200).

The number of trophectoderm cells removed was counted by fixing and staining the biopsies. The biopsied blastocysts were monitored for morphological changes by phase-contrast microscopy, and hCG secretion in the culture supernatants was measured up to day 14 of culture. These results were compared with those of non-manipulated human blastocysts [42].

Effect of Trophectoderm Biopsy on Morphological Development of Human Blastocysts

In a preliminary set of experiments, 47 human blastocysts were manipulated to assess the feasibility of the technique [4]. The results showed that blastocysts formed on days 5 and 6 after insemination were successfully manipulated as compared with those formed on days 7 and 8. Although there was no impairment of subsequent development as assessed by the rate of hatching, it was essential to monitor further developmental changes before considering the blastocysts to be suitable for transfer.

A second set of experiments was therefore performed and the in vitro development and hCG secretion of manipulated blastocysts were monitored up to day 14 of culture. Fourteen blastocysts formed on days 5 and 6 were slit and biopsy of the herniating trophectoderm cells was performed (Table 21.2). After biopsy, hatching was observed in six blastocysts (43%), and overall there was evidence of growth in five hatched and zona-intact blastocysts. The sequential morphological changes observed up to day 14 in non-manipulated controls [42] were also documented in this group of biopsied blastocysts.

The approximate size of the biopsy was varied randomly: a large biopsy was performed in six and in the remaining eight a small biopsy was performed. All of the six large biopsies had ten or more cells (mean = 18.3, range 10–30) and the eight small biopsies had less than ten cells (mean = 5.6, range 1–9).

Table 21.2. Comparison of morphological changes and hCG secretion by non-manipulated controls, biopsied and zona-slit blastocysts

Blastocyst	*n*	Hatching	Adherent/ growth	hCG secretion[a] (*N*)
Non-manipulated control	9	4 (44%)	3 (33.3%)	223.1 ± 62.5 (9)
Total biopsied	14	6 (43%)	5 (35.7%)	81.1 ± 21.3 (12)
Biopsy size				
up to 8 cells	5	2 (40%)	2 (40%)	124.6 ± 38.7* (5)
up to 10 cells	8	3 (37.5%)	3 (37.5%)	104.4 ± 28.1** (8)
more than 10 cells	6	3 (50%)	2 (33.3%)	34.6 ± 15.2*** (4)
Zona-slit	18	18 (100%)	10 (55.5%)	193 ± 51.3 (15)

[a]Mean ± SEM cumulative hCG secretion (mIU/ml) from day 3–14 in culture. N = number in which secretion was detected.
hCG secretion in the different biopsy sizes compared with the non-manipulated controls: *no significant difference; **$P < 0.05$, ***$P < 0.01$.

Effect of Trophectoderm Biopsy on hCG Secretion by Human Blastocysts

No hCG secretion was detected in two of the 14 biopsied blastocysts. There was a significant decrease in the mean cumulative secretion ($P < 0.01$, Table 21.2) after biopsy, though the trend in secretion was similar to that of the non-manipulated controls over the given time period. There was a correlation between the morphological changes reported above and the hCG secretion in

Table 21.3. Correlation between morphological changes and cumulative hCG secretion (mean ± SEM mIU/ml) by zona-slit and biopsied blastocysts

Blastocyst morphology	Controls (n)	Zona-slit (n)	Biopsied (n)
Hatching	175.4 ± 67.0 (4)	264.6 ± 73.8[a] (9)	57.3 ± 21 (5)
Adherent	225.2 ± 63.2 (3)	313.2 ± 86.9 (7)	62.8 ± 26.1 (4)
Non-adherent	25.8 (1)	94.5 ± 11.3 (2)	35.2 (1)
Intact	261.2 ± 102.8 (5)	85.7 ± 38.8 (6)	98.2 ± 33.1 (7)
Growth	312.3 ± 115.2 (4)	129 ± 69.7 (3)	114.2 ± 37.3 (2)
No growth	57.0 (1)	42.4 ± 28.2 (3)	91.7 ± 46.1 (5)
Total	223.1 ± 62.5 (9)	193 ± 51.3 (15)[b]	81.1 ± 21.3 (12)[c]

[a]In the zona-slit blastocysts, a comparison has been made between those blastocysts that hatched completely and those which hatched partially and then retracted into the zona. The difference in the cumulative secretion in these two groups was significant ($P < 0.01$).
[b]No hCG was detected in three zona-slit blastocysts.
[c]No hCG was detected in two biopsied blastocysts.
n = number in which hCG secretion was detected.

that the hatched adherent embryos, and the intact embryos which showed evidence of growth, secreted higher levels of hCG (Table 21.3).

The amount of hCG secreted by biopsied blastocysts varied inversely with the size of the biopsy: when a large biopsy was performed (> 10 cells) the hCG secretion was significantly less ($P < 0.01$) compared with when a small biopsy was performed (< 10 cells, Table 21.2). In addition the secretion was inversely proportional to the number of cells removed (Fig. 21.2), though the size of the biopsy did not affect the time course of hCG secretion.

Trophectoderm Biopsy: a Possible Approach?

The in vitro morphological development of human blastocysts after trophectoderm biopsy was similar to that of non-manipulated blastocysts. As would be expected, on removing trophectoderm cells there was a decrease in hCG secretion and this correlated with the number of cells biopsied. This in vitro observation is similar to the in vivo observations made in marmosets [11]. Pseudopregnant marmosets did not require hCG supplementation after transfer of blastocysts biopsied on day 9, and gave birth to normal offspring. However, only when 75 IU hCG were administered on days 13 and 17 of the pregnancy did transfer of blastocysts biopsied on day 10 result in four live offspring (50%). In four other recipients no hCG was administered and no pregnancies occurred.

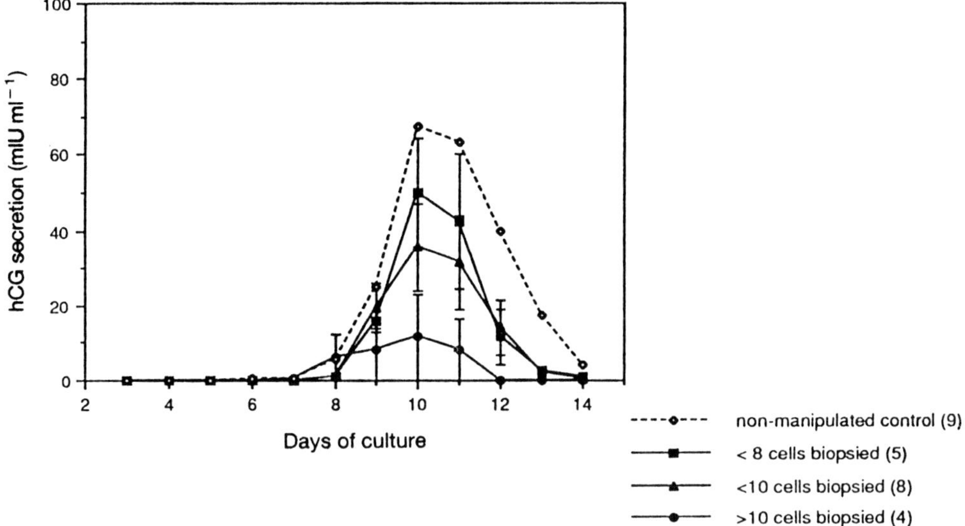

Fig. 21.2. Variation in mean ± SE daily hCG secretion with the size of the trophoectoderm biopsy. (The mean cumulative hCG secretion for these groups is shown in Table 21.2.) ◇, Non-manipulated control (9); ■, < 8 cells biopsied (5); ▲, < 10 cells biopsied (8); ●, > 10 cells biopsied (4).

The difference in day 9 and day 10 biopsies was in the number of cells removed: 10% of the total on day 9 and 25%–40% on day 10. The removal of eight cells from the human blastocyst on day 5 or 6 would decrease the total cell number by approximately 10% [40]. It appears that if the number of cells removed can be carefully controlled the remaining trophectoderm can compensate in vitro without significantly reducing hCG secretion. The ability to compensate would probably be greater in vivo, although this can only be reliably assessed by transferring manipulated embryos in the uterus.

However, the main drawback of this approach is that currently IVF embryos are transferred to the uterus on day 2 or 3 after insemination. This would not allow the transfer of biopsied blastocysts in the same IVF cycle, and would therefore require cryopreservation. There have been few reports on blastocyst transfer after IVF [43] despite the high success rates in farm animals [44]. Therefore it would be important to establish routine blastocyst transfers before a trial of biopsied blastocyst transfer can be undertaken.

Assisted Hatching

Hatching is a process during which the blastocyst expands and the zona pellucida becomes very thin until a wide opening develops which allows the blastocyst to emerge partially or completely out of its zona. It enables the blastocyst to come in contact with the endometrium for implantation. Cohen et al. [45]

suggested that suboptimal culture conditions could cause hardening of the zona which results in impaired or delayed hatching in vitro. This could be one of the factors contributing to the low implantation rate following IVF. The same authors observed that cleaved embryos with thin areas in their zonae implanted more frequently than embryos with uniformly thick zonae. Following these observations, they transferred 144 "zona dissected" embryos (four–six-cell stage) and reported an increase in implantation rates (from 11% to 23%). This technique has been termed "assisted hatching" [6]. Recently an increase in the implantation rates after transfer of "zona-dissected" frozen–thawed human embryos has also been reported [46]. These observations suggest that a number of IVF embryos are unable to breach the zona at the time of hatching in vivo.

Effect of Zona-Slitting on Morphological Changes up to Day 14

In order to assess the effects of zona slitting on in vitro development and hCG secretion we monitored 18 zona-slit blastocysts up to day 14 (Table 21.2). Hatching was initiated in all slit blastocysts (100%) and occurred completely in nine (50%). This was statistically significant compared to non-manipulated controls ($P < 0.01$). Overall, hatching occurred earlier (day 6–7) and thinning of the zona prior to hatching was less obvious in this group than in the non-manipulated controls. A total of ten blastocysts showed evidence of further growth in this zona-slit group. The range of morphological changes observed up to day 14 was very similar to those previously documented in non-manipulated controls [42].

Effect of Zona-Slitting on hCG Secretion up to Day 14

Of the 18 zona-slit blastocysts, no hCG secretion was detected in three. The pattern of hCG secretion after slitting the zona showed a similar trend to that of the non-manipulated controls (Fig. 21.3) and there was no significant difference in the mean cumulative secretion over days 3–14 by zona-slit controls compared to the non-manipulated controls (Table 21.3). The nine zona-slit blastocysts that hatched completely had a significantly higher mean cumulative secretion ($P < 0.01$) than that of the six blastocysts that herniated but failed to hatch completely (Table 21.3). This is in contrast to the non-manipulated controls where hatching did not appear to affect hCG secretion. Also, of the zona-slit blastocysts, those which adhered and showed evidence of growth secreted higher levels of hCG (Table 21.3). These in vitro observations suggest that artificial openings at the blastocyst stage facilitate hatching without impairing further development.

However, the consequences of zona-slitting early cleavage stage embryos need to be discussed. First, the intact zona serves as a protective barrier in the early embryo against microorganisms and viruses, which could gain access to the embryo through the slit. Second, individual blastomeres could escape through large slits in the zona and third, if the slit is narrow, trapping of part of the blastocyst within the zona to give a "figure of eight" configuration could lead to an increased incidence of monozygotic twinning and blighted ovum formation [13]. Alternatively, a slit made at the blastocyst stage coincides with

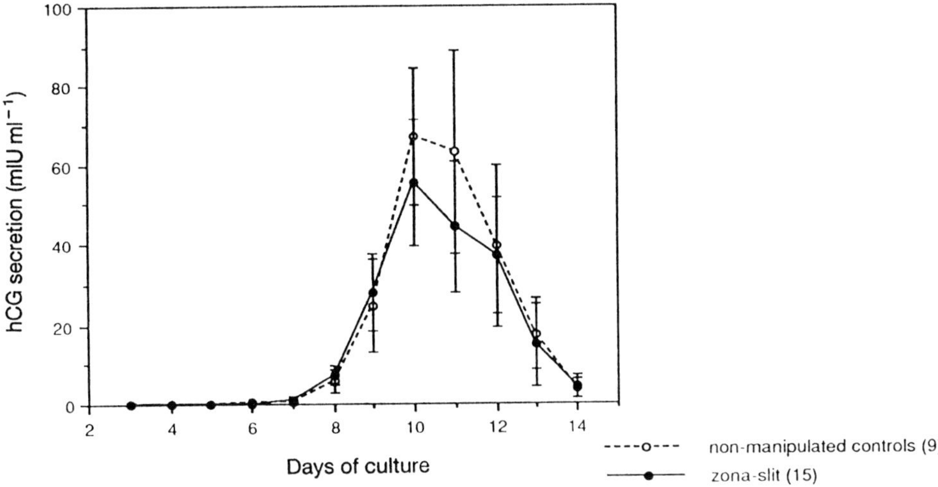

Fig. 21.3. A comparison of the daily hCG secretion (mean ± SE) in non-manipulated blastocysts and zona-slit blastocysts. There was no significant difference in secretion on any of days 3–14. ○, non-manipulated controls (9); ●, zona-slit (15).

the time when the blastocyst would have initiated hatching in vivo. At this stage there would be no loss of individual blastomeres as compaction and differentiation have already occurred resulting in formation of tight junctions, and the phenomenon of blastocyst trapping could be avoided by standardising the size of the slit. Trapping was not observed in any of the 18 zona-slit blastocysts in this study though it was seen in the controls. This could be due to the size of the slit being large enough to facilitate complete hatching or complete retraction. Also, initiation of blastocyst hatching soon after slitting the zona, in contrast to the early cleavage stages, might maintain the size of the slit. As discussed above, transfer of blastocysts after IVF would allow zona slitting at this stage.

Correction of Polyspermy

The incidence of triploidy following routine IVF is 4%–5%, and usually occurs due to polyspermic fertilisation. The introduction of microsurgical techniques to enhance fertilisation (partial zona dissection (PZD), subzonal insemination (SUZI)), has increased the rate of polyploidy significantly [47]. If the extra male pronucleus could be removed and the genetic composition of the zygote restored, more embryos would be available for transfer. Malter and Cohen [8] described a technique for enucleation of polyploid zygotes and subsequently reported an increased zygote survival rate (87%) and cleavage rate (73%) following the technique [48]. However, the main drawback of this approach is the inability to identify accurately the male pronucleus, as its size and association with sperm tails are not reliable criteria. Currently, the pronucleus farthest from the second polar body is removed. If the extra male pronucleus can be reliably identified, then this approach of embryo micromanipulation could be applied to multipronucleate embryos.

Gene Therapy

It seems appropriate to discuss briefly the micromanipulation of human embryos to insert genes into the human germ line. With the steady advances in germ line gene transfer technology using various animal models [49], its application to humans appears feasible. The different approaches developed include the microinjection of purified DNA into the pronucleus of the fertilised egg, retrovirus-mediated gene transfer and the injection of genetically altered ES (embryonic stem) cells into the blastocoelic cavity with subsequent incorporation into the inner cell mass. The major drawback of the first two strategies is the formation of insertional mutations in the host genome. As regards the third approach, human ES cells have not been generated as yet, and the therapy can lead to the formation of genetic mosaics. If these problems can be overcome and appropriate genes are selected for insertion into embryos, the application of this technology in humans is a possibility.

Conclusion

It is essential to stress the need to evaluate critically the current and new micromanipulation techniques. Both the technique and the artificial environment that the embryo is subjected to as a consequence of manipulation can affect embryo viability and its further development. The appropriate selection of techniques and preliminary experiments in animal models will assist in evaluating the safety of the technique. Improvements in culture conditions and precautions in maintaining the physicochemical environment during manipulation can help minimise any adverse effects. As exciting years lie ahead in this field, the development of less invasive techniques to obtain human embryos will also be of immense benefit.

References

1. Steptoe PC, Edwards RG. Birth after the reimplantation of a human embryo. Lancet 1978; ii:366.
2. Barlow DH, Egan D, Ross C. The outcome of IVF pregnancy. In: Matson PL, Lieberman BA, eds. Clinical IVF Forum, Current views in assisted reproduction. 1990; 63–9.
3. Handyside AH, Pattison JK, Penketh RJA, Delhanty JD, Winston RML, Tuddenham EDG. Biopsy of human preimplantation embryos and sexing by DNA amplification. Lancet 1989; i:347–9.

4. Dokras A, Sargent IL, Ross C, Gardner RL, Barlow DH. Trophectoderm biopsy in human blastocysts. Hum Reprod 1990; 5:821–5.
5. Dokras A, Sargent IL, Gardner RL, Barlow DH. Human trophectoderm biopsy and secretion of chorionic gonadotropin. Hum Reprod 1991; 6:1453–9.
6. Cohen J, Elsner C, Kort H, Malter HJ, Massey J, Mayer MP, Wiemer K. Impairment of the hatching process following IVF in the human and improvement of implantation by assisting hatching using micromanipulation. Hum Reprod 1990; 5:7–13.
7. Gordon JW, Grunfeld L, Garrisi GJ, Navot D, Laufer N. Successful microsurgical removal of pronucleus from tripronuclear human zygotes. Fertil Steril 1989; 52:367.
8. Malter HE, Cohen J. Embryonic development after microsurgical repair of polyspermic human zygotes. Fertil Steril 1989; 52:373–80.
9. Handyside AH, Kontogianni EH, Hardy K, Winston RML. Pregnancies from biopsied human preimplantation embryos sexed by Y-specific DNA amplification. Nature 1990; 344:768–70.
10. Monk M, Muggleton-Harris A, Rawlings E, Whittingham DG. Pre-implantation diagnosis of HPRT-deficient male, and carrier female mouse embryos by trophectoderm biopsy. Hum Reprod 1988; 3:377–81.
11. Summers PM, Campbell JM, Miller MW. Normal in vivo development of marmoset monkey embryos after trophectoderm biopsy. Hum Reprod 1988; 3:389–93.
12. Krzyminska UB, Lutjen J, O'Neill C. Assessment of the viability and pregnancy potential of mouse embryos biopsied at different preimplantation stages of development. Hum Reprod 1990; 5:203–8.
13. Malter H, Cohen J. Blastocyst formation and hatching in vitro following zona drilling of mouse and human embryos. Gamete Res 1989; 24:67–80.
14. Whittingham D, Penketh R. Preimplantation diagnosis in the human pre-implantation period. Hum Reprod 1987; 2:267–70.
15. Edwards RG, Gardner RL. Sexing of live rabbit blastocysts. Nature 1967; 214:576–7.
16. Gardner RL, Edwards RG. Control of the sex ratio at full term in the rabbit by transferring sexed blastocysts. Nature 1968; 218:346–9.
17. Saiki RK, Scharf S, Faloona F, Mullis KB, Horn GT, Erlich HA, Arnheim N. Enzymatic amplification of β-globin genomic sequences and restriction site analysis for diagnosis of sickle cell anaemia. Science 1985; 230:1350–4.
18. Leese H. Non-invasive methods for assessing embryos. Hum Reprod 1987; 2:435–8.
19. Leese HJ, Humpherson PG, Hardy K, Hooper MA, Winston RML, Handyside AH. Profiles of hypoxanthine guanine phosphoribosyl transferase and adenine phosphoribosyl transferase activities measured in single preimplantation human embryos by high-performance liquid chromatography. J Reprod Fertil 1991; 91:197–202.
20. Gordon JW, Gang I. Use of zona drilling for safe and effective biopsy of murine oocytes and embryos. Biol Reprod 1990; 42:869–76.
21. Monk M, Holding C. Amplification of a β-haemoglobin sequence in individual human oocytes and polar bodies. Lancet 1990; 335:985–8.
22. Verlinsky Y, Ginsberg N, Lifchez A, Valle J, Moise J, Strom C. Analysis of the first polar body: preconception genetic diagnosis. Hum Reprod 1990; 5:826–9.
23. Tarkowski A, Wroblewska J. Development of blastomeres of mouse eggs isolated at the 4- and 8-cell stage. J Embryol Exp Morphol 1967; 18:155–80.
24. Epstein CJ, Smith S, Travis B, Tucker G. Both X chromosomes function before visible X-chromosome inactivation in female mouse embryos. Nature 1978; 274:421–4.
25. Willadsen SM. A method for culture of micromanipulated sheep embryos and its use to produce monozygotic twins. Nature 1979; 227:298–300.
26. Nijs M, Camus M, Van Steirteghem AC. Evaluation of different biopsy methods of blastomeres from 2-cell mouse embryos. Hum Reprod 1988; 3:999–1003.
27. Monk M, Handyside AH. Sexing of preimplantation mouse embryos by measurement of X-linked gene dosage in a single blastomere. J Reprod Fertil 1987; 82:365–8.
28. Monk M, Handyside AH, Hardy K, Whittingham D. Preimplantation diagnosis of deficiency of hypoxanthine phosphoribosyl transferase in a mouse model for Lesch–Nyhan syndrome. Lancet 1987; ii:423.
29. Wilton LJ, Trounson AO. Biopsy of preimplantation mouse embryos: development of micromanipulated embryos and proliferation of single blastomeres in vitro. Biol Reprod 1989; 40:145–52.
30. Hardy K, Martin KL, Leese HJ, Winston RML, Handyside AH. Human preimplantation development in vitro is not adversely affected by biopsy at the 8-cell stage. Hum Reprod 1990; 5:708–14.

31. Navidi W, Arnheim N. Using PCR in preimplantation genetic disease diagnosis. Hum Reprod 1991; 6:836–49.
32. Ozil JP, Heyman Y, Renard JP. Production of monozygotic twins by micromanipulation and cervical transfer in the cow. Vet Rec 1982; 110:126–7.
33. Ozil JP. Production of identical twins by bisection of blastocysts in the cow. J Reprod Fertil 1983; 69:463–8.
34. Wang ZJ, Trounson A, Dziadek M. Developmental capacity of mechanically bisected mouse morulae and blastocysts. Reprod Fertil Dev 1990; 2:683–91.
35. Van Blerk M, Nijs M, Van Steirteghem AC. Decompaction and biopsy of late mouse morulae: assessment of in vitro and in vivo developmental potential. Hum Reprod 1991; 6:1298–304.
36. Nijs M, Van Steirteghem A. Developmental potential of biopsied mouse blastocysts. J Exp Zool 1990; 256:232–6.
37. Muggleton-Harris A, Findley I. In vitro studies on spare human preimplantation embryos in culture. Hum Reprod 1991; 6:85–93.
38. Hartshorne GM, Avery S. Effect of biopsy on subsequent viability of mouse and human embryos in vitro. Abstract 11–20. Symposium on preimplantation embryo development, Serono Symposia, 1991.
39. Winston NJ, Braude PR, Pickering SJ, George MA, Cant A, Currie J, Johnson MH. The incidence of abnormal morphology and nucleocytoplasmic ratios in 2, 3 and 5 day human pre-embryos. Hum Reprod 1991; 6:17–24.
40. Hardy K, Handyside AH, Winston RML. The human blastocyst: cell number, death and allocation during late preimplantation development in vitro. Development 1989; 107:597–604.
41. Summers PM, Taylor CT, Hearn JP. Characteristics of trophoblastic tissue derived from in vitro culture of preimplantation embryos of the common marmoset monkey. Placenta 1987; 8:411–22.
42. Dokras A, Sargent IL, Ross C, Gardner RL, Barlow DH. The human blastocyst: its morphology and hCG secretion in vitro. Hum Reprod 1991; 6:1143–51.
43. Bolton VN, Wren ME, Parsons JH. Pregnancies after in vitro fertilisation and transfer of human blastocysts. Fertil Steril 1991; 55:830–2.
44. Iritani A. Current status of biotechnological studies in mammalian reproduction. Fertil Steril 1988; 50:543–51.
45. Cohen J, Wiemer KE, Wright G. Prognostic value of morphologic characteristics of cryopreserved embryos: a study using videocinematography. Fertil Steril 1988; 49:827–34.
46. Tucker MJ, Cohen J, Massey JB, Mayer MP, Wiker SR, Wright G. Partial dissection of the zona pellucida of frozen–thawed human embryos may enhance blastocyst hatching, implantation and pregnancy rates. Am J Obstet Gynecol 1991; 165:341–5.
47. Garrisi GJ, Talansky BE, Grunfeld L, Sapira V, Navot D, Gordon JW. Clinical evaluation of three approaches to micromanipulation-assisted fertilisation. Fertil Steril 1991; 54:671–7.
48. Malter HE, Cohen J, Grifo J, Tang YX. An enucleation technique for the correction of polyspermy in the human and genetic analysis for the determination of pronuclear "gender". Abstract at 7th World Congress on IVF and Assisted Procreation, 1991; 605.
49. Gordon J. Strategies for human germ line gene therapy. In: Verlinsky Y, Kuliev A, eds, Preimplantation genetics. Plenum Press, New York, 1991; 221–32.

Chapter 22

Preimplantation Diagnosis of Genetic Defects

A. H. Handyside

Introduction

Human genetic disease, including genetic contributions to congenital abnormalities and some common diseases, has been estimated to affect 1 in 20 births and is a significant cause of illness and mortality in infants [1]. Since treatment after birth and in adulthood is only possible in a few conditions and with variable success, the emphasis remains on prevention through prenatal diagnosis. In this context, the incidence of genetic diseases resulting from identifiable genetic defects is of the order of 1%–3% with chromosomal abnormalities predominating over defects affecting single genes of which close to 5000 have now been identified [2]. How this incidence of affected individuals translates into couples seeking prenatal diagnosis is a complex equation. Often couples will only become aware of the risk through diagnosis of an affected relative or after having an affected child themselves. In future, however, population screening and identification of carriers may alert couples to the risk before they have children.

Current methods of prenatal diagnosis involve sampling cells of fetal origin, for example, by amniocentesis in the second, or chorion villus sampling (CVS) in the first trimester of pregnancy and using cytogenetic, biochemical or DNA methods to detect the genetic defect. If the pregnancy is affected, however, couples face the difficult decision of whether or not to terminate the pregnancy and some have repeated terminations before establishing a normal pregnancy. In vitro fertilisation (IVF) and diagnosis at preimplantation stages of embryonic development in vitro, or preimplantation diagnosis, would allow only unaffected embryos to be returned to the uterus. Any pregnancy should, therefore, be unaffected by the disease and the possibility of a termination following diagnosis at later stages of pregnancy avoided.

Preimplantation diagnosis, using established methods for IVF, potentially has several other advantages. First, several embryos can be screened in a single reproductive cycle since women are routinely superovulated to control the timing of ovulation and increase the numbers of oocytes which can be collected. This increases the probability of identifying unaffected embryos in couples at high risk of transmitting a genetic defect. Second, although pregnancy rates among infertile couples treated by IVF have in general remained low, averaging 13%–21% clinical pregnancies per embryo transfer in the UK in 1989 [3], larger centres often achieve higher rates. For example, at Hammersmith Hospital, in an 18-month period to March 1991, the clinical pregnancy rate as assessed by the detection of a fetal heart by ultrasound was 34% per embryo transfer [4]. Thus, even though several attempts may be necessary, the time taken to establish a normal pregnancy is likely to be relatively short. Finally, preimplantation diagnosis may be more acceptable to some couples at risk of relatively mild or late onset diseases, for example Huntington's chorea, since they may not feel justified in terminating a pregnancy diagnosed as affected later in gestation if there is the prospect of only mild impairment or an initial period of normal life.

Preimplantation Embryo Biopsy

For detection of genetic defects, one or more cells must be removed or biopsied from each embryo. Biopsy of some of the outer trophectoderm cells at the blastocyst stage has a number of advantages. Primarily, a maximum number of cells can be recovered at this advanced preimplantation stage maximising the chances of an accurate diagnosis. However, only about half of normally fertilised embryos reach the blastocyst stage in vitro [4] and only about half of these could be successfully biopsied [5] which would significantly reduce the likelihood of identifying unaffected embryos for transfer. Also, pregnancy rates after blastocyst transfer have been inconsistent and, at best, no more successful than earlier transfers at cleavage stages [6,7].

The alternative is to biopsy embryos at earlier cleavage stages when pregnancy rates after transfer are more consistent even though this restricts the number of cells which can be biopsied. During cleavage, each division subdivides the zygote into successively smaller cells. To minimise the reduction in cellular mass involved in removing a single cell, therefore, embryos have been biopsied as late as possible, at the 8-cell stage on the morning of the third day post-insemination (day 3), which then leaves only 8–12 h for genetic analysis before transferring selected embryos later on the same day. Biopsy of one or two cells at the 8-cell stage does not adversely affect preimplantation development [8] and several girls have now been born following identification of sex in couples at risk of X-linked disease [9].

Detection of Genetic Defects

The time limitation following cleavage stage biopsy and access to one or only a few cells biopsied from each embryo severely restricts the methods which can be used for the detection of genetic defects. However, several recent developments have allowed analysis of both chromosomal abnormalities and single gene defects. For example, fluorescent detection of in situ hybridisation (FISH) with chromosome-specific DNA probes [10] is rapid and efficient compared with other in situ methods and can be used to detect abnormal numbers of chromosomes. For single gene defects, the polymerase chain reaction (PCR) enables amplification of short fragments of DNA over a million-fold [11] within a few hours making it possible to detect even single base changes in the DNA of single cells.

Chromosomal Abnormalities

Chromosomal abnormalities especially abnormalities of chromosome number are a major cause of genetic disease. Several trisomies, for example, are compatible with development to term but cause congenital abnormalities of varying severity, the most well-known of which is trisomy 21 which causes Down's syndrome. Autosomal monosomies are generally lethal early in development, however sex monosomy for the X chromosome results in Turner's syndrome. Although older women are at increased risk of chromosomal abnormalities especially trisomies, the risk is still very low (of the order of 1%) compared with the high risk often associated with single gene defects. Preimplantation diagnosis, therefore, may not be worthwhile in these circumstances. Couples in which one partner is carrying a balanced translocation, however, and who have a history of miscarriage with a trisomic conceptus as a consequence are known to be at very high risk, in some cases 50%, of having another affected pregnancy. In these cases, preimplantation diagnosis has the positive advantage of screening several embryos in a single reproductive cycle.

For the detection of chromosomal abnormalities, cytogenetic analysis of banded metaphase chromosomes would be ideal. However, problems with spreading of chromosomes and a tendency for the chromosomes to be too short for banding have so far prevented reliable karyotyping of human embryo nuclei by standard procedures. FISH with chromosome-specific DNA probes has several advantages [10]. First, fluorescent detection of hybridisation is both sensitive and efficient, and allows detection in both metaphase and interphase nuclei. It also requires less time than conventional autoradiographic methods for detection of radiolabelled probes. Second, by conjugating probes to different haptens, different coloured detection systems can be used for the simultaneous detection of each probe. Finally, by using a mixture of short DNA probes to a particular chromosome or chromosomal region, "chromosomal painting" is possible which should allow the detection of various translocations and other structural abnormalities.

FISH to interphase nuclei of human preimplantation embryos has been demonstrated for X- and Y-specific probes [12] and for a chromosome 18 specific probe [13]. With each of these probes, the majority of interphase nuclei had the appropriate number of hybridisation signals indicating that the efficiency of hybridisation was high. However, hybridisation failure and conversely artifactual signals in some nuclei together with a relatively high incidence of tetraploid nuclei indicate that several nuclei must be examined for a reliable result with a single probe. For example, in attempting to identify the sex of an embryo by FISH with a Y-specific probe to a single nucleus, hybridisation failure would lead to the misidentification of a male as a female, and similarly, with an X-specific probe hybridisation to a tetraploid nucleus would also lead to the misidentification of a male as female. Recently, this problem has been overcome by exploiting the potential of FISH for the simultaneous detection of both X- and Y-specific probes [14]. The use of two probes and the identification of either one X and one Y signals in males or two X signals in females allows the sex of embryos to be accurately identified from a single nucleus in the majority of nuclei analysed.

The potential of FISH for the simultaneous detection of probes specific for different chromosomes has led to the suggestion that the embryos of older women undergoing IVF could be screened for the common trisomies [12]. As outlined by Feldberg et al. [15], the abortion rate in IVF pregnancies for women over 40 years of age is 2.5 times more frequent than for women around the age of 30 years, probably due to a combination of chromosomal abnormalities in embryos and lower uterine receptivity. The most frequent autosomal trisomies found during prenatal screening are 21, 18 and 13 and their incidence rises significantly in women over the age of 35 years. Each of these trisomies is compatible with development to term but may result in later abortions or perinatal death. If these could be screened in combination with trisomy 16, the most frequent autosomal trisomy in abortuses, not only would this prevent the birth of trisomic individuals but also reduce the risk of a miscarriage. Together these trisomies account for about 50% of trisomic abortions [16]. However, the technical problems, particularly of overlapping chromosomes, are likely to be exacerbated as the number of probes is increased and it seems unlikely that this would be feasible with the number of cells available from a cleavage stage embryo.

Single Gene Defects

Approaching 5000 conditions caused by single gene defects have been described [2]. Although many of these are rare, in couples known to be at risk, the chance of having an affected child is often as high as 1 in 4 or 1 in 2 depending on whether the condition is dominant or recessive.

Pregnancies from Embryos Sexed by DNA Amplification

X-linked recessive diseases are transmitted by women carrying a defect in a gene on the female X chromosome. The most common of these is Duchenne muscular dystrophy (DMD) affecting 1 in 3500 male births. With these diseases, boys are affected if they inherit the defect on the X chromosome from their mother because the genes involved are not duplicated on the male Y chromosome whereas girls inheriting the defect are unaffected because they inherit the normal gene on the X chromosome from their father. The probability that a male embryo is affected is, therefore, 50% and the overall probability of having an affected child is 1 in 4. An increasing number of X-linked diseases, notably DMD [17], have been mapped and extensively characterised at the molecular level and prenatal diagnosis of normal or carrier females as well as normal or affected males is now possible by DNA analysis. Others are not so well characterised and all that can be offered is to diagnose the sex of the fetus giving the option of terminating males with a high probability of being affected. Similarly for preimplantation diagnosis, a number of approaches are being developed for specific diagnosis of a few X-linked diseases, but initial efforts have concentrated on identifying the sex of embryos so that normal or carrier females can be selected for transfer in any of these recessive disorders [18].

Using PCR for DNA amplification of Y-specific repeat sequences, the sex of cleavage stage embryos can be accurately identified from single biopsied cells [19]. The identification and transfer of normal or carrier female embryos was attempted in eight couples known to be at risk of transmitting various X-linked diseases [9]. These included X-linked mental retardation, Lesch–Nyhan syndrome, adrenoleukodystrophy, retinitis pigmentosa and hereditary sensory motor neurone disease type II. Also, two couples at risk of Duchenne muscular dystrophy opted for this approach even though a specific DNA diagnosis by conventional methods may have been possible after CVS.

After routine assessment of each couple for IVF, women were induced to superovulate using an established protocol involving an initial period of suppression of ovarian function with an LHRH agonist followed by stimulation of folliculogenesis and administration of human chorionic gonadotrophin to trigger ovulation [20]. The numbers of oocytes recovered and normally fertilised in these predominantly fertile couples were similar to those obtained with infertile couples. Normally fertilised embryos developing to the 6- to 10-cell stage by the morning of day 3 were biopsied and a single cell removed (in one cycle two cells were removed). The single cells were then lysed and the Y-specific fragment (if present) amplified by PCR while the biopsied embryos were returned to culture. After gel electrophoresis of the amplification products, the sex of the embryos was identified on the basis of the presence of the Y-specific fragment and up to two of the best female embryos were selected for transfer in the evening of the same day.

Five out of the eight women became pregnant after a total of 13 treatment cycles, three after one and one each after two and three treatment cycles. The first two were both twin pregnancies and the other three singletons. The sex of each of the seven fetuses was examined by CVS and karyotyping at about 10 to 11 weeks. All were female except for one singleton pregnancy in which the karyotype indicated that the fetus was male. Since this couple is at risk of transmitting type II hereditary sensory motor neurone disease and in their case a specific diagnosis to determine whether a male fetus is affected is not possible, the couple took the decision to terminate the pregnancy. Both of the twin pregnancies and the remaining two singleton pregnancies have now gone to term. Apart from the second of the second set of twins which was stillborn, all are apparently normal healthy girls. Detailed post-mortem examination failed to reveal any gross abnormality in the stillborn twin and the cause was probably intrapartum hypoxia prior to Caesarean delivery.

Table 22.1. Single gene defects detected by DNA amplification from single cells

Affected gene	Defect	Disease/ mouse mutant	Detection of defect	Reference
Mouse single gene defects				
1. β-major haemoglobin	Complete deletion	β-thalassaemia	Absence of amplification	[22]
2. Myelin basic protein	Partial deletion	Shiverer	Absence of amplification	[23]
Human single gene defects				
1. CFTR	Closely linked RFLP	Cystic fibrosis	Restriction digest	[24]
2. Dystrophin	Partial deletion	Duchenne muscular dystrophy	Absence of amplification	[24]
3. β-globin	Point mutation	Sickle cell disease	Restriction digest	[25]
4. CFTR	3 bp deletion (ΔF 508)	Cystic fibrosis	Heteroduplex formation	[26], [27]

From [21] with permission.

The misidentification of the sex in one case is now known almost certainly to have been because of amplification failure from a single cell. Amplification from each cell of disaggregated male embryos has subsequently demonstrated that amplification failure occurs with a frequency of about 15% so that at least two cells amplified independently would be necessary for an acceptable level of accuracy. Although biopsy of two cells from 8-cell stage embryos does not appear to harm their development, the use of dual FISH with X- and Y-specific probes is currently being tested as an alternative approach.

Specific Diagnosis of Single Gene Defects

Initially, attempts at specific diagnosis are being directed at prevalent single gene defects and especially those which are predominantly caused by one or a limited number of mutations in the genes involved using PCR to amplify a fragment of DNA containing the defective sequence [21] (Table 22.1). Examples include sickle cell anaemia which is caused by a single base change in the β-globin gene [25] and cystic fibrosis which is carried by 1 in 20 of the Caucasian population and is caused in a majority of cases by a three base pair deletion at position ΔF508 of the CFTR gene [28]. The reason for this is that many other single gene defects, notably the haemoglobinopathies, are caused by heterogeneous mutations and these would have to be first identified in each family before attempting to amplify the affected region of the gene.

A fragment of the CFTR gene including the ΔF508 region has been successfully amplified from first polar bodies biopsied from oocytes [26] and the deletion detected by mixing the amplified DNA with DNA amplified from cells known to be either homozygous normal or homozygous deleted. The mixtures are denatured and cooled to allow heteroduplex formation, i.e. the formation of double-stranded DNA from a normal single strand and a deleted single strand (if these are present in the mixture), prior to gel electrophoresis. Since migration of the heteroduplex is significantly retarded, the genotype of the cell can then be deduced from the presence or absence of the heteroduplex bands in the various mixtures. If there has been no recombination between ΔF508 in the CFTR gene and the centromere during meiosis, the presence of two copies of either the normal or deletion affected allele in the polar body implies that the other allele is present in the oocyte itself. If on the other hand, there has been recombination the first polar body and oocyte will both have copies of both alleles and the eventual genotype of the oocyte which is determined by the allele segregated to the second polar body at fertilisation cannot be predicted. Apart from recombination, the main drawback of this approach is that only the maternal defect is screened and in this case as cystic fibrosis is autosomal recessive, the presence of the father's allele in carrier embryos cannot be detected.

Verlinsky et al. [29] have attempted diagnosis by polar body analysis combined with analysis of a single cell biopsied at cleavage stages following fertilisation but with no ongoing pregnancy success. The only pregnancy which has gone to term, however, resulted from transfer of an embryo diagnosed as unaffected by blastomere biopsy alone and was later diagnosed conventionally as affected. It is difficult to account for this misdiagnosis. Certainly contamination with a sperm detached from the zona during biopsy is a possibility since they break off the pipette tip containing the biopsied cell for PCR. However, this should theoretically have resulted in a carrier or affected diagnosis unless there was amplification from the sperm but not the cell which seems unlikely.

Recently, a more reliable PCR protocol has been described using a second round of amplification with "nested" primers, i.e. primers annealing within the sequence of the first amplified fragment [27]. This results in a significant improvement in the yield of amplified product without exceeding the time available after cleavage stage biopsy. Furthermore, amplification from single cells is efficient and tested blind has always accurately detected the correct alleles. Unlike amplification failure with Y-specific sequences, the minority of cases in which amplification of this CFTR fragment fails would not result in a misdiagnosis. Diagnosis would simply not be possible for that embryo and it would not be transferred.

Preimplantation diagnosis was attempted in three couples who all had previously given birth to a child suffering from cystic fibrosis. In each case, both parents carried the predominant ΔF508 deletion affecting 70%–75% of cystic fibrosis carriers. IVF was used to recover several oocytes and these were fertilised with the husband's sperm. Normally fertilised cleavage stage embryos were biopsied on the day 3 post-insemination and one cell was removed from each and the region containing the deletion amplified using this nested amplification protocol. With two couples in which unaffected and carrier as well as affected embryos had been identified, they chose to have one unaffected and one carrier embryo transferred as these appeared morphologically to be the best embryos. After uterine transfer on the same day as the biopsy, one of these patients became pregnant and has subsequently delivered a healthy normal girl, free of both alleles with the deletion [30].

Similarly efficient PCR protocols for two individual-specific mutations of the hypoxanthine phosphoribosyl transferase gene causing Lesch–Nyhan syndrome have also been developed and preimplantation diagnosis attempted, so far without pregnancy success (Lesko et al., unpublished results). The prospects for the detection of single gene defects in general, therefore, are optimistic for those in which, either the defect itself has been sequenced allowing the design of the relevant primers for PCR, or there is a closely linked marker which has been similarly characterised.

References

1. Weatherall DJ. The new genetics and clinical practice, 3rd edn. Oxford: Oxford University Press, 1991.
2. McKusick VA. Mendelian inheritance in man, 8th edn. Baltimore, MD: Johns Hopkins University Press, 1988.
3. Interim Licensing Authority. The fifth report of the Interim Licensing Authority for Human in Vitro Fertilization and Embryology. ILA, Clements House, Gresham St, London EC2, 1990.
4. Hardy K. Development of the human blastocysts in vitro. In: Bavister B, ed. Preimplantation embryo development. New York: Springer-Verlag, 1991.
5. Dokras A, Sargent IL, Ross C et al. Trophectoderm biopsy in human blastocysts. Hum Reprod 1990; 5:821–5.

6. Dawson KJ, Rutherford AJ, Winston NJ et al. Human blastocyst transfer, is it a feasible proposition? Hum Reprod 1988; suppl 145:44–5.
7. Bolton VN, Wren ME, Parsons JH. Pregnancies following in vitro fertilization and transfer of human blastocysts. Fertil Steril 1991; 55:830–2.
8. Hardy K, Martin KL, Leese HJ et al. Human preimplantation development in vitro is not adversely affected by biopsy at the 8-cell stage. Hum Reprod 1990; 5:708–14.
9. Handyside AH, Kontogianni EH, Hardy K, Winston RML. Pregnancies from biopsied human preimplantation embryos sexed by Y-specific DNA amplification. Nature 1990; 344:768–70.
10. Trask BJ. Fluorescence in situ hybridisation: applications in cytogenetics and gene mapping. Trends Genet 1991; 7:149–54.
11. White TJ, Arnheim N, Erlich HA. The polymerase chain reaction. Trends Genet 1989; 5:185–9.
12. Griffin DK, Handyside AH, Penketh RJA et al. Fluorescent in situ hybridisation to interphase nuclei of human preimplantation embryos with X and Y chromosome specific probes. Hum Reprod 1991; 6:101–5.
13. Schrurs B, Winston RML, Handyside AH. Preimplantation diagnosis of aneuploidy by fluorescent in situ hybridization: evaluation using a chromosome 18 specific probe. Hum Genet 1992; submitted.
14. Griffin DK, Wilton L, Handyside AH et al. Dual fluorescent in situ hybridization for simultaneous detection of X and Y chromosome-specific probes for the sexing of human preimplantation embryonic nuclei. Hum Genet 1991; in press.
15. Feldberg D, Farhi J, Dicker D, Ashkenazi J, Shelef M, Goldman JA. The impact of embryo quality on pregnancy outcome in older women underging in vitro fertilization-embryo transfer (IVF-ET). J In Vitro Fert Embryo Transf 1990; 7:257–61.
16. Boué A, Boué J, Gropp A. Cytogenetics of pregnancy wastage. Adv Hum Genet 1985; 14:1–58.
17. Koenig M, Hoffman EP, Bertelson CJ et al. Complete cloning of the Duchenne muscular dystrophy (DMD) cDNA and preliminary genomic organization of the DMD gene in normal and affected individuals. Cell 1987; 50:509–17.
18. Handyside AH, Delhanty JDA. Cleavage stage biopsy and diagnosis of X-linked disease. In: Edwards RG, ed. Preimplantation diagnosis of human genetic disease. Cambridge: Cambridge University Press, 1992.
19. Handyside AH, Pattinson JK, Penketh RJA et al. Biopsy of human preimplantation embryos and sexing by DNA amplification. Lancet 1989; i:347–9.
20. Rutherford AJ, Subak-Sharpe RJ, Dawson KJ et al. Improvement of in vitro fertilisation after treatment with buserelin, an agonist of luteinising hormone releasing hormone. Br Med J 1988; 296:1765–8.
21. Hardy K, Handyside AH. Biopsy of cleavage stage human embryos and diagnosis of single gene defects by DNA amplification. Arch Pathol Lab Med 1992; 116:388–92.
22. Holding C, Monk M. Diagnosis of β-thalassaemia by DNA amplification in single blastomeres from mouse preimplantation embryos. Lancet 1989; ii:532–5.
23. Gomez CM, Muggleton-Harris AL, Whittingham DG, Hood LE, Readhead C. Rapid preimplantation detection of mutant (shiverer) and normal alleles of the mouse myelin basic protein gene allowing selective implantation and birth of live young. Proc Natl Acad Sci USA 1990; 87:4481–4.
24. Coutelle C, Williams C, Handyside A, Hardy K, Winston R, Williamson R. Genetic analysis of DNA from single human oocytes – a model for pre-implantation diagnosis of cystic fibrosis. Br Med J 1989; 299:22–4.
25. Monk M, Holding C. Amplification of a β-haemoglobin sequence in individual human oocytes and polar bodies. Lancet 1990; 335:985–8.
26. Strom CM, Verlinsky Y, Milayeva S et al. Preconception genetic diagnosis of cystic fibrosis. Lancet 1990; 336:306–7.
27. Lesko J, Snabes M, Handyside A, Hughes M. Amplification of the cystic fibrosis DF508 mutation from single cells: applications toward genetic diagnosis of the preimplantation embryo. Am J Hum Genet 1991; in press.
28. Riordan J, Rommen JM, Kerem B-S et al. Identification of the cystic fibrosis gene: cloning and characterisation of complementary DNA. Science 1989; 245:1066–73.
29. Verlinsky Y, Rechitsky S, Cieslak J et al. Reliability of preconception and preimplantation genetic diagnosis. Am J Hum Genet 1991; 49 suppl: 22.
30. Handyside AH, Lesko J, Tarin J, Winston RML, Hughes M. Birth of a normal girl following preimplantation diagnosis for cystic fibrosis. N Engl J Med submitted.

Discussion

Templeton: In the slides of hCG secretion, when did the biopsy actually take place?

Dokras: We performed the biopsy on day 5 or day 6, depending on blastocyst formation. The earliest that we detect hCG is day 7. hCG is detected in culture supernatants only after the biopsy.

Fishel: The second biopsy, which was for an assisted hatching, seemed to show an hCG level the same as that of the control, which was higher than the previous biopsy. Is that because the biopsy was more into the blastocyst? Was there no herniation? In the first series of experiments there was a reduced hCG and a herniation, but there was the same level of hCG doing an assisted hatching biopsy.

Dokras: Because they are two entirely different techniques. For a trophectoderm biopsy we make a slit that allows cells to come out and the cells that come out are then biopsied. That is the basic technique of preimplantation diagnosis. For the second technique we made a similar slit and did nothing else. We did not remove any of the cells, they remained intact as part of the entire blastocyst and the blastocyst hatched out and adhered. And in those we found no differences compared to controls.

Shaw: And that was done on the grounds that it may ultimately assist implantation.

Dokras: That is right. They are completely different.

Leese: The rates of implantation were so good following biopsy at the eight-cell stage. Presumably a slit had to be made to get the biopsy. Was that a factor in contributing to the success on transfer?

Handyside: It is an interesting question and one that I discussed with Jacques Cohen. We dissolve away part of the zona and we are putting quite a large hole in the zona pellucida. He felt that our increased implantation rates might be because we are putting a larger hole in the zona at a later stage of development than he was doing in his series. But in fact, those implantation rates are similar to the normal IVF implantation rates that we are currently getting in the clinic and there does not seem to be any evidence of a particularly beneficial effect in our hands.

Dokras: Jacques Cohen was at the RCOG in March 1992 and he has changed his technique. He is now making a slit. He made a mechanical slit the way I showed initially. The results I showed were of his first series where he got a significant increase, and as the numbers increased the significance was not that obvious. He has begun to use Acid Tyrodes and is making a large slit just the way Dr Handyside's group is doing for the biopsy. With his new series using acid tyrodes he is getting very high implantation rates, as high as 45%. We shall have to see the data in print but he is using that technique.

Leese: Are we assuming that effective zona lysins in vivo, or putative zona lysins, are reduced or have limited effect because of the zona hardening in vitro?

Dokras: That is the assumption.

Baird: Both chapters have pointed out the advantage for the geneticist of biopsying at blastocyst stage. Most geneticists are horrified at the thought of putting their money on the line on a single cell and in the near future one will need more than one cell. It makes the work being done with blastocyst biopsy even more important.

I wonder whether we should now be looking at the age at which embryos are transferred, rather than having to wait until it becomes necessary to do so. In the history of IVF, some 10 years ago there was the hope that blastocysts would have a higher implantation rate and that if an embryo survived to blastocyst it would have a better chance of implanting and going on. But people by and large gave that up because of the relatively few embryos that went to blastocyst.

With current technology, what is the experience of keeping embryos? What percentage of embryos do we start off with and get to, say, five cells, when that would be a realistic time to biopsy them.

Dokras: In our hands approximately 30%–35%, go on to form blastocysts.

The premise that an embryo that reaches the blastocyst stage will definitely implant on transfer is not true. That conclusion is supported by the data that show that embryos that look like blastocysts apparently have vacuoles. When we do nuclei counts they have anything from 18 nuclei right up to 80 or 100 nuclei, which makes us wonder if it is true that anything that reaches the blastocyst stage will actually implant. What we have to try and do is to assess blastocyst transfers based on two criteria, increase the number of embryos reaching the blastocyst stage in vitro, and also give some thought to the quality of the blastocyst, which will probably have to be taken into consideration. For every embryo to have a vacuole-like appearance is not the answer.

Handyside: It is important that we do not get over simplistic about it. It is quite a complex thing. Technically speaking there is now no doubt that we will have very efficient ways of detecting defects in the single cell. So, that is not the level of inaccuracy. But it has been implied that the single biopsied cell might not be representative of the embryo. Preimplantation development tends to be clonal, so that even if we take a patch of blastocyst we may be getting the evidence of a single cell. It may not be something that we can avoid at the preimplantation stages – those kinds of errors of aneuploid mosaics, for example. We might have to look for it but currently I do not think we have enough information to say how frequent those are.

Winston: We are starting to re-look at later transfer and perhaps we shall be able to report on that later this year or next year. When we previously transferred blastocysts in our programme we were certainly unlucky and we discontinued it, and it will be interesting to review that again.

Interestingly, we found this delayed rise in hCG in vivo and I have just had news of two pregnancies where the hCG has now reached, after biopsy, normal levels, but delayed over a week or two. So that is something which continues.

We have confidence limits for what we would expect the hCG to be in the blood after embryo transfer, and two current newly pregnant patients have just reached normal hCG values. So there is a delay which we are seeing in vivo after transfer. There is obviously some slowing down of development, presumably because of fewer cells in the embryo.

I certainly would not be dismissive of previous stage biopsy. The fact is we have demonstrated that it works.

Handyside: But you got it wrong once.

Winston: It was, to be fair, a very early technique in the very early stages of doing single cell PCR.

Handyside: That is a unique situation. It would give the wrong impression if people went away feeling that DNA amplification could cause errors like that from single cells in the future. Normally if the amplification fails, we simply do not get a diagnosis on that embryo.

Clearly with an average of only five or six embryos, we have to have a very efficient technique. We have to have a diagnosis on the majority of those to end up with two embryos to transfer in order to have a reasonable chance of pregnancy success. But having said that, the single cell failures are now down to about 5% under the right conditions, and it is 100% accurate, when it does work. When we were doing the Y sequences, it was probably the wrong strategy to use, as the target sequence is only there in the normal cells.

Winston: It is quite different for cystic fibrosis and we have the evidence to demonstrate that. There is a high degree of reliability for single cells.

Templeton: I would agree with David Baird, but nonetheless, in the blastocyst biopsies it has to be demonstrated that the trophectoderm biopsy is representative, in genetic and chromosomal terms, of the rest of the embryo. Presumably that still has to be done.

Dokras: One would assume on genetic terms it would be. But chromosomal? We do not know how soon mosaicism would develop. That is what we want to look at and that is why we started karyotyping.

Baird: But the difference is that one is likely to get ten shots at it. So that even if there is mosaicism and two of the 10 cells are aneuploid, one can be pretty confident.

Dokras: Even if one uses fluorescent in situ hybridisation, the way Dr Handyside showed 80% efficiency, if we have eight to ten interphases there, we shall definitely be able to give a more accurate diagnosis.

Shaw: We can see that while the techniques are being developed, there is still some controversy about the timing of the stage at which these biopsies should be performed and perhaps their predictability. But obviously great progress is being made.

Section V

Training in Infertility

Chapter 23

Training in Infertility: the Role of the Royal College of Obstetricians and Gynaecologists

I. D. Cooke

Although the role of the Royal College of Obstetricians and Gynaecologists (RCOG) is mainly concerned with the training of clinicians, it is important to recognise that there are other individuals, particularly scientists and technicians working the the field of infertility, who could appropriately be grouped under the College umbrella. Various forms of associate membership could be envisaged as described in the Futures Committee report recently presented to Council.

Current Mechanisms

There are a number of mechanisms for training in infertility that currently exist within the framework of the College. General training for the Membership of the RCOG comprises four six-month modules with a 12-month elective. The Part I examination encompasses basic science and Part II clinical aspects. Experience is gained in Senior House Office posts in District General as well as Teaching Hospitals where units frequently provide basic secondary level investigation and treatment without a special interest. The training therefore is uneven and it is possible for trainees to have no exposure to the more sophisticated aspects of reproductive medicine during the whole of their preparation for the MRCOG.

After passing the MRCOG, general accreditation is obtained by approval of a programme of work by the Higher Training Committee. One post of post-MRCOG registrar work may be taken into account but the emphasis on the training is generalist in character. Trainees are encouraged to develop a special interest and this is unlikely to happen in reproductive medicine unless the trainee is working in a tertiary unit. The number of trainees therefore who are accredited and have a special interest in reproductive medicine is few. This means that future Consultants who could be described as having a special interest are also few. The definition of "special interest" is an individual who spends less than 50% of his/her time in a particular subspecialty. A subspecialist would spend greater than 50% of his or her time in the subspecialty. The Subspecialty Board supervises the development of a syllabus/training programme for centres and for trainees in each of the subspecialties. When subspecialisation was established, it was envisaged that no more than 6% of trainees would ultimately qualify and function as subspecialists but it will take many years before even this small proportion is qualified and functioning. The training programme lasts three years, two years of training with agreed modules and one year of research although the latter may be recognised retrospectively. There is in addition the requirement to do one year as a "generalist" senior registrar.

The Working Party on Continuing Medical Education has recently reported and a number of mechanisms has been enumerated. An Education Board following the Futures Committee recommendations will be established. An agreement must still be reached about the allocation of cognate points for different types of education such as hospital meetings, educational courses, national and international meetings, audit, self-examination systems such as Logic, even the writing of papers or giving of lectures.

The Scientific Advisory Committee (SAC) convenes study groups such as the one from which this book is published. Subsequent scientific meetings are held as a digest of the proceedings. Other meetings are arranged on a regular basis on agreed topics and Consultants and Senior Registrars' Annual Conferences provide further updates. The Fertility Subcommittee of the SAC has compiled a list of assisted reproduction training centres willing to offer training throughout the UK and has made that available, particularly to enquirers from abroad.

Of more relevance perhaps is the recent completion of the volume "Infertility: Guidelines for Practice" which has been agreed by the Scientific Advisory Committee and Council and should shortly be available for distribution. It addresses the issues of the type of investigation and treatment that should be offered at primary, secondary and tertiary level with detailed descriptions of assisted reproduction techniques. More controversially perhaps it has addressed the issues of counselling, of audit and research attempting to influence practice positively. There will doubtless be scope for future preparation of documents that can have useful roles in training in specific areas as well as infertility.

Recent Changes

It is appropriate to review training in the context of the major changes that have occurred in the field in the last few years. There has been an explosion of knowledge and technical development. There has been a shift from experimental and research procedures to routine clinical practice, usually in academic or tertiary centres. Microsurgery has become established, but in few centres. In vitro fertilisation and its plethora of associated techniques with suitable acronyms, computer assisted semen analysis and the techniques of minimally invasive surgery with the use of laser have all arriveed. There is a major problem in evaluating these and extending the appropriate use of these techniques on a national basis.

IVF has developed to an unprecedented extent in the private sector which contributes the dominant proportion of services. The largest clinics are also in the private sector and it is clear from data of the Interim Licensing Authority that better results are obtained in larger centres.

The RCOG contributed 50% of the funding of the Voluntary Licensing Authority (VLA) together with the Medical Research Council. Subsequently the VLA changed its name to the Interim Licensing Authority to create pressure for the establishment of a statutory Licensing Authority and for financial reasons the RCOG considerably reduced its contribution at that time. With the passage of the Human Fertilisation and Embryology Authority Act in October 1990 the HFE Authority came into being and has subsequently promulgated a Code of Practice and has developed licensing procedures, schedules of fees and regulations about donors that impact on centres providing assisted conception procedures.

There are very few clinics offering the full spectrum of infertility investigations and treatment. They are not adequately staffed, training facilities are not available and it would be reasonable to conclude that in general the National Health Service has poor facilities for these procedures. There is limited counselling and indeed exposure of clinicians to counselling in general. There is poor audit of practice and there are significant misleading claims as to the efficacy of assisted conception procedures. Little of the practice is standardised although the HFEA Code of Practice and more recently the document "Infertility: Guidelines for Practice" have addressed these problems. There is no national and therefore no regional strategy relating to infertility and one of its significant components, assisted conception. Although the latter is demonstrably "high tech", it is frequently not seen in the context of comprehensive investigation and service delivery. Considerable resources are already expended on gynaecological services that encompass many aspects of infertility but this fact is not recognised and there seems to be an artificial block between what is seen as more conventional treatment such as tubal surgery and the high technology of in vitro fertilisation.

There are only six accredited training programmes in the UK in reproductive medicine. This is too few to have a major impact on the future provision of competent subspecialists who could provide effective regional services. Of even more importance is the fact that manpower approval for trainees is limited although at Senior Registrar level a small number has been top sliced from the Joint Professional Advisory Committee's ceiling number. Funding, although agreed centrally, is frequently delayed before being implemented regionally and the constraints on funding inhibit the development of further training centres. Indeed the RCOG has needed to support a few subspecialty training fellowships, a significant drain on its resources. If trainees, who complete their training in the subspecialties, have difficulty obtaining Consultant posts in their subspecialty, this information is likely to feed back and inhibit applications for further subspecialty traineeships. It is clear that the provision of high quality regional services cannot be accomplished until there is an adequate number of individuals well trained in the difficult techniques of assisted conception.

One of the problems of the restriction to tertiary centres of assisted conception techniques is that the more sophisticated investigations as well as the treatments are suboptimally employed in secondary centres. This means that those training in the secondary centres have little exposure to modern techniques or to the emergent philosophy of up to date reproductive medicine. This deficit reaches down to the most junior training levels and if not remedied will ensure that the next generation of trainees will have had virtually no exposure to this major branch of obstetrics and gynaecology. Indeed the present licensing procedures laid down by the HFEA help to ensure that there is no communication between licensed and other centres so that the transmission of information, even about a secondary centre's own patients, will be zero, inhibiting every educational process inherent in clinical practice.

There is also little staff rotation through subspecialist and other programmes. Thus training grade staff have little opportunity for rotation and exposure to newer techniques in this discipline. The future Consultant practising as a generalist will have no understanding of the sophisticated nature of assisted conception and is likely therefore to develop inappropriate investigation routines and waste resources, ultimately referring patients inappropriately and too late to the tertiary centres. As it is clear that the outcome of all procedures is related to maternal age this is a fact of the utmost public health importance.

Recommendations

Examiners for the MRCOG tend to be generalists but it is important that future trainees become aware of the subspecialist aspects of the subject at a theoretical as well as practical level so that two standards of practice, those provided by the generalist and those provided by the subspecialist, do not develop. Changes introduced by the development of reproductive medicine must be incorporated into the theoretical and clinical aspects of the MRCOG examination.

One of the best ways to improve the understanding of the subject is to encourage Registrar rotations between generalist and subspecialist units. This will be difficult as the staff numbers, particularly at intermediate level in the subspecialist units, are so small that rotation of a skilled member out may seriously compromise the technical capacities of the unit. Rotations therefore may need to be over and above the basic staffing. This will also add to the educational potential.

A problem arises, however, because there are more reproductive medicine units, particularly in assisted conception, outwith the NHS and the concept of rotation between the NHS and the private sector is one that will need to be addressed. Although there would be enthusiasm from the private sector for exposure to comprehensive services available within an NHS unit, the problems of minimal staffing and their cost to the private clinic are inhibitory factors and part-time rotations, for example on some days per week, may need to be pursued.

Regional centres of excellence need to be developed so that a critical mass of experts and skills is readily achieved. Training programmes can then be developed on the basis of quality in clinical service and research. It is easier to train when there are larger numbers of principals in a group than when the numbers are unitary. Subspecialty training is optimally done at Senior Registrar level by training fellowships. The number of funded posts must be increased and allocation made as of right to regional training centres. The current uncertain security of funding and unpredictability of long-term allocation militate against effective training. It also operates against trainees being incorporated into a programme and being allowed to develop appropriate responsibility.

In addition to the activities of tertiary centres it is important that second care services develop within a comprehensive framework. Only then can appropriate judgements be made about the ideal treatment for a couple. At present treatments are more often determined by what is available than by what is appropriate for the patients. Training cannot be effective if biased judgements are made for this reason.

There will always need to be regular clinical and scientific updating meetings for trainees and these should become the responsibilities of regional centres. To this end they should have a training budget.

Even though future Consultants may not be working in the subspecialist unit, it is imperative that couples are suitable and efficiently investigated and treated prior to referral to a reproductive medicine subspecialty centre. To do that, the latest thinking and data supporting efficacy of techniques must be presented regularly to those in secondary and indeed primary centres. In the latter, contributions by regional centres to DRCOG training for general practitioners should also be given regularly. Teaching of generalists working in secondary centres should keep in mind the need for a broad approach and not concentrate excessively on technical details.

There is also a need to present a strong image to the community. To this end, factsheets should be written and produced for public distribution using the scientific skills of College Fellows and Members and Birthright skills for design and publicity. These could become a fundraising instrument through a Publications Department. There would also be the advantage of high quality and consistent information being given to the public. Regular audit of regional centres should be co-ordinated with international figures. Although the Human Fertilisation and Embryology Authority has responsibility for collection of statistics for IVF and donor insemination, there is a need for much wider data collection and presentation, not least to recipients of treatment but also to those working in primary and secondary care. Interpretation of such data by commentaries in quality leaflets could help to inform the public debate in the area of infertility. The acquisition of quality data would spur the development of policies for the advancement of reproductive medicine and its services and would provide a stimulus for proactive contact with the media and consumer organisations such as ISSUE.

To manage or oversee these activities there would need to be a strong reproductive medicine group created within the RCOG. Currently the Fertility Subcommittee is less effective than it might be because of a long chain of reporting through to Council. There may be a case for developing subspecialty committees in general but there is certainly one for developing a Reproductive

Medicine Group under the umbrella of the Educational Board. Such a group would be able to develop policy, suggest and take initiatives which would allow the College to liaise with the Department of Health more effectively. It would also allow the RCOG to build stronger links with scientific societies such as the British Fertility Society and to make more effective contact with the Society for the Study of Fertility which is more strongly scientific and less clinically based. There would then be a cohesive force for the development of reproductive medicine and sciences for the benefit of the population.

Chapter 24

Infertility Nurses and Counselling

M. M. Inglis and J. Denton

Introduction

Couples and individuals diagnosed as infertile present with special psychological needs. Appropriate recognition and support of these needs is the responsibility of the caring team and different members can contribute in different but overlapping ways throughout the whole spectrum of the infertility experience. There is a requirement for the development and exercise of counselling skills by nurses in this field. These skills are concerned principally with the management of care, the giving of information and support in ensuring that the couple make informed consent. In addition there must be access to the services of specially trained counsellors whose role is different. This counselling is concerned with helping couples to explore problems, gain insight into themselves and their situation, examine implications of treatments offered and make critical choices.

There is evidence that an early, routine offer of counselling helps in the course of subsequent treatment particularly if it becomes complex and protracted. The Human Fertilisation and Embryology (HFE) Act 1990 states that people seeking licensed treatment (i.e. in vitro fertilisation or treatments using donated gametes) or consenting to the use or storage of embryos, or to the donation or storage of gametes must be given "a suitable opportunity to receive 'proper counselling' about the implications of taking the proposed steps" before they consent [1]. Ideally counselling should be available for all infertile couples before, during and after treatment.

It is against this backdrop that this chapter seeks to explore the special needs of infertile couples and the difference between the use of counselling skills and the contribution of "proper counselling". It also seeks to explore the role of the infertility nurse who may enhance her nursing contribution by gaining counselling skills. Some nurses may decide to undertake the longer training to be able to take a proper counselling role.

The Current Situation

The rapid progress in reproductive technology over the last 15 years has demanded the development of new roles involved with providing infertility services. Nurses are caring for infertility patients in general infertility clinics, family planning clinics, gynaecological wards and general practice as well as licensed treatment centres. As the specialty evolves it is necessary to review the role of these nurses and to define and implement appropriate training and education.

It has become apparent that care of the infertile couple extends beyond the management of organic disorders. It must also address underlying psychological problems resulting from diagnosis and treatment. A survey conducted by the Royal College of Nursing (RCN) Fertility Nurses Group (FNG) revealed that nurses working in infertility centres have exceptionally diverse roles and diverse professional training. Many feel inadequately trained to undertake some aspects of the work involved; 93% felt they did not have the counselling skills to deal with the degree of distress experienced by the couple and individuals. There is little peer group support within the nursing structure in this field, particularly in the private sector and this perception often gives rise to a feeling of isolation among nursing staff [2].

The information received by the RCN FNG and feedback from nurses' conferences reveals that nurses feel strongly that they do not have the skills to deal adequately with the degree of stress, anxiety and complexity of feelings thay they encounter daily in infertility clinics and licensed treatment centres. There is not enough time, space or skill allocated to this important area of care. Nurses are troubled that they may even be increasing distress by giving inappropriate responses to requests for help.

Counselling should also be made available to couples whose infertility treatment does not include licensed treatment. Some of these couples will not have access to this scarce resource. For others, it may not be considered a suitable treatment, or they may decide for personal reasons that they do not wish to pursue licensed treatments.

Special Needs of Infertile Couples

The diagnosis of infertility undermines fundamental assumptions that most couples/individuals carry with them from early developing life, the assumption that they will become parents. Strong links between expectations about parenthood and infertility set the foundation for what has been described in the literature as the crisis of infertility. Marital, sexual, social and career relationships are all put under extreme pressure with resulting tensions and conflicts. For many, becoming a parent is closely associated with the primary source of their identity. Coughlan [3] reported a high incidence of divorce in childless couples and Schellon [4] noted the frequency of suicide in childless couples.

Religious and cultural attitudes towards parenthood and childlessness impose pressures on infertile couples which can confuse the person's perceived need and attitude towards childlessness [5]. Parenting is viewed as an important adult goal which if impeded will have serious implications in later life [6]. If the reassuring rhythm of expectation of family life is disrupted a feeling of disequilibrium and disorientation can result. This feeling of being out of control of their lives together with the uncertainty of outcome of treatment results in a high level of anxiety and depression for many couples. Of 94 couples questioned in a study carried out in Houston, Texas, 41% said they would have liked one to one counselling, and 72% said they would like contact with other patients [7].

Infertile couples have an added difficulty in getting the support they need because they still tend to be an invisible group. Partners find it difficult to help each other and often become isolated within the relationship and within the community. Because of the protracted nature of infertility, couples sometimes stay in treatment for as long as 15 years, and even up to the menopause.

Couples who fail to become pregnant describe the experience as a slow death of hope; hope of becoming pregnant, hope of giving birth and becoming parents. There needs to be a grieving process, a letting go of loss, the loss of the hoped for child. To adapt to loss, it must be perceived and accepted as such. Where there is no visible loss to focus on, and it is rarely acknowledged or spoken about, resolution of loss can be a slow process. Couples often say they come to terms with childlessness eventually but that feeling infertile goes on indefinitely.

New technology has provided hope for some who could not previously conceive but it has also brought added pressure. With improved techniques constantly emerging it is never possible for the couple to say they have explored every solution. This makes it more difficult to come to terms with their predicament, to grieve the loss of the potential child and to consider other options or get on with their lives.

The Need for a Combined Medical, Social and Psychological Model of Care

Some of the difficulty in recognising and making provision for counselling needs may be that historically involuntary childlessness has been viewed only as a medical problem with only medical treatments as a solution. Diagnosis and investigation have involved only the exploration of physical aspects, the so called "medical model". In the absence of sufficient time and training it is easy for the nurse to concentrate mainly on tests and investigations. Dependence on medical specialists to regulate internal bodily functions re-enforces lack of autonomy and places the patient in a non-adult position [8].

In a retrospective survey of 843 couples who had attended an infertility clinic it was found that the emotional well being of couples was affected by prolonged periods of clinical investigation. Greater emotional and marital difficulties were reported by both men and women when the cause of infertility lay with the man [9].

Research is beginning to indicate that "pre-treatment counselling to help couples identify concerns and develop coping strategies" is likely to be an effective way to reach the vulnerable who most need help [10–12]. Dennerstein and Morse [13] found that when couples had prior contact with a counsellor before treatment, counselling was much more likely to be taken up and be effective. Infertile couples who have not had the opportunity to reflect at length on all the options available and the implications of certain treatments may accept new techniques as an inevitable next step, rather than making an informed choice.

There is a need for separate space, time and skill to explore the broader issues of how infertile couples cope in a pronatalist society where most married couples have children or seem to be pregnant. Couples also often describe a need for time to deal with fears about the future without children and whether the relationship will survive in the absence of children.

Multidisciplinary Approach to Psychological Care of the Infertile

"The range of tasks contained within this generic term (counselling) should be seen as a continuum with considerable overlaps between the tasks" [14]. The RCN would agree with this statement and also with the statement that medical and counselling expertise are complementary and would not want to draw rigid or bureaucratic boundaries between the disciplines. A difficulty in defining counselling needs and tasks is knowing where the provision of information and support spills over into counselling. A multidisciplinary approach to support and counselling where all members of the team have an opportunity to obtain skills in this field would be in the best interest of the couple. It is clearly impossible and uneconomic for any one profession or discipline to possess all the knowledge, skill and resources to meet the needs of the patient.

Dealing with overlaps and managing blurred boundaries is a difficult task and requires maturity and tolerance. If the nurse keeps as a guide the United Kingdom Central Council for Nursing, Midwifery and Health Visiting (UKCC) [15] guidelines which put the best interests of the patient first it should be possible for him or her to meet the needs of the couple and the team and still work within the limits of competence.

The Nurse's Role in the Continuum of Care

The British Association of Counselling (BAC) Code of Practice makes a distinction between listening skills, counselling skills and counselling. One important distinction between the exercise of counselling skills and counselling is the intention of the user. Counselling skills should enhance the performance of the functional role as a nurse [16]. The patient in turn will perceive her in the role of the nurse rather than counsellor.

Theoretical skills are relatively easy to master but it is the work needed to achieve personal development which takes much longer [17]. The nurse's own anxieties, doubts and fears can get in the way of being able to help the couple with their own particular problem. The aim of counselling skills is to develop the role of the health care professional, enhancing performance by improving communication and self awareness, not to create a different role. It is not a preparation for the individual wishing to work as a counsellor.

The Nurse's Role in the Giving of Information

Information

Nurses often have a large part to play in information giving which is clearly distinguished from counselling but requires its own skills. The way in which this process is conducted may affect the way the whole infertility treatment is experienced. Many couples come to counselling because they feel they have been unable to get adequate information or the time to understand it, to ask questions and to process the information. Studies reveal that patients forget as much as 50% of information presented as soon as five minutes after a consultation and that forgetting is random so there is no greater likelihood of recall for important items. Listening and communication skills facilitate the giving and receiving of information.

The main aim in information giving is to lower the anxiety level and enable the couple to make informed choices for which they can give consent where appropriate. To do this the nurse needs to keep in mind:

1. An awareness of a wide range of complex feelings;
2. The need to allow these feelings to be expressed;
3. The need to acknowledge the feelings as a normal reaction to the diagnosis of infertility;
4. The need to treat the partners as a couple with a shared problem;
5. The need to ensure that the couple have accurate and complete information.

The degree and complexity of information giving in licensed treatments is daunting. Time, space and skill are required and couples will vary enormously in their level of knowledge and their ability to receive and process it.

The HFE Act stipulates that information should be given on the limitation, outcome and side effects of treatments. The techniques, psychological effect of treatment and counselling availability are also described. The client should be informed about what information will be recorded and who has access. Details about who will be the child's parents, the child's need to know about his or her origin and the centre's statutory duty to take account of the welfare of any resulting or affected child are also important issues about which the couple should be informed where relevant [1]. The nurse may not be the major disseminator of this information but being one of the most visible and accessible members of the team, she will inevitably be asked to clarify many of the issues.

Donor Gametes

Information about consent to the use or storage of embryos or to the donation or storage of gametes is also extensive. It involves procedures, discomfort and risks in collecting gametes. The screening tests, utility, parentage, anonymity, identity, stored information are all issues which need careful discussion [1]. Such discussion will inevitably require nurses to reflect on the ethical, legal and kinship issues involved so that they can respond in an informed way to points of clarification presented by the couple. Complete and accurate information is a major factor in ensuring that the couple can make informed consent.

Nurse's Role in Giving Support Around Treatment

The uncertainty involved in every treatment is a difficult issue for the couple and the team. Freeman et al. [11] emphasise the importance of counselling couples in ways which help them to recognise the outcome statistics but at the same time retain enough hope to engage in treatment.

Dealing with loss of control over their lives is a major stress for the couple. At this time they are likely to be very susceptible to influence by those from whom they seek help. Allowing them to move at their own pace through treatment gives back a degree of feeling of control and avoids their becoming too dependent on the team. Couples are often at different stages of their acceptance of childlessness and badly timed suggestions about other options can increase anger and stress. Careful listening is required to hear the views of each partner. Even within one couple the understanding and acceptance of the problem may be quite different.

Aims in Giving Support

These can be usefully summarised as follows:

1. To enable people to retain a feeling of control;
2. To include them in decision making and treatment plan;
3. To acknowledge the sensitive nature of the investigations and treatment;
4. To acknowledge the invasiveness of treatments;
5. To acknowledge the success, failure rates and uncertainty of outcome;
6. At the end of treatment to facilitate the grieving process if it seems appropriate;
7. To offer counselling if it seems appropriate.

The time when couples finally decide to end treatment may be the first time they allow themselves to experience feelings of loss and grief. It is not possible to prepare people adequately for these feelings as hope needs to be retained to engage in treatment. Individuals experience all or some of the components of grief differently. The skill is again to listen, observe, allow and acknowledge their feelings as a normal reaction to loss and failure and support the couple through these feelings.

The Counsellor's Role in the Continuum of Care

Definition of Counselling

Counselling in whatever context is the process by which people are enabled to explore and understand the experiences which affect their lives and relationships. Having understood and gained insight into their situation they should be able to make the necessary adjustments and progress towards the resolution of their problems. The aim is also to lower anxiety and stress so that the couple are in a better position to make more informed decisions.

Counsellors choose and are trained to work mainly with emotions. Their experience, their work setting, personal supervision and closely defined task enables them to work in considerable depth [18]. The counsellor also provides non-judgemental acceptance, respect and a confidential relationship.

Counselling should provide a comfortable and private setting where sessions can take place undisturbed. If a working alliance is to be achieved counselling must be taken up voluntarily and not imposed.

Infertility Counselling

Infertility counselling requires specialist knowledge of the issues and needs of the infertile couple as well as some clinical insight into the management of infertility.

Proactive counselling as well as having an educative role in preparing couples for a range of feelings, can have a preventive function and promote health by anticipating problems in advance. The counsellor can help the couple assess themselves. As a result they may decide to postpone or withdraw from treatment, a decision which may be in the best interests all round.

The repeated trauma of prolonged failure arouses self doubt, and erodes the couple's sexual and emotional resources. Couples may resort to blaming each other for the inability to conceive each disclaiming their share of the failure. Old assumptions may have to be given up so that, while inevitably grieving the loss of spontaneous conception, they can begin accepting other ways of becoming pregnant, other ways of parenting or adjusting to childlessness [19]. It is important at this stage to allow enough time for reflection so that adjustment to the situation occurs before other options are taken up.

Counselling for Assisted Conception

Couples contemplating licensed treatments such as in vitro fertilisation (IVF), gamete intra-fallopian transfer (GIFT), donor insemination (DI) and others need time to reflect on the implications of treatment success or failure on themselves, their family and any resulting child.

The HFEA Code of Practice identifies three different areas of counselling.

1. *Implications counselling* explores the personal and family implications of infertility treatments.

2. *Support counselling* should be available to help people experiencing stress before, during and after treatment. Support can be provided by nurses and other members of the team as well as counsellors.
3. *Therapeutic counselling* focuses on resolving issues and adapting to changed circumstances whether treatment succeeds or fails [20].

Nurses and other members of the team are concurrently involved in information giving and support while in the counselling process there is a complex interplay of the three types of counselling described above (Fig. 24.1).

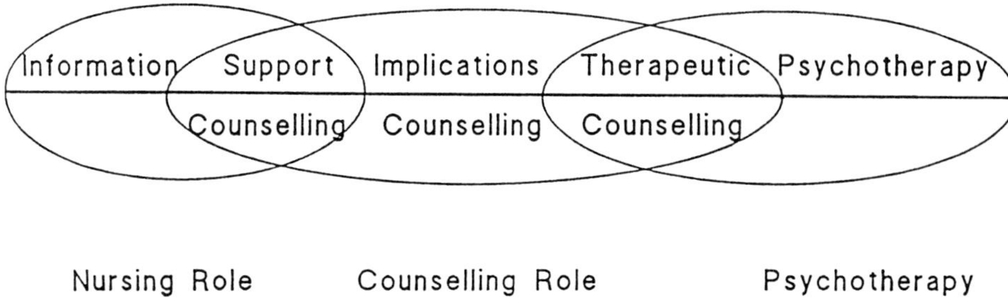

Fig. 24.1. Functions within overlapping boundaries in psychological care.

Training Pathway of the Infertility Nurse

It has been noted that management of the infertile is a rapidly developing field and that there are three principal areas in which the nurse may seek further training. These are:

1. Special nursing support in infertility treatment programmes
2. Counselling skills development
3. Formal counselling training

Special Nursing Support in Infertility Management

In the responses to the RCN FNG professional survey there was a request for guidance over the extended role of nurses in fertility care and for the promotion of professional guidance through written statements of practice. There was also identification of an overwhelming need for specific fertility nurse training at all levels.

The current changes in the National Health Service highlight the need for assessment of the quality of health care. The RCN have established a Standards of Care Project to help nurses and other health care professionals develop their own standards within a multidisciplinary framework. The RCN FNG have produced Standards of Care for Infertility Nurses which will be published shortly. These will provide a framework which nurses can use to establish and maintain high standards of care.

It is important that training for infertility nurses complies with the requirement of postregistration education. It is essential that the nursing role is identified and that the skills and experience gained are assessed so that work in this specialised field is valued as part of career development. The RCN Institute of Advanced Nursing Education (IANE) are in the process of developing a programme of training for infertility nurses. This will encompass all aspects of infertility and participants will be assessed.

Counselling Skills Training

The RCN recommends that in addition to the above training course all nurses working in the field of infertility should have the opportunity of obtaining counselling skills and other specialised training. There are many counselling courses already established. The RCN IANE organises a one year certificated course in counselling skills.

This course uses a theoretical framework to increase communication and self awareness and to extend knowledge and understanding of ways in which theories from the field of developmental and social psychology can inform the principles and practice of counselling in health care.

Formal Counselling Training

Counsellors working in the field of infertility at the moment come from many different professional backgrounds such as nursing, social work, psychotherapy, psychology, theology and teaching. There are a number of nurses who have had counselling training as distinct from counselling skills training. They may be taking a full counselling role and undertaking implications and therapeutic counselling. It may be necessary for them to supplement their training with a specialist course such as the five modules course described in KFCC Report [14].

Nurses who plan to undertake a counselling training in order to take a "proper" counselling role need to be clear about the level of training provided by the course they choose. They must be aware of the tasks involved in infertility counselling and at which level they wish to work to select appropriate training.

In addition to adequate training, before a nurse undertakes a counselling responsibility, there is a need to ensure that sufficient time and an appropriate counselling setting is available. Ideally the counselling task should be the whole or major part of the role undertaken. It should not be possible for other members of the team to make demands on counselling hours. If roles are mixed it may be difficult to maintain professional boundaries, gain the trust of the client and maintain confidentiality.

Conclusions

Infertility nursing and counselling should be seen as an evolutionary process, adapting and changing as needs develop and are identified.

It is common experience that counselling provision is inadequate. The length of time required to train a counsellor means that it will be some time before there are sufficient trained individuals available, while those already trained may need specialist training in infertility. The limited research available seems to indicate that not withstanding the recognised need counselling uptake may be low. If provision and uptake continue to be low, and the need remains high, then the nursing and medical staff will continue to be the emotional as well as the medical caretakers. There is an urgent need to provide counselling skills in this area.

The RCN would wish to encourage developments in this area, but not at the risk of producing rigid or bureaucratic guidelines. Those who wish to work in an extended role are recommended to have appropriate training and accept accountability for that role. Nurses who wish to make a transition to a counselling role are advised to embark on a course of training which will ultimately lead to accreditation. They should abide by a clearly defined code of ethical practice and identify to whom they are accountable for the counselling process.

The larger group of nurses working in the field of infertility should be given an opportunity to acquire the appropriate counselling skills to enhance their performance in their role as soon as possible. It is abundantly clear that no one discipline can fulfil all the needs of the patients. Medical reproductive technology has raced ahead and counselling provision needs to catch up.

References

1. Human Fertilisation and Embryology Act 1990; 13(6); Schedule 3 para 3(1)(a), London: HMSO.
2. Royal College of Nursing Fertility Nurses Group. Report of Professional Survey, 1990.
3. Coughlan WC. Marital breakdown. New York: Columbia University Press, 1965.
4. Schellon AMCM. Artificial insemination in the human. New York: Columbia University Press, 1957.
5. British Infertility Counselling Association. Guidelines for counsellors, 1992.
6. Menning B. Infertility: a guide for the childless couple. Englewood Cliffs, NJ: Prentice Hall, 1977.
7. Mahlstedt P, Macduff S, Bernstein J. Emotional factors and the in vitro fertilisation and embryo transfer process. J In Vitro Fert Embryo Trans 1987; 4:232–6.
8. Raphael-Leff J. Infertility: diagnosis on life sentence. Br J Sexu Med 1986; 13:129–32.
9. Connolly KJ, Edelman R, Cooke ID. Distress and marital problems associated with infertility. J Reprod Inf Psychol 1987; 5:49–57.
10. Bresnick E, Taymor ML. The role of counselling in infertility. Fertil Steril 1979; 32:154–6.
11. Freeman EW, Boxer AS, Rickels K, Tureck R, Mastroianni L. Psychological evaluation and support in a programme of in vitro fertilisation and embryo transfer. Fertil Steril 1985; 43:48–53.
12. Edleman R, Connolly K. The impact of infertility and infertility investigations: four case illustrations. J Reprod Inf Psychol 1989; 7:113–19.
13. Dennerstein L, Morse C. Psychological Issues in IVF. Clin Obstet Gynaecol 1985; 56:316–22.

14. The Report of the King's Fund Centre Counselling Committee (KFCC). Introduction. Counselling for Regulated Infertility Treatments, 1991.
15. United Kingdom Central Council for Nursing, Midwifery and Health Visiting (UKCC). Code of Professional Conduct for Nurse Midwife and Health Visitor, 2nd edn, 1984.
16. British Association of Counselling (BAC). Code of Ethics and Practice for Counselling Skills.
17. Buckroyd J, Smith E. Learning to help. Nursing Times 1990; 86: No 35.
18. Segal J. Counselling people with multiple sclerosis and their families. Counselling and Communication in Health Care, Chichester: Wiley, 1991.
19. Raphael-Leef J. Psychological processes of childbearing. London: Chapman and Hall, 1991.
20. Code of Practice. Human Fertilisation and Embryology Authority (HFEA).

Chapter 25

Training and Accreditation of IVF Embryologists

S. G. Hillier

Introduction

Embryology, as a clinical laboratory activity, had no practical relevance to reproductive medicine before the late 1970s, when in vitro fertilisation (IVF) and embryo transfer were introduced to treat tubal infertility. Nowadays, the embryology laboratory is each IVF centre's linchpin, and the scientist in charge of it carries responsibilities that crucially affect the patient's wellbeing and the outcome of treatment.

In this chapter I briefly survey the events that have led to the emergence of IVF embryology as a specialised clinical laboratory function during the past decade or so. I also assess current options for the training and accreditation of this new breed of clinical scientist. Finally, I conclude with a personal recommendation that an "Association of Clinical Embryologists" should now be constituted to promote the advancement of IVF embryology in the UK.

The Pioneering Phase of IVF Embryology

The scientists who pioneered IVF embryology were mainly gamete biologists who earned their academic spurs through laboratory-based studies of fertilisation and early development in mammals. Pre-eminently, it was the work of the Cambridge physiologist R. G. Edwards, in collaboration with his late colleague, the gynaecologist P. C. Steptoe, that led to the birth of the world's first "test tube" baby in 1978 [1]. Edwards' research during the twenty years or so leading up to that signal event included far-reaching experimental studies of ovarian follicular development, oocyte maturation, fertilisation and preimplantation in mice, and latterly human beings. Those fundamental research contributions and his subsequent innovative application of them to reproductive medicine qualify Edwards as the original IVF embryologist [2].

With the demonstration that IVF followed by embryo transfer was feasible as a means of treating tubal infertility, the new challenge was to increase its effectivess and practicability. Among those IVF scientists who made particularly important contributions at this stage was A. O. Trounson, an Australian reproductive biologist who had worked in the mid-1970s as a postdoctoral fellow with R. M. Moor at the Agriculture and Food Research Council Institute of Animal Physiology, Cambridge. Trounson's work focused mainly on the relation between oocyte developmental potential and follicular maturity in domestic animal species [3], allowing him opportunities to hone basic techniques of gamete-handling and embryo culture that were ultimately applicable to human IVF. Back in Australia, Trounson teamed up with clinical colleagues in the Department of Obstetrics and Gynaecology at Monash University, Melbourne, and set about exploiting those skills to establish what was one of the world's most influential and successful IVF clinics by the end of the 1970s [4].

Gamete-handling and embryo culture techniques initially employed by the Monash team rapidly entered into widespread clinical use [5], due to their efficacy as well as the unit's open policy towards training other IVF scientists and clinicians. Several other IVF centres also greatly influenced contemporary IVF laboratory practice through their published work and/or provision of training facilities – notably Bourn Hall (Cambridge, UK) [6], the Howard and Georgeanna Jones Institute (Norfolk VA, USA) [7], and the Royal Women's Hospital (Melbourne, Australia) [8]. However, many, if not most, of the IVF clinics opened at the beginning of the 1980s owed initial successes to their use of working practices that originated at Monash [9].

Thus IVF embryology was developed into a clinical laboratory activity by adapting techniques designed initially for basic experimental research. Even culture media originally formulated for mouse embryos [10] were used empirically to culture human embryos [11]. Many IVF clinicians of the day treated their patients in the same vein, leading to various sensational "breakthroughs" that heightened public awareness of IVF and, with hindsight, highlighted the need for legislation to regulate both laboratory and clinical practice in this new field of reproductive medicine (Table 25.1).

IVF embryology continued to evolve in this piecemeal fashion throughout most of the 1980s. During much of this period, standards of laboratory practice varied considerably from centre to centre and there was no generally agreed way of working. The situation began to improve as opportunities for contact between centres at workshops and meetings increased. Eventually, "state-of-the-art" procedures became more widely adopted, and it came to be generally accepted that minimum standards of training and practice for IVF embryologists were desirable, as proposed initially by the Interim Licensing Authority for Human Fertilisation and Embryology (ILA) (formerly the Voluntary Licensing Authority or VLA) [12], and subsequently endorsed by the Human Fertilisation and Embryology Authority (HFEA) [13] (Table 25.2).

Table 25.1. Events leading up to the enacting of legislation covering IVF research and practice in the UK

Year	Event	Country
1982	Warnock Commission set up to inquire into human fertilisation and embryology	UK
1982	Britain's first IVF twins born	UK
1983	World's first IVF baby from a frozen-thawed human embryo	Netherlands
1984	Britain's first IVF triplets born	UK
1984	Britain's first IVF quads born	UK
1984	Birth of world's first GIFT baby reported	USA
1984	World's first IVF baby born following egg donation	Australia
1984	Warnock Report issued, recommending establishment of statutory licensing authority to regulate IVF and embryo research	UK
1985	Voluntary Licensing Authority (VLA) is set up by the Medical Research Council and RCOG	UK
1985	Britain's first IVF baby born from a frozen–thawed human embryo	UK
1986	Britain's first baby born following GIFT	UK
1987	Britain's first IVF baby born following egg donation	UK
1987	White paper published - "Human Fertilisation and Embryology: A Framework for Legislation"	UK
1989	The VLA becomes the ILA (Interim Licensing Authority)	UK
1989	"Human Fertilisation and Embryology Bill" published and introduced to the House of Lords	UK
1990	First baby born following preimplantation genetic diagnosis (gender selection to avoid X-linked genetic abnormality)	UK
1990	First IVF baby born following sperm microinjection	Italy
1991	HFEA takes over from the ILA	UK

Contemporary IVF Embryology

In a modern IVF centre, the embryologist carries out multiple day-to-day functions that crucially affect the quality of service provided. As well as being responsible for the IVF laboratory (Table 25.3), the IVF embryologist must

Table 25.2. Extracts from the HFEA Code of Practice dealing with IVF embryology

Qualifications

Paragraph 1.11 "The person in charge of an embryology laboratory should have an appropriate scientific or medical degree, plus a period of experience in an embryology laboratory sufficient to quality the person to take full charge of the laboratory."

In-Service Training

Paragraph 1.14 "Centres should arrange relevant staff training for all staff taking part in specialist scientific, clinical and counselling activities for which existing formal qualifications are not entirely sufficient. Centres with too few staff to provide adequate training themselves should make arrangements for staff to be trained where there are such facilities. All staff taking part in specialist activities should also receive regular updating."

Laboratory Facilities

Paragraph 2.7 "It is essential that centres follow good laboratory practice, whether their laboratories are used for research or for clinical services."

Paragraph 2.8 "All blood products, other than those of the woman receiving treatment, with which gametes or embryos might come into contact should be pre-tested for HIV and hepatitis-B."

Paragraph 2.9 "The room where eggs are collected for in vitro fertilisation should be as close as practicable to the laboratory where fertilisation is to take place."

Maintaining and Improving Standards

Paragraph 2.17 "Centres should have procedures for improving and updating laboratory, clinical and counselling practice, so that every effort is made to achieve optimum procedures and outcomes by the standards of professional colleagues elsewhere. . . "

co-ordinate laboratory function with other elements of service provision, as illustrated in Fig. 25.1. This includes liaison with other departments and with clinicians and nurses on the IVF team, and it may also extend to contact with the patients themselves. In many centres the embryologist confronts patients personally with information about the progress of their treatment or its outcome. This requires communication skills and levels of sensitivity to patient needs not usually required of a laboratory scientist.

IVF laboratory practice also impacts directly on patient wellbeing, independent of nursing and clinical input. An example of this is the serum supplement added to embryo culture medium: patient's own serum, donor serum, pooled donor serum, or a commercially available human serum albumin preparation? The embryologist decides, ensuring that the material is processed aseptically, not damaged during storage, and if necessary pretested for HIV and hepatitis-B. In the Netherlands, 4 years ago a situation came to light in which several IVF patients had contracted hepatitis-B through receiving embryos transferred in medium supplemented with a batch of hepatitis-B-positive donor serum, providing a vivid example of the potential hazards of deficient laboratory practice [14].

Table 25.3. Expertise in the IVF laboratory

• Basic tissue culture technique
• Oocyte assessment and handling technique
• Semen assessment and sperm handling technique
• Embryo assessment and culture technique
• Documentation of working laboratory procedures
• Maintenance of accurate patient records
• Operation of quality assurance procedures
• Gamete and embryo cryopreservation
• Familiarity with follicular monitoring protocols (endocrine and ultrasound)
• Liaison with clinical, nursing and clerical staff
• Sensitivity to patients' needs
• Familiarity with relevant scientific, legal and ethical issues

A familiar problem faced by IVF embryologists is that attempts at treatment are all too frequently unsuccessful. Even when morphologically "normal" embryos are transferred to a patient but pregnancy does not occur, it is commonly assumed that deficient practice in the embryology laboratory is somehow to blame. It is therefore incumbent on the embryologist to adopt, and be seen to adopt, exacting standards of practice that are not the limiting factor in treatment outcome. This extends to running appropriate quality assurance programmes (e.g. weekly mouse IVF, or mouse embryo cultures [15]) and participating in in-service training, similar to scientists in longer-established branches of laboratory medicine such as chemical pathology, haematology and cytogenetics etc.

Thus, IVF embryology has come of age and its exponents should reasonably be regarded as true paramedical professionals. The question therefore arises as to who is now responsible for overseeing their training and accreditation, and for defining the standards of practice they should follow.

Professional Standards

In August 1991, the HFEA brought into force a statutory licensing system covering all centres engaged in human embryo research and infertility treatment

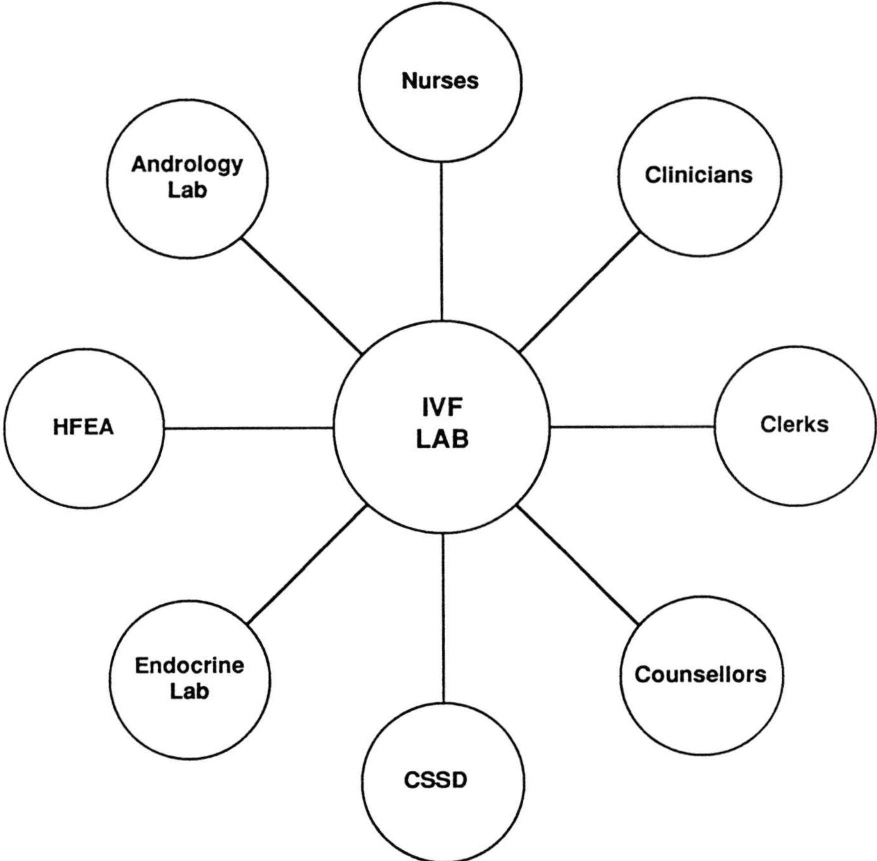

Fig. 25.1. The IVF embryologist's sphere of interaction. CSSD, Central sterile supplier department.

involving in vitro fertilisation or donor insemination. A Code of Practice (COP) was published setting out standards to which IVF centres are expected to operate in order to be licensed. Sections of the COP dealing with staff engaged in scientific services are given in Table 25.2 [13].

The HFEA COP exhorts IVF laboratories to follow good laboratory practice (GLP), leaving it to embryologists themselves to decide what GLP is. Most IVF embryologists use facilities, techniques and equipment that are already proven and in widespread use, being known to be safe for patients and their embryos and to be capable of sustaining acceptable levels of treatment success. Thus to a considerable extent, maintenance of acceptable standards of work practice in IVF laboratories is automatically achieved through professional peer pressure and the drive to treat patients successfully.

So should GLP in the IVF laboratory be defined any further, and if so, by whom? Those best placed to answer these questions are IVF embryologists themselves, many of whom feel that minimum guidelines are desirable but ought not to be overly prescriptive. Guidelines that have been offered frequently fail in this regard. For example, preliminary guidelines published by the Safety and Standards Committee of the European Society of Human Reproduction and Embryology (ESHRE) [16] cover the minutest details of laboratory procedure, including water purification systems, osmolarity and pH optima of embryo culture media, the frequency at which culture incubators should be cleaned, and even advice on labelling culture tubes.

Training and Accreditation

Unlike the other professionals (clinicians and nurses) involved in the provision of IVF services, embryologists have no college or society to regulate training and accreditation, oversee maintenance of acceptable standards of practice, provide representation in the event of litigation, and perhaps above all else, give them professional credibility.

Currently, the standard of practice followed in individual IVF laboratories is variable, and is established according to the senior scientist's own level of experience and expertise. Among the 72 or so IVF centres currently registered with the HFEA, several large and particularly well-established ones have sufficient staff and expertise to be able to offer training facilities to staff from other IVF centres. Some even run workshops and residential training programmes. Thereby, most IVF scientists become familiar with state-of-the-art techniques and procedures sucessfully practised elsewhere. However, whether they choose to adopt such practices themselves, and whether they teach them to their own juniors, are matters of individual judgement.

The HFEA COP currently recommends that the director of an IVF embryology laboratory should have an appropriate scientific or medical degree plus "a period of experience in an embryology laboratory sufficient to qualify the person to take full charge of the laboratory" (Table 25.2) [13]. Neither the duration nor the nature of that experience is specified. What are these minimal criteria, who should define them, and who should see that they are followed?

If pressed on these issues, most IVF embryologists in the UK would agree that even a junior embryologist should at least have earned a degree or equivalent in a relevant subject, followed by a minimum period (say 6–12 months) of hands-on experience in an established (licensed) IVF centre, before being allowed to carry out solo IVF procedures. A laboratory director, i.e. the senior embryologist, would usually be expected to hold a higher degree such as an MSc or PhD. However, further prescription of training requirements beyond an agreed minimum level is generally felt to be unnecessary.

A potential difficulty that might arise in prescribing in any detail the training that IVF embryologists should receive is illustrated by developments in the USA. In the absence of formal proposed criteria for training and accreditation by a recognised professional body such as the American College of Pathologists, the American Fertility Society has published detailed guidelines covering aspects of professional staffing and the integration of clinical and laboratory procedures in IVF [17]. The section dealing with the qualifications of the embryo laboratory director is typical of the heavily prescriptive tone of the document as a whole, proposing among other things that such a person "should have had a period of training of at least 6 months and at least 60 completed ART [Assisted Reproduction Technology] procedures in a program that performs at least 100 IVF procedures per year with a minimum annual 10% IVF live birth rate per retrieval cycle".

Few IVF embryologists in the UK are members of any of the professional bodies (e.g. Royal College of Pathologists or Association of Clinical Biochemists) that oversee the training and accreditation of scientists in other clinical laboratory specialities, and currently there is no equivalent organisation with direct relevance to embryology. Some belong to the Institute of Biologists or the Institute of Medical Laboratory Scientists, but these organisations currently do not make special provisions for IVF embryology. Many more IVF embryologists are members of learned scientific or clinical societies such as the Society for the Study of Fertility, the British Fertility Society or ESHRE. However, none of these bodies is constituted to oversee the professional interests of any particular specialist group from within its wide-ranging scientific and clinical membership. Accordingly, it would seem that a new forum is now required, constituted by IVF embryologists themselves to oversee training, accreditation and practice in this unique area of laboratory medicine.

Summary and Recommendations

1. IVF embryology evolved as a specialist clinical laboratory activity through piecemeal adaptation of techniques initially developed for research purposes from experimental studies of animal gametes and embryos in vitro.
2. The pioneering phase in the establishment of IVF embryology was concluded in the late 1980s, when IVF success rates reached a plateau and contemporary technique became more or less standardised.
3. The sphere of influence and the responsibility of the IVF embryologist extends beyond the laboratory itself, and impacts crucially on patients' wellbeing and other aspects of service provision.
4. Professional peer pressure and the drive to achieve acceptable rates of clinical success are major motives for IVF embryologists to use state-of-the-art procedures and follow GLP.
5. Agreed minimum levels of training, and guidelines covering acceptable work practices would probably be welcomed by most IVF embryologists. However, any such standards should not be over-prescriptive. Moreover, they should be drawn up by IVF embryologists for IVF embryologists; ideally with the support and encouragement of the HFEA.
6. This move towards the professionalisation of IVF embryology should also include the formation of an Association of Clinical Embryologists. This association should have as its major objective the advancement of IVF embryology in the UK through promoting appropriate standards for training, accreditation and practice, and providing a corporate voice for those scientists engaged in this newly established paramedical activity.

References

1. Edwards RG, Steptoe PC, Purdy JM. Establishing full term pregnancies using cleaving embryos in vitro. Br J Obstet Gynaecol 1980; 87:737–56.
2. Edwards RG. Conception in the human female. London: Academic Press, 1980.
3. Moor RM, Trounson AO. Hormonal and follicular factors affecting maturation of sheep oocytes in vitro and their subsequent developmental capacity. J Reprod Fertil 1977; 49:101–9.
4. Trounson AO, Wood C. In vitro fertilization results, 1979–1982, at Monash University, Queen Victoria and Epworth Medical Centres. J InVitro Fert Embryo Transfer 1982; 1:42–7.
5. Trounson AO, Mohr LR, Wood C, Leeton JF. Effect of delayed insemination on in vitro fertilization, culture and transfer of human embryos. J Reprod Fertil 1982; 64:285–94.
6. Edwards RG, Fishel SB, Cohen J et al. Factors influencing the success of in vitro fertilization for aleviating human infertility. J In Vitro Fert Embryo Transfer 1984; 1:3–23.
7. Jones HW Jr, Jones GS, Acosta AA et al. The programme for in vitro fertilization at Norfolk. Fertil Steril 1982; 38:14–21.
8. Lopata A. Concepts in human in vitro fertilization and embryo transfer. Fertil Steril 1983; 40:289–301.
9. Wood C, Trounson AO, comps. Clinical in vitro fertilization. Berlin: Springer Verlag, 1984.
10. Whittingham D. Culture of mouse ova. J Reprod Fertil Suppl 1971; 14:7–21.
11. Trounson A, Hoppen HO, Lutjen PJ, Mohr LR, Rogers PAW, Sathananthan AH. In vitro fertilization: the state of the art. In: Rolland R, Heineman MJ, Hillier SG, Vemer V, eds. Gamete quality and fertility regulation. Amsterdam: Excerpta Medica, 1984; 325–43.
12. The Fifth Report of the Interim Licensing Authority for Human In Vitro Fertilization and embryology. London: Interim Licensing Authority, 1990; ISBN 0951-7480.
13. Human Fertilisation and Embryology Authority Code of Practice. London: HFEA, 1991.
14. Staatstoezicht op de Volksgezondheid. Onderzoek Hepatitis B besmetting bij In Vitro Fertilisatie. Rijswijk, The Netherlands: Staatstoezicht op de Volksgezondheid, 1988.
15. Hillier SG, Dawson KJ, Afnan M, Margara RA, Ryder TA, Wickings EJ, Winston RML. Embryo culture: quality control in clinical in vitro fertilisation. In: Thompson W, Joyce DN, Newton JR, eds. In vitro fertilisation and donor insemination (Proceedings of the Twelfth Study Group of the Royal College of Obstetricians and Gynaecologists). London: Royal College of Obstetricians and Gynaecologists, 1985; 125–40.
16. European Society of Human Reproduction and Embryology. Safety and Standards Committee: Preliminary Recommendations. ESHRE Newsletter No. 9, 1989; 14–20.
17. The American Fertility Society. Revised minimum standards for in vitro fertilization, gamete intrafallopian transfer, and related procedures. Fertil Steril 1990; 53:225–6.

Discussion

Fishel: This week, interestingly enough, in the Nottingham Region there was a public "information awareness" meeting, on infertility in general. It was very revealing, but there was something that I wanted to pick up on Ms Inglis's presentation and relate it to Professor Cooke's material. Her comment was that it is quite clear that the emotional wellbeing of the couple is affected drastically by protracted investigations. It was quite clear at the meeting that we had, from the comments that people were making, that these couples may do best after preliminary investigations to see an infertility specialist – somebody who spends a lot of their obstetrics and gynaecology time in infertility – as opposed to the obstetrician and gynaecologist who hangs on to his infertility patients. Yet that angered the obstetricians and gynaecologists enormously; they were tremendously offended by such a suggestion. But then we get the patients' experiences, and most of us know of those experiences: they go on and on with inappropriate therapies.

The second point relates to the embryologist. A lot of what Dr Hillier had to say really touched my heart. Another question we might ask is: who are the embryologists today? It is wrong to assume that the majority of them are professionally well qualified. I do not mean to knock them; many of them are technically very good, but nonetheless, as a profession we are not protecting ourselves. Somebody, who has been doing hormone tests in the laboratory may become an embryologist, having never seen a sperm or an egg in his life. That person will struggle along doing the embryology. Dr Hillier made the point but I want to stress it. The embryologist has paternity in his or her hands, and that is a really onerous task apart from all the other tasks as well. It is something that the RCOG should take on board, and move the standards much further forward.

Braude: I should like to endorse that. People do not even know what to call themselves when they are embryologists. We see them call themselves senior scientists, or consultant embryologists, and there needs to be some kind of professional credibility.

The point that Dr Hillier did bring up is that there is no other body which it impinges on solely apart from the RCOG. Might it be possible to have some category such as Licentiate of the RCOG for those who are clinical embryologists, a qualification that can be licensed, and the training that goes with it, to provide that kind of credibility that perhaps embryologists are seeking.

Hull: There is a precedent for emerging specialties. Take anaesthesia, for example. For many years there was a faculty of anaesthesia within the Royal College of Surgeons. They developed their own degrees and a fellowship for the Faculty of Anaesthesia, and now they have their own college.

There is no reason why there should not within this college be developed a faculty of fertility specialists, or clinical embryologists, which could act in that very way and possibly blossom in the future as the College of Anaesthesia has.

Hillier: It was an afterthought, but it is so obvious. I could kick myself for not promoting it more visibly earlier. If we are recommending that the embryologists should get their house in order, that is all well and good. We all accept that. But it has to be done against the backdrop and within the accommodation of a far more authoritative and experienced group of people.

In struggling to think who that might be, and in considering the Royal College of Pathologists, the Institute of Biologists and other bodies, I had overlooked the obvious fact that we are part of the Royal College. Most of us here have been here so many times that we almost feel a part of it, and it would be a logical move to regularise that.

Watson: At the end of his talk, Professor Cooke spoke of various scientific societies and I was disappointed that he did not include the British Andrology Society. They are a group that have some relevance to these discussions.

Ms Inglis included counselling and nursing, and that was her brief. But would she comment on the conflict of roles in nursing and counselling. Is this an argument for independent counselling?

Inglis: My heart sank when I read the title because I felt it was too difficult a brief. Obviously there are two disciplines here. Was there some particular problem?

Watson: No. Just the general overall message that came across. I wondered if there was a case for independent counsellors in the infertility setting.

Inglis: This has been discussed. I always felt independent. I do not know how an independent counsellor would be defined but to me it is someone who is autonomous within the team and has their separate consultative support. I always had that in the department in which I worked. But is it to be someone who geographically is outside of the unit?

Watson: I was not thinking geographically. I was thinking in terms of relationships with the patient, and not having a confusion of roles.

Inglis: That was what I was suggesting, that the nursing and counselling roles should be separate. I started off having a slightly mixed role and it did have conflicts, because people did tend to think that if your patient was late then you could help out. But gradually I became more and more independent. If the nurses want to make a transition to a counselling role, they have to have the same conditions as any other counsellor would have, and that is to be separate. They should not be expected to have a nursing role, or at least the major component of their work should be a counselling one.

It is difficult to be absolutely bureaucratic about it and say they will not do any nursing work, but ideally they will be seen as separate and independent.

Nieschlag: As the only participant from mainland Europe I feel obliged to comment on Professor Cooke's presentation and his recommendations.

We are on the edge of a new phase, where doctors and specialists will move easily across borders. At that point in time it may be useful to look cross border and to establish contacts with societies and bodies in other countries, and maybe European organisations. On a society level I should like to mention the European Society for Human Reproduction, and on the organisational level, the European Community in Brussels. At this time, if it is not already too late, it should be a commitment to look for European connections. They have become matters of fact and we have to live with them.

Cooke: I agree absolutely. The poem written by the Poet Laureate on the occasion of the election commented on the declining sperm count of men in Western Europe due to toxins, and therefore a vote for the environment and andrology obviously has a great future!

Abdalla: A dichotomy exists between the private sector and the National Health Service in training in infertility. That dichotomy is partly because the private sector is not monitored. Part of the development of our practice has happened in the private sector, but it is autonomous, and there is no relationship between the College, the Health Service side and the private sector, at least in some units. I have a feeling that it is a two-way street. There is a need to train infertility specialists in the private clinics, but at the same time these private clinics should have, in my opinion, the good fortune to be audited, and each one of their practices should be supervised.

We need to have some kind of a working party where this can be discussed.

One other point. I hope we are practising counselling, but I have the feeling that the rôle of the doctor in counselling is never discussed – as if the doctor had no role in counselling patients but was confined to acting as a technician who would sit and say, "Let us try this", and so on. I believe that the doctors have an immensely important role in spending time on explanations and on support counselling as well. That is not a substitute for the independent counsellor or for the nurse, but I look on it as a team role. If we put all the emphasis on the role of the independent counsellor or of the nurse, we shall be delegating the role of the doctor to that of surgeon or technician.

Inglis: I thought I had explained that.

Drife: My comment was stimulated by a question addressed to Ms Inglis and it happens to echo Mr Abdalla's.

I could not believe my ears when I heard someone suggest that there is a conflict of roles between a counselling role and a nursing role. I have been wondering if I am the only member of the group to have been distressed by the idea that either one is a doctor or one is a counsellor. I feel I am losing touch with reality in my perception of what doctoring is all about, and what we try to teach that doctoring is all about. If I have to employ someone else to speak to the patient, it beggar's belief.

On a recent visit to London a piece in the *Evening Standard* caught my eye – a colour feature on the cover naming "London's 100 top specialists". Not being a London specialist, I bought a copy and turned to the gynaecology section. This was prefaced by a paragraph which said that gynaecologists by and large are people who do not form good relationships with their patients and who cannot talk to their patients. Ours was the only specialty within medicine where they had elected to include a paragraph criticising the relationship the gynaecologists have with their patients. If we draw this sharp comparison between doctoring and counselling, we are acquiescing at this process that has already gone too far.

Winston: I must take strong issue with Professor Drife. I do not know that publication and I am surprised he should pay so much attention to the press.

The issue is an important one and there is a confusion here, and as a medical practitioner perhaps I might speak for the counsellors.

There is an independence about counselling, which is the exquisite thing. It is not that we do not talk to our patients, but if the patient perceives that the person counselling them is involved in some way with the delivery of treatment, it alters the relationship. The counsellor, in the role that I believe is important, is truly independent of that treatment, and therefore is not influencing that treatment. In my clinic where there are three counsellors who are truly independent, they do not influence the treatment, they do not necessarily tell me what they discussed with the patient, they may get quite critical and analytical about our treatment and I believe that is very much for the health of my patients.

Hull: To address a number of the issues raised by Professor Cooke. The College is prepared to recognise private facilities for training. For example, in Bristol our private service (perforce we have had to go private in IVF) is recognised by the RCOG as part of our subspecialty training centre. So that mechanism already exists. The bigger problem is funding for trainees. Whilst there is manpower approval for a senior registrar for that post, there are no funds.

I am more concerned, however, about manpower approval for any other trainees within the subspecialty. Whilst there are theoretical mechanisms for increasing the number of consultants on the basis of arguments of workload and so on, there is no such theoretical argument for the juniors. If services and training are to improve without any question there will have to be an increase in the number of training posts. And I do not know how we argue for them.

I should like to raise the issue of how much training is it appropriate to give people at this level, not for subspecialty training, but as part of their general pre-MRCOG experience. Anybody at registrar level has got to be in the reproductive medicine unit for at least six months to achieve reasonable competence, and a year to be of any use whatsoever. If that is so, it is not possible to work rotating programmes for the present numbers of juniors, and we would probably have to make an arrangement whereby the rotation involved a supernumerary attachment to such subspecialist units for periods that are probably 6 months at the most. One way or another there has got to be an increase in training posts. How do we argue for that? And could I ask for some views about the length of time that trainees should spend in such a unit?

Cooke: I agree absolutely. I was not addressing the mechanism, I was saying that this was what needs to be done. I agree that it would take six to 12 months, preferably 12, to induce a level of knowledge. My fear is that unless we do that, there will be a more and more exclusive enclave of people who understand quite a lot, and the rest, who will form the vast majority, will know nothing about assisted reproduction. That will be a major problem for the future when they move through to consultant posts.

How we do that requires a great deal of argument, with JPAC in particular, and through them the Department of Health.

Hull: I wanted to take the argument beyond the principles, to spelling out some of the detailed requirements that the College will have to argue for in the future.

Cooke: This is not the place. We have to do what is possible: we have to have long-term objectives and we have to address what is possible through the Department of Health.

There should be an increase in number and they should rotate on the general basis. I do not think that everyone ought to rotate, but say a third of all registrars ought to have an opportunity to rotate for a year through a reproductive medicine unit. But unless we get the distribution, the staffing, and the economic support for the units and the trainees, that is absolutely impossible. We cannot look at the training programme in isolation from all the others.

Hull: But it would be wrong for this meeting to accept that there is an impossible limitation. Surely we should use this opportunity to spell out what is required.

Baird: In Lothian we have faced this issue. All the registrars in the region do rotate through reproductive medicine in its widest sense, not just IVF technology, but also medical abortion and all the other things.

But let us assume that eventually the National Health Service does take on some responsibility for assisted conception, which it will have to do eventually. There will be some kind of limitation on the degree of resources. What impact is that likely to have on the necessity for prescribed professional accreditation for non-medical staff, that is, for clinical gamete biologists, or clinical embryologists?

Hillier: It would clearly make it more pressing that the system became available and the mechanism was enacted. But to some extent it would tend to stratify the problem, which has two levels. I did not want to confuse the topic by speaking of senior embryologists and laboratory technicans, but there is no question in my own mind that the Institute of Laboratory Scientists, which is responsible for state registration of technicians, will have to take on board the subspecialisation of laboratory activities. That is part of the answer to the question.

But at the other level it would be nice to be able to say that the people who work in these laboratories in an academic scientific capacity would become like the clinical biochemists, the bacteriologists, and so forth, and that they would be recognised as one of the clinical scientific fraternity. But I would again caution against the notion that it will be possible to prescribe and regulate the training of these individuals in quite the same way as the other laboratory technicians.

Baird: I am not suggesting it will. I am suggesting that that will be required by an employer: there is no question about it. The reason that it has been allowed to persist unrecognised, as at the moment, it because 95% of this activity is confined to the private sector. But the minute the NHS becomes the employer, the employee will be required to show proof of a certain level of qualification and professional expertise, which would not be tolerated in the private sector.

Unless this issue is faced fairly soon, we shall get prescription from the existing bodies, such as the Institute of Medical Laboratory Technicians and the Royal College of Pathologists, which may prescribe an inappropriate level of training and qualification.

Hillier: If I may reinforce that, I agree entirely. It is extremely helpful to have the point made so eloquently and strongly. It is now the last and the most realistic opportunity that we have to put our house in order. There has not been the pressure to do this in the past, but now is the time we must do it.

Campbell: I found the session very interesting. When the HFEA began, we were a rather large body, hardly knew each other and did not really have a staff. And we had seven months to design a full national regulatory system. Admittedly we were able to look at some of the examples of the Interim Licensing Authority (ILA), but we had a much tougher and broader job to do.

I am not passing judgement on clinical practices – I am a layman though there are inspectors to advise – but what has struck me since we have begun our work is the range of standards in the field. It has rather surprised me. I thought there might be a range of "1 to 3" but there is a range of "1 to 8" or"1 to 9". I would have thought that one of our tasks, working with you, would be to try to raise those at the bottom to higher standards, and we are already asking a number of you for permission – where we see you have done something particularly good or impressive, can we borrow it to show your best trick to someone else and thus raise standards?

The areas we have been discussing – training, counselling and embryologists – are all areas that so far are only dealt with very slightly by the authority. We will have to turn to them in more detail in the future, and this sort of discussion, and any conclusions you come to when you draft your recommendations, will be very helpful.

Section VI

Service Provision

Chapter 26

The Patients' Viewpoint

J. Dickson

There are many myths and misconceptions about infertility and its treatment. The view of the general public appears to be that infertility is not a problem. This springs first from the idea that it is a 1:1000 risk, when we know that 1 in 6 couples will need to seek help with their fertility at some time in their fertile lives (Chapters 2 and 3). This is further compounded by those who can have children seeing the problem as something that the infertile should be able to handle. Where do these prejudices and misunderstandings come from?

People's expectations of health and health care are very high. We are a very healthy nation compared to a couple of generations ago, as well as being more health conscious than in the past. Infertility treatment has come a long way in the last 15 years. To most people treatment of the infertile is in vitro fertilisation (IVF), or, as I am sure they would call it, "Test Tube Babies". The public would expect that the few people who need this treatment would be able to get it on the NHS. The point that I am trying to make is that there is a great deal of difference between the world of infertility and its treatment and the public's expectations. Yet purchasing managers in the Health Service, MPs, and even Government Ministers are only reflecting the public attitude that infertility should not be given a high priority.

I have been told, when campaigning for better treatment in the NHS, that treatments like gamete intra-fallopian transfer (GIFT) and IVF are experimental and unproven, and that they lead to unacceptable levels of multiple births. The problem is that the person with these views manages a District Health Authority (DHA), and uses these arguments to cut back on the care of the infertile. Those involved in treatment have a very big job in front of them in educating those involved in health care, both about infertility, and how successful some treatment can be.

Family planning has led to control of fertility. Young people learn about fertility and how to control it, but do they learn about infertility at the same time? Fertility is the other side of infertility. There are many myths about fertility and infertility and we need to dispel them. The view that we now control our fertility needs to be mitigated by two important additions to the public health message. First, that with 1:6 couples needing help at some time in their fertile lives, infertility will strike them or somebody close to them, a brother, sister or friend. So we do need to bring infertility into sex education. Second, that women's fertility decreases. That they should be aware of both declining fertility and infertility when making their decisions as to when to start their family. There are many angry women who leave childbearing until their mid-thirties and then find they have a fertility problem. They wished they had been aware of the risks of their decision to leave having children so late in their lives.

Introducing infertility into sex education and family planning could then lead on to caring for fertility and stopping preconceptual problems that cause infertility. Sexually transmitted diseases cause some infertility, and infection like Chlamydia can cause pelvic inflammation leading to blocked tubes. The coil can exacerbate infection and, thereby, lead to infertility. The condom, being a barrier method, helps to stop the transmission of infection. How much should we be involved in warning people of the risk of becoming infertile in the first place, because much of it could be avoided if people were more aware of the chances they are taking with their own fertility.

Patients' Feelings and Loss

People are unprepared for their own infertility. If they though about it at all, they thought it was not going to happen to them. Most people find the slow realisation that they have a problem is a great blow to their self-confidence. The year of trying is a period when they feel very alone, often not able to share their problem with anyone. This is when self-help groups like Issue can give much-needed information and support.

When the patient starts having diagnostic tests, most often at a local District Hospital, these take a year. There is now no real excuse for this. The treatments are available and we should be using a standard diagnostic protocol to test aggressively for the reason for the infertility. The present system can lead to people taking two years in discovering and testing for the problem, which is very destructive to both the emotional and psychological aspects of the patients.

Infertility can be a devasting blow to a man or a woman. It strikes at who they are, and want to be, and can lead to broken relationships and mental and physical ill health. For animals, it is said that their primary function is to project their genes into future generations, so it would be surprising if failure to do so in humans did not have very dramatic costs. There is something about being human that can only be learned in families. It is fine if you choose not to be a parent, but the depression and sense of failure that can descend on those who are infertile is more than many can bear alone.

There are other types of infertility, like secondary infertility, miscarriage and ectopic pregnancy, where people feel trivialised. Each of these needs a different kind of care and attention. If you are treating any of these patients, remember their emotional needs are in many ways more important than their medical ones, but so often these are missed. These are the patients who feel angry, let down and hurt by current attitudes towards their treatment and conditions.

There is an insensitivity in producing any treatment that feels to the patient that they are on a conveyor belt. Reactions from Doctors like, "Not pregnant yet!" or the cheery "Back again?", do not square with how that patient is feeling. They are not helped by false hope, contained in, "You will get pregnant with this", or, "We will have you pregnant by Christmas". You can never be sure that you will get a patient pregnant, so do not say so.

Problems within the National Health Service (NHS)

Waiting lists have become a part of the service. The problem is that what we have in the United Kingdom looks like a system that withholds from people the treatment that they need, rather than one that rigorously tests to obtain a diagnosis, and then makes sure the patient is hastened to the best source of that treatment. The Government talks as if they would like it to work that way, but in reality, we know this must require more resources.

People are given hope that they will be treated. The long wait for a full diagnosis is tolerated on the basis that it will lead to treatment, and that the treatment will succeed. But if the final diagnosis is blocked tubes, for most couples this means they will have to go to the private sector. These people are isolated by their experience. They feel that they pay their National Insurance contributions, but that somewhere in the small print is the get-out clause which says that certain types of treatment are not provided in the NHS. So much of their time has been wasted. They scrape together the money they need (if they are able) and seek private help. This does not mean they are happy with the situation. They are emasculated and broken people and the hurt of their infertility is compounded by a society which does not care.

There must be a standard diagnostic protocol for treatment of infertility. Ideally this would be national, but to start with I would go along with a regional one like that of the Yorkshire Fertility Group. So much time, effort and money is wasted at the moment, through the time taken to complete a diagnosis. The number of times that tests are repeated is unforgivable in a service that is so short of funding. In an under-resourced area like infertility this is the one thing we could correct now to make more of the current funding.

Sometimes the testing is inadequate. The number of patients diagnosed as having unexplained infertility is around 25%, and yet I know specialists who claim that this should be as low as 10% or 5%. In some cases this poor testing is followed by the use of drugs, without any clear idea of how they will help the problem. We still hear of people being prescribed clomiphene, when no tests had been undertaken to see the state of the woman's ovulation.

Patients still complain to us that every time they visit a NHS Unit they see a new doctor. The patient has plucked up courage to seek treatment but feels she is not being taken seriously. A great deal of each visit is taken up with the new doctor getting to know what stage the treatment has reached, and there is no way that the patient can feel any confidence in what is happening to her. She becomes yet more isolated and withdrawn.

Inadequate discussion of results, either of tests or of treatment, is another common problem. This is sometimes not directly the fault of the person giving the news; for example, if you are telling the couple the results of the man's sperm tests. Once the news has sunk in that he is infertile, both of them may be so shocked by that news that they hear nothing else that is said to them. However, there are also many occasions when people have been given results in writing, where they do need the chance to discuss what they are being told. It is not just about giving information, but also being sure that the couple understand the implications of what they are being told.

Because infertility treatment has grown out of obstetrics and gynaecology, there are too few specialists in the treatment of the male. We need more andrologists in this country working as part of large specialist centres, not practising separately from the rest of infertility treatment.

The feelings of those who go on with the treatment but still fail to have a child are akin to bereavement. The loss can never be filled. We help with coping strategies but all these tend, by their nature, to be partial, needing to be rebuilt from time to time. The loss of the children they hoped to have must be faced. There is sadness for what will not now be. The anger of why this should happen has to be worked through as will the immense frustration that so much of what was planned will have to be rewritten and rebuilt. To start with, there can be a terribly frightening lack of purpose about their lives, but working through all this with those who have already travelled this road can lead to a strategy to cope with infertility.

Counselling

I find counselling one of the most difficult areas in the Human Fertilisation and Embryology Authority's (HFEA) Code of Practice. As I travel around the country and speak to licensed providers of infertility treatment, I find they are all giving counselling, but almost every provision is different. There is still a great feeling that counselling is not necessary, and that if it is, it can be done by the consultant or a member of the staff in the last five minutes of a visit.

I am sure the consultant would be horrified if the counsellor said they were going to provide infertility treatment. So why do some in the infertility field continue to deny the value of trained counsellors? When you look at the areas that counselling can be asked to cover – medical, information, supportive, therapeutic, implications, and crisis – do you really feel that you can cover all these properly?

To start with medical counselling, an expert in an area might feel that they can and should cover this, but is the person who gives a treatment the best person to describe it, and tell you about the benefits, drawbacks and risks, and what other options there may be? I have heard a professor say that a laparoscopy is a minor surgical procedure. That may be the case from the other end of the laparoscope, but it is an operation under general anaesthetic, and the gas used leaves the woman bloated and uncomfortable. If you say this is a minor surgical procedure and nothing else, then you have not prepared the patient for what is going to happen to them.

Information counselling also sounds easy. All you have to do is give information in a clear and understandable form. This we do all the time on our Fertility Helpline, which is one of the services we offer. This gives people support, information and counselling. We carry an insurance policy, because we can be sued if the information or advice we give is wrong. If you are not prepared to take information counselling seriously then you should not be doing it. Best advice may not be enough. You need to accept that bad advice may lead to somebody taking action against you.

Supportive counselling in the past has often been left to nurses. As British people find dealing with feelings very difficult, that may well mean it was not done at all. To be supportive you have to be in touch with the patients' feelings. That means dealing professionally with hurt, anger, fear, isolation, loss and broken or damaged relationships and a whole lot more. If sometimes this does not hurt you, then you are not doing it very well. It is difficult and requires a high level of personal input, and this may very well leave you feeling emotionally bruised and flat.

Therapeutic counselling is using the skills of the counsellor to make the patient better. It is often about making the patient value themselves and can come at any time in treatment though the most obvious value would come at the end of treatment. If the couple were very upset and low, the role of the counsellor would be to get that couple to view what had happened to them through their many years of treatment – get them to face the hurt, pain and frustration they have endured. They would look at the failure of the treatment, and then grieve for the child they have failed to create. This grieving is an important part of the therapy. Many people can describe in great detail the child they never had, so to talk of the death of that child is not too strong. The child that never was must die so that the parents might live. The child dies, and is grieved over, and then maybe there is hope. The psychology is clear, but this is going to be a long emotional path. Would you rather the counsellor did this?

Implications counselling is very difficult. When there is more than one option and they all have different implications, how do you help people to look at the choices that confront them? The counselling that most people do not get, and most need, is help at the start. As an infertile couple they can seek treatment, adopt either here or from overseas, or cope with their infertility. To make the correct choice they would need to know the likelihood of success or failure in each case, the costs and the risks as well as some of the stresses and strains that may be part of that choice. I think that many people, if they knew how they would feel at the end of treatment, might well choose not to undertake this in the first place.

Crisis counselling is dealing with the patient when they cannot cope. Often the need for this kind of counselling is not when the clinic is open, and it could in theory be 24 hours a day and seven days a week. Even where the response is immediate there is still going to be a great deal of time spent before the patient is over whatever is causing this cry for help. Some patients need very large amounts of time and support at these times. Can any medical professional give that sort of input, and still be available to provide treatment for their other patients?

Most of you will know that there are two major ways that the counsellor can relate to their client; these are termed directive and non-directive. My reading of the HFEA Code of Practice leads me to say that there are strands of both in their thinking.

The directive approach assumes that a particular form of action is correct and seeks to persuade the client that they should agree to that course of action. The non-directive does not seek to guide but only points up options and helps the client to look at the pros and cons of different choices that may be open to them.

To give an example, I would not usually be directive but can think of an area in which I would feel this could be justified. This is where couples are using donor sperm to have a child. The history of this treatment is that most couples, have, in the past, not told the child about the special nature of their background.

In part this is to do with the difficulty that many men have had with accepting a permanent reminder of their infertility, but I have spoken to many women who have conceived through Donor Insemination (DI) and they were also sure that this was the right action.

Implications counselling would take them through the options of telling and not telling the child. With openness the crunch comes at the beginning, as you tell people close to you, and the child when he or she is old enough. If you decide not to tell them everything is easy at the beginning, but the problem comes later as the child starts to receive sex education. I understand this can start as early as seven years old. The child may then come home and ask questions that make the parents uncomfortable. The most difficult time being as an adolescent, when the truth may come out but not at a time, or in the way, that parents would have planned it. The problem with counselling is that a couple using DI to solve their infertility are focused to their need for a child and problems seem so far off.

Maybe it would help by taking a case and looking at how the two types of counselling would approach the same problem. A couple are looking at using DI, and the counsellor is exploring what they plan to tell any children that may be produced.

In the non-directive case the counsellor would explore the two options: keeping the treatment secret from the child, or telling the child at the appropriate time. There is no right or wrong answer, but it is important that the opinions are properly explored, and all the implications are looked at carefully.

In the directive case the counsellor may firmly believe that not to tell the child is letting the family live a lie. The value of openness is stressed and the couple are encouraged to agree to tell the child. I would argue that non-directive counselling best suits infertility patients, but have in the past argued for openness myself, and I think it is too easy to choose not to tell, only to find later how very difficult that is to maintain.

Problems Caused by the New Human Fertilisation and Embryology Authority (HFEA)

As the new Authority has only been running since 1 August 1991, it may seem a little early to start to make any comment, but some effects are clear. First let me say that, in the main, I am in favour of the Human Fertilisation and Embryology Act, and the HFEA that it created. I was far more worried by the lack of legislation, but there are some things that still trouble me.

I had a telephone call recently from a woman who had a two-year-old child from DI. She had just returned from an unnamed clinic, where she had become pregnant for the first time. She said that she received treatment without any counselling, and then afterwards had been given a leaflet explaining the new HFEA Code of Practice, relevant to her as a patient.

She was upset to find that any child born as a result of the treatment she had just received would have a right to ask the HFEA if it was the child of donation. She did not want this to happen, and felt that because of this she would opt to have no further treatment. There was also the aspect that she would be asked to treat the two children differently. The earlier child would not have the same rights, and yet the desire of the couple to treat children born to them, as if they were their own genetic children. Was it really intended that this law would lead to some choosing not to have treatment, because its terms were unacceptable? The argument was that a child had a right to know its origins, but here the child will not be born. I think the price is too high, and we have a bad piece of legislation. Was any research done into the feelings of those who have to use DI? My experience is that most would choose not to tell the child. I do not think we should be using anything stronger than directive counselling to persuade the prospective parents.

Success Rates

I understand that 90% of the IVF children in this country are the result of treatment in private clinics. People who are paying for treatment want the best, but where can they find independent information about success rates? We receive many calls from those who want to know the best clinic to attend. I understand that success rates are not the only way to judge treatment, but I will continue to put pressure on the HFEA to provide this form of information. The most recent data published in 1991 by the Interim Licensing Authority (relating to 1989) showed that at least three clinics have a live birth success rate of 2.7% or less. Should we accept that patients can be treated in a clinic where the failure rate is greater than 97%?

Donors

We have noticed a disturbing trend that some women have resorted to finding their own sperm donors. They find a man prepared to give them some of his sperm and then using crude techniques they self-inseminate using fresh sperm. This raises all kind of health, moral and legal issues. We were not even sure if we should take calls from people who were thinking of this kind of action, but felt that we had to make them aware of the risks they were running.

We have also found that some women, who need egg donors, have been asked by the clinics to find their own donor. We are not sure if the intention is that the donor would then be used for the woman who had recruited her or if the donor would be swapped to keep the anonymity that the HFEA says is important. I do not think that women who need donors should be asked to help find donors in this way.

Conclusion

There are many ways that those working with the infertile can help to make their experience better than it is at present. Most of all we need to work together, the patients and all those involved in their treatment, to make the rest of the population understand infertility – how common it is, and what it feels like for those unfortunate enough to experience it. This is vital if more resources are to be put into this area of treatment.

Chapter 27

Service Provision: A Tertiary Clinic Director's View

D. H. Barlow

This book explores a wide range of topics within the field of infertility, indicating current research-based developments. The theme of this chapter is service provision. From the couple's standpoint this might be paraphrased "How do we provide couples with the most sensitive and effective service, preferably without cost?" From the health manager's viewpoint it might be paraphrased as "How do we provide the couples in our Health District with the most cost-effective service?" In giving the tertiary clinic director's view the question might be "How do we give infertile couples the best possible service and how do we translate that into a form of service provision which someone will purchase, hopefully the health authority or board?

Gynaecologists and Infertility

The investigation of infertility at the level of secondary care is usually initiated by gynaecologists although this might be provided by urologists, andrologists or endocrinologists. In British practice the vast majority of gynaecologists are generalists who provide a broad spectrum of obstetric and gynaecological care and in many centres, particularly in England, their workload greatly exceeds recommended RCOG workload targets [1]. As a result what can be offered in the investigation of infertility will inevitably be limited but will vary with the consultant's view of the relative importance of infertility problems. In many gynaecology clinics there is insufficient time and resources to support the introduction of new developments in infertility management.

Those innovations which are compatible with the basic model of gynaecological practice: the outpatient clinic and the operating list will be easier to introduce. For example, the current interest in hysteroscopic surgery and transcervical resection of the endometrium may lead to a greater interest in the hysteroscopic investigation of infertile women. New developments demanding care on an irregular pattern such as ovulation tracking, ovulation induction or those which require laboratory developments are less likely to be delivered by a secondary care gynaecology clinic unless the consultant is particularly motivated by an interest in infertility. Similarly the routine gynaecology service is unlikely to have infertility counselling facilities.

Subspecialisation

In Britain there has not been any co-ordinated planning of NHS tertiary infertility centres so that the position is patchy and in general tertiary centres tend to be in teaching hospitals. Within these there is now a group of centres which carry RCOG recognition as Subspecialty Training Centres in Reproductive Medicine. These provide an appropriate range of specialist facilities and workload to permit training in reproductive medicine at the Senior Registrar level [2]. A major component of that training is infertility work. The centres currently recognised are Aberdeen, Bristol, Edinburgh, Glasgow, Oxford and Sheffield. These centres are accredited to provide training which carries joint specialist accreditation in Obstetrics and Gynaecology and in Reproductive Medicine.

Private Clinics

The other element in British practice is the wholly private hospital clinic. These have flourished since the successful development of assisted reproduction and vary in the quality and breadth of the services provided. The best private clinics involve senior staff with international reputations but some others have senior clinical staff whose training would not be accredited for providing an equivalent service in the NHS. Few of the private clinics function as tertiary referral centres ouside the context of providing assisted reproduction.

Contracting

Tertiary referral centres in the NHS will themselves be limited by the available resources. In the "reorganised" NHS this means that the tertiary centre's NHS work is limited by what Health Authorities are prepared to "purchase" from the clinical service [3]. Generally this is negotiated as part of the NHS Gynaecology "contract" and is not negotiated separately. It is uncertain if infertility would fare better or worse if it was provided under a separate contract. There is little doubt that Health Administrators who are perennially short of funds may well opt to limit spending on what some view as "less essential" services such as infertility [4]. With the increased role of clinical budgeting [5] in the provision of clinical services the increasing disparity between the nature of a clinical gynaecology service and the range of services involved in a tertiary infertility service introduces significant complexities in meeting the needs of the two types of service. Although some form of budgeting separation of general gynaecology and tertiary infertility work might help resolve any such problems it would render the infertility services more separate and more vulnerable to a "low priority" in Health Authority purchasing decisions.

The Role of the Tertiary Clinic

A tertiary clinic will have a variety of service and research and development roles. At the service level the tertiary clinic will provide a fuller range of relevant investigations than is usually provided by generalist gynaecologists. Similarly the range of treatment options will be wider and it is in tertiary clinics that there is the first possibility of applying the findings emerging from current research on the causes of infertility.

Investigation

At best, current investigations provide many couples with only a vague guide as to their fertility prognosis unless there is an outstandingly severe problem such as gross pelvic damage or azoospermia. Often it is the couples whose "routine" basic investigations at a gynaecology clinic lead to the "unexplained infertility" label who are referred for tertiary clinic advice. What is then offered will vary considerably from clinic to clinic and will be greatly influenced by any research bias of the clinic since few can offer the whole spectrum of "more sophisticated" investigations. In addition, since studies on new investigative methods will usually be reporting on intermediate endpoints rather than the fertility prognosis, it is often years later before it is clear whether the innovation adds significantly to the diagnostic and prognostic assessment of the couple. If, however, the range of investigations used in clinical practice is to be developed and improved, it is tertiary centres which will provide this translating research studies into clinical practice. Similarly the importance of audit in determining the utility of new practices cannot be overemphasised. This model for innovation may well be different in health care systems where the work is mainly private.

Treatment

The tertiary clinics have an important role in providing more specialised treatments and in introducing treatment innovation to clinical practice. The specialised treatments will probably include more specialised surgical techniques such as microsurgery, operative laparoscopy and hysteroscopy. There may be a broader experience of specialised drug therapies, often at the clinical trial level. For example GnRH agonists have been used by some tertiary specialists for more than a decade and are only recently becoming more widely used.

Assisted Reproduction

The tertiary clinic will generally have facilities for assisted reproductive techniques. Traditionally the first of these was therapeutic donor insemination (TDI) which was applied to those with azoospermia, infertility and severely impaired spermatogenesis as well as for rarer problems such as some genetic conditions. More recently in the 1980s IVF has assumed an important place in the therapeutic ladder. IVF has not only opened up new treatment possibilities where previously there was no good option – as with damaged obstructed tubes – but has altered the pattern of treatment for other groups so that couples with impaired spermatogenesis will often try IVF rather than TDI. Similarly, the results of IVF in patients with unexplained infertility in the more successful centres, justify its use although the most appropriate stage at which to introduce this remains uncertain.

The other assisted reproductive techniques offered will vary from centre to centre and may include stimulated ovulation with intrauterine insemination (IUI) GIFT, or intratubal insemination.

The Funding of Assisted Reproduction

Since few heath authorities are prepared to fund assisted reproduction it is common for this aspect of the tertiary service to be outside the NHS provision with the assisted reproduction funded by charges to patients. These charges must be in the context of "private patient status" if they are to be legitimate since there cannot be charges for NHS treatment. If the assisted reproduction is not subsidised then the least expensive method of charging is to charge patients as users of an "NHS hospital private patient facility" so that the clinic space, infrastructure, equipment, consumables and staff are covered but leave out private fees for the consultant. This model has worked well in Oxford for many years [6] and provides successful IVF, without subsidy, for a cost of less than £1100 per cycle (excluding drugs).

Continuity of Care

Another important role of the tertiary clinic is, where possible, the continued care of the patient during pregnancy and delivery. The overall care of the patient requires continued management, not terminated when the couple achieve pregnancy. There are a number of reasons for this.

1. There may be an accompanying recurrent miscarriage problem requiring special support.
2. There may be a multiple pregnancy to be managed.
3. There may be the particular anxieties of an infertile couple in pregnancy.
4. There is the need to avoid an overinterventionist approach to the management of a pregnancy following infertility.

Paediatric Consequences

The effect of infertility treatments in increasing the rate of multiple births with its impact on neonatal services is a matter of concern. If a tertiary infertility clinic tends to feed infertility pregnancies to a linked antenatal clinic it is vital that neonatal paediatric liaison is maintained, since such a clinic risks attracting a disproportionate proportion of multiple pregnancies to its hospital.

The pragmatic approach we have reached in Oxford is that when women from outside the District become pregnant and request antenatal care from the same team of clinicians, such referrals are only accepted if they are singleton. Our Oxford paediatricians recognise that a majority of multiple pregnancies (usually twins) would not require transfer to our regional neonatal intensive care unit facilities if they were born in their own Districts. However, of those multiple pregnancies delivered in their own districts, a proportion will require some level of care because of a degree of prematurity short of requiring regional neonatal intensive care. If these latter cases were delivered in Oxford they would swamp the neonatal facilities and distort the local service.

Counselling

Finally, a tertiary centre has a role in developing infertility counselling facilities. These are difficult to organise because funding is limited and specifically trained counsellors are few. With the emphasis placed on independent counselling by the Human Fertilisation and Embryology Act 1990 [7] it is to be hoped that this area will be developed more generally. As the technical possibilities continue to expand, the importance of the couple being able to stand back from the "conveyer belt" of treatments becomes ever more important [8].

The Integration of Tertiary Centres Within the NHS

The role of the independent private clinics tends to be mainly in assisted reproduction. It is important to emphasis that any development of independent private hospital clinics should not be at the expense of the established range of infertility investigation and treatment available in an NHS Regional tertiary clinic (accepting that some of the integrated service may have to be charged for as an "NHS hospital private patient facility" as discussed above).

The advantages of an integrated service are numerous:

1. The specialist and his/her team can maintain an overall perspective of all the elements of the service.
2. The relative merits of different options can be weighed critically since the options are all within the control of the one team.
3. The flow of patients from GP or secondary referral clinic to the service can be as seamless as possible since the referral opens all of the range of "doors" into different investigations and management.
4. Trainee clinicians within the service gain a broad and comprehensive experience of all aspects of infertility provision within a single service and subsequently training is facilitated.
5. The issue of general practitioner co-operation and collaboration in prescribing "fertility drugs" is complex and difficult because of the issues of responsibility for prescribing and also the cost. With an integrated tertiary clinic there is a greater opportunity to develop agreed District and Regional protocols in collaboration with pharmacists and medical committees.

An Integrated Service

The Oxford service at the John Radcliffe Hospital is integrated within an obstetric and gynaecological service in an NHS teaching hospital and is run from the Academic Department. As described above some elements of the service are not funded by the Health Authority and must be run as an "NHS private practice facility" [6]. Those are the TDI service, the IVF Unit and the Andrology Laboratory. The basic fertility clinic and its associated basic investigations and treatments, the operating theatre work and ovulation induction are NHS facilities. Where a couple require the other facilities they may opt to be charged as private patients, without private doctor's fees being levied. Thus the staff providing the service are NHS staff paid from the budget generated by the charges to the patients for the use of the "NHS hospital private patient facility".

Within the range of facilities the pattern of referral can be complex. Referrals from general practitioners and other consultants are assessed and appointments usually arranged for the fertility clinic but many couples will be more appropriate for direct referral to the TDI or IVF programmes. At the point of referral there may be clear evidence to suggest a potential male problem so that direct referral to the andrology clinic with its andrological investigations and links with our urological colleagues is the best plan. Since the same group of clinical staff are involved in all of these services the couples can pass between different components of the service without disruption of the clinical involvement. As stated above in such an integrated service it is important to guard against the assumption that couples automatically want to "progress" through the facilities simply because they are available.

Providing a Tertiary Service

Each tertiary clinic has its own flavour and will face service demand influenced by what the clinic offers. The expansion of treatment possibilities over the past few years will have influenced the pattern and scale of referral. I have illustrated the trends from our experience in Oxford. In 1985 the fertility clinic offered investigation and had an associated TDI clinic. By 1991 the fertility clinic had expanded to provide a more comprehensive tertiary referral service including IVF, and an andrology facility. Comparison of the two years shows an increase in new couples seen from 168 to 375. This is the work load seen in the fertility clinic. In 1985 it was possible to see all referrals (147 for further investigation and 21 for TDI) at the fertility clinic whereas by 1991 the role of the clinic was to see couples requiring further investigation (301 couples) or specialised discussion (74 couples). Couples whose referral letters clearly indicate that particular services such as TDI, IVF or andrology facilities are appropriate now bypass the fertility clinic. A better comparison between the 168 referrals of 1985 is the overall referrals into the whole fertility service. Extrapolating the overall new referrals surveyed for the first three months of 1992 to annual statistics indicates 1048 referrals, a more than sixfold increase. The distribution of the referrals is 516 to the fertility clinic, 48 to TDI, 92 to andrology and 392 to IVF.

With the increased tertiary role the outside-District proportion of the work increased from 20.2% (34 couples) in 1985 to 31.2% (117 couples) in 1991. This outside-District role is particularly marked in the more specialised services such as gonadotrophin ovulation induction (50% of 1991 referrals outside-District) and in the IVF workload (69.9% of 1991 treatment cycles outside-District).

Analysing the whole infertility service provision for one year (1991) the team managed 375 new and 648 return visits to the fertility clinic, 259 andrology assessments, 510 TDI cycles, approximately 200 gonadotrophin ovulation induction cycles, 174 IUI cycles for antibody or male problems and 568 IVF cycles as well as the associated work load of infertility investigations and operations. All those working in infertility will be well aware that each of these various types of "cycle" involves several hospital consultations.

Staffing

The team required for this scale of service provision can be illustrated by the Oxford service. The team involves a nurse manager, three secretaries (fertility clinic, andrology/TDI unit, IVF unit), a clerical assistant (IVF unit), five scientists (andrology/TDI unit, 2, IVF unit, 3), two MLSOs (IVF unit), six nurses (IVF unit, 4, andrology/TDI unit, 2) and four medical staff (IVF unit, 3, andrology/TDI unit, 1). These staff are wholly involved in infertility work. In addition there is my own commitment as Director of the Service and the involvement of my clinical lecturer in the work with both of us maintaining the link into the antenatal service for the successful patients. Counselling support is provided in collaboration with our medical social work team.

Legislation

Much of the activity of a tertiary infertility centre is covered by the remit of the Human Fertilisation and Embryology Authority (HFEA) as determined by the Human Fertilisation and Embryology Act 1990. It is accepted that the work of the HFEA will tend to protect standards in the field of assisted reproduction where the activity is covered by the HFEA's remit but it is of concern that some important closely related treatments within assisted reproduction, such as GIFT, are not covered. The monitoring and licensing process of the HFEA places a significant administrative burden on the relevant parts of the fertility service but in some ways the most difficult burden relates to the very difficult confidentiality provisions of the Act [7]. These are out of step with normal methods of ethical medical communication between doctors and are a continuing problem for those providing an integrated service of non-HFEA clinical work as well as HFEA clinical work.

The Future

The developments in service provision in recent years have been substantial. The tendency has been for the developments to increase the gulf between general gynaecology and infertility practice. The major change noted in the Oxford service between 1985 and 1991 is noted above. The change is one of increasing demand, increasing outside-District referrals, an increase in the diversity of the service and a significant increase in the size of the team involved. In the tertiary centre the infertility work is no longer simply the province of a consultant and his or her junior medical staff with some nursing support. The team of about 25 people involved in our clinical service (wholly research staff have been excluded) demand a significant amount of management to ensure the smooth running of the activity. In addition that volume of more than a thousand new couples referred per year plus those already within the fertility service creates a considerable consultant management workload because of the range of complex queries to be dealt with, some of which undoubtedly require thoughtful and sensitive handling and can be very time consuming [9, 10].

In conclusion I suggest that with the aim of providing the most effective and wide-ranging infertility care the evolution of tertiary infertility work should be in Regional centres. In those centres the infertility service should be positioned alongside gynaecology rather than enmeshed within it since so much of the work is outside the scope of the gynaecology services as they are currently organised.

References

1. Royal College of Obstetricians and Gynaecologists. Workload for consultant obstetricians and gynaecologists. London: RCOG 1987.
2. Royal College of Obstetricians and Gynaecologists. Report of the RCOG working party on further specialisation within obstetrics and gynaecology. London: RCOG 1982.
3. Liddell A. Developing the purchaser role. In: Beck EJ, Adam SA, eds. The white paper and beyond. Oxford: Oxford Medical Publications, Oxford University Press, 1990; 70–7.
4. Levitt R, Wall A. The reorganised national health service, 4th edn. London: Chapman and Hall, 1992; 168–72.
5. Coles J. Direction and purpose in clinical budgeting. In: Brooks R, ed. Management budgeting in the N.H.S. Keele: Health Services Manpower Review, 1986; 31–50.
6. Barlow DH. The organization of in vitro fertilization in the hospital service. In: Bonnar J, ed. Recent advances in obstetrics and gynaecology – 17. Edinburgh: Churchill Livingstone, 1992; 139–48.
7. Her Majesty's Stationery Office. Human Fertilisation and Embryology Act, 1990.
8. Appleton, T. Caring for the IVF patient – counselling. In: Fishel S, Symond EM. eds. In vitro fertilisation – past, present, future. Oxford: IRL Press, 1986; 161–9.
9. Franklin S. Deconstructing "desparateness": the social construction of infertility in popular representations of new reproductive technologies. In: McNeil M, Varcoe I, Yearley S, eds. The new reproductive technologies. London: Macmillan, 1990; 200–29.
10. McWhinnie AM. Test tube babies. . . the child, the family and society. In: Fishel S, Symonds EM, eds. In vitro fertilisation – past, present, future. Oxford: IRL Press, 1986; 215–27.

Chapter 28

Public Health Aspects of Subfertility

T. A. Sheldon and S. L. Ibbotson

There is a vast literature describing advances in the techniques of diagnosis and treatment of subfertility. This chapter will consider the broader perspective provided by public health, which examines how the efforts of society may best be organised to produce the maximum improvement in health and welfare. This necessitates taking a few steps back to gain a broad view of the demand for diagnosis and treatment in relation to available knowledge of the effectiveness of technologies, and the supply of skills and resources, in order to produce an optimum pattern of health care to maximise what is now called "health gain" which embraces both length and quality of life.

This chapter will consider information requirements necessary to carry out this task, the need for evidence of effectiveness, and aspects of the efficiency, equity and quality of services provided.

The Size of the Problem

A starting point for discussing subfertility is to consider the size of the problem and the burden of need for services it generates. The question "how common is subfertility?" is not easily answered and there is much confusion over the interpretation of prevalence data [1]. (Prevalence is a measure of the number of cases of subfertility in a given population at a specified time.) Part of the problem is that there are several definitions of subfertility/infertility and the appropriate definition depends on the purpose to which it is to be put and who is asking the question.

In aetiological research it is sensible to talk about the distribution of time to conception and birth or, if using cut-off periods for categorisation, the percentage of couples not conceiving within one, two or more years of having regular unprotected sexual intercourse. On the other hand when they are used for planning health services the definitions should reflect knowledge of the natural fecundability of the subfertile populations and of the timing (in relation to the duration of non-conception) and the demand for intervention. In this context it is more appropriate to consider involuntary subfertility which is unlikely to resolve spontaneously within a reasonable period of time. The issue then becomes the length of time during which it is reasonable to follow an expectant course of management.

Using a definition of subfertility which includes all couples who have failed to achieve a desired pregnancy after engaging in unprotected sexual intercourse for at least a year needlessly inflates estimates of potential demand, since around 60% of these will conceive within the next three months. These individuals may be considered to be in the right hand tail of the distribution of conception times of the "normal" population. The probability of spontaneous conception decreases with the duration of involuntary non-conception, as the proportion of couples in this group who are "not normal" increases.

The most cost-effective use of resources may be achieved if couples were to be defined as subfertile only when conception has failed to occur after two or three years of unprotected intercourse. This must, however, be qualified by the need to consider also the age of the woman and likely years of reproductive life remaining. In view of both the natural decline of fecundability and the decrease in effectiveness of many treatment strategies with increasing age of the woman, it may be appropriate to define as "subfertile" a woman, for example, in her late thirties at an earlier stage than a woman in her twenties. Thus the public health (rther than purely epidemiological) definition, relates to the social context of likely demand and probability of treatment success. This consideration is likely to assume greater importance if the current trend of postponing the age at which a woman has her first child continues [2].

Data from prevalence studies, then, can be misleading due to variation in terminology and definitions as above (including whether resolved subfertility episodes are included), and sampling biases (e.g. including only referrals to hospital). Estimates of combined primary and secondary subfertility rates range from as low as 4% to over 15% of women of reproductive age [1]. More effort is needed to collect reliable data on a national scale in a way which will allow valid comparisons over time. This is important in terms of planning health service provision, especially in the face of evidence which suggests that the incidence of subfertility is increasing independently of any increase in demand for services, though care must be taken to identify how much of this is due to an increase in voluntary childlessness [3]. In this context, the suggestion that questions about fertility be added to the General Household Survey, which already contains questions on contraception, is worth exploring in more detail. Health authorities are giving more attention to epidemiologically based health needs assessment (stimulated by the Department of Health's Project 26) and a substantial number of conducting health and lifestyle surveys. It may be possible to include surveys of fertility as part of such work, as was shown by Page [4] using FHSA population registers.

Despite the conflicting estimates from the literature what is clear is that, using any of the usual definitions, a significant proportion of couples experience subfertility at some time. Estimates of incidence are also important (The incidence is a measure of the number of new cases of subfertility occurring in a defined population during a given period of time.) The number of new "subfertile" couples may significantly influence annual demand for related health services especially in situations where a high proportion of demand is met. There is a dearth of studies examining demand for these services. Estimates of demand for services are difficult to interpret since they will vary according to the availability of subfertility services (and according to associated factors such as criteria for acceptance and the attitudes and practices of referring doctors). It is hard to disentangle demand from met demand. Page [5] found that approximately 0.5% of women between the ages of 20 and 44 years were demanding these health services annually, a figure which agrees with that found by Hull et al. [6]

Morbidity and Need: Should Society's Resources Be Used to Manage Subfertility?

Although we know something about prevalence, incidence and demand, much less is known about the psychosocial morbidity associated with subfertility and thus it is hard to estimate the burden of suffering represented by the prevalence.

While it is clear that there is a complex interplay between psychopathology and subfertility, a review of the literature reveals many methodological problems which make the interpretation of seemingly contradictory results and the determination of causal associations difficult [7]. Many studies are retrospective or examine women already receiving treatment and thus it is difficult to distinguish previous existing morbidity from that associated with the processes of diagnosis and treatment. The interpretation of case–control studies, involving the comparison of infertile couples with those who have no difficulty in conceiving, is also problematical. Infertile couples often perceive their infertility as a complex life-crisis, which may remain unresolved. Couples who experience no difficulty in conceiving are perhaps unlikely to have experienced a similar life-crisis. Because of this, finding a suitable control group may be impossible. The small sample sizes, heterogeneity of the sample populations, reporting biases, lack of standardised and relevant outcome measures and the variability in timing of studies in relation to the duration of infertility and the stage of the diagnosis and treatment process at which studies are undertaken, all compromise the validity of studies and make comparison of results very difficult. There is a need for workers in this area to explore more rigorously, probably by the use of carefully conducted longitudinal studies, the morbidity associated with subfertility and the impact on this of diagnosis, counselling and treatment [8].

Traditional measures of morbidity (such as health service use and self-reported sickness) are not necessarily relevant in that childlessness or subfertility are often not considered to be diseases in the usually accepted sense (although they might be the result of a disease process). Subfertility may be considered to be a symptom rather than a disease [9]. Accordingly the use of assisted conception techniques may be justified as symptomatic treatment akin to other medical treatments which do not cure the underlying disease state, but merely alleviate its expression.

The treatment of subfertile couples may be justified in other ways. For example, there is the notion that the resources of society should be used to increase individual and social welfare. Thus the change in utility or welfare associated with treatment or counselling is of more relevance [10]. In society there is a growing recognition that priority should be given in the use of resources towards optimising the satisfaction of needs, although there may be disagreement as to the definition and ranking of these needs and the most effective ways of achieving the desired ends. In this context there are disputes over the way in which assessment of "need" should be undertaken and over the resulting criteria which should be applied to individuals and couples who seek help to overcome subfertility.

A more general justification could be found using Doyal and Gough's [11] general theory of human need. They argue that autonomy and physical health constitute basic needs which are important conditions which facilitate social participation. Whilst traditional measures of morbidity reflect more the dimensions of physical health, social participation and autonomy may be more relevant to the area of reproductive control. "To be denied the capacity for potentially successful social participation is to be denied one's humanity" [11]. Peoples' ability to develop a social role is an attribute of human autonomy.

It has been argued that for many couples, and more especially for women, parenthood (motherhood) is a socially significant activity which if denied as a result of subfertility, leads to a questioning of self-worth [12] and to a feeling of being devalued. In our society reproduction and the care of children are valued social roles. The denial of such a role by subfertility or other causes of childlessness or the conflict of this role with other valued roles, such as that of undertaking paid work, can significantly reduce personal autonomy.

It can be argued that control over reproduction is one of the more important dimensions of autonomy in the sense that individuals have control over themselves and their environment. Thus just as access to contraception and termination of pregnancy can help couples (in particular women) maintain autonomy, facilitate fulfilment of the social role of work, or ensure that the care of children and their socialisation occurs under optimal conditions, so treatment of subfertility helps couples who want to achieve autonomy or feelings of self-determination by fulfilling the desirable social role of parenthood. In this context it could be argued that subfertility corresponds to a handicap in the WHO International Classification of Impairments, Disabilities and Handicaps [5,13]. Handicaps are the impact that impairment or disabilities may have on a person's role. For infertile couples "biological parenthood seems to be a crucial factor in their sense of control and fulfilment in their lives" [14].

In rooting the justification for resourcing subfertility treatments in the lost reproductive social role of subfertile women (couples), however, there is a danger that other important and at times competing social roles (e.g. as paid workers) may be denied so limiting womens' opportunities in society. Thus it is important to acknowledge the social context of reproductive technologies, in particular the way they may reinforce stereotypical attitudes to women, motherhood and childbirth. There is a risk that the increased availability of more effective treatments may result in childless women coming under greater pressure to reproduce. The criteria for selecting women for assisted conception (for example) are often based on conventional assumptions about what constitutes a suitable environment for child rearing and discriminates against single women, lesbians and unmarried couples [15].

Assessing Effectiveness

Even if it is accepted that subfertility is relatively common and that it represents a legitimate claim on societal resources, this is not sufficient for the purposes of health service planning. The effectiveness of alternative strategies must also be examined. As in the discussion of prevalence, there is a variety of informational and methodological issues which need highlighting.

Ever since the publication of Cochrane's lecture "Effectiveness and Efficiency" [16] in 1972 we have been increasingly made aware of the need for properly conducted randomised controlled trials (RCTs) to demonstrate the effectiveness of interventions. In order to establish whether subfertility treatments are effective, better designed randomised controlled trials are needed. Early in the innovative phase of any treatment development it is necessary to publish short case reports and small before–after studies to generate hypotheses which will be tested by subsequent research. As the subdiscipline matures, one expects an increase in the standards of evidence by which effectiveness is judged. This phase has been reached for many subfertility treatments and it is no longer acceptable to see comparisons between non-randomised (and non-comparable groups) receiving different treatments used as evidence of effectiveness.

Unfortunately, such is the variability in reporting methods that comparison of the results between trials and centres is often difficult. In order to aid the research community and provide a reliable guide for those who purchase or commission health care services, national (and international) criteria and standards for the recording and presentation of results are vital. From the public health point of view the most useful reproductive outcome is successful maternities or motherhood (parenthood). The use of "babies born" as the numerator in describing the outcome of assisted conception techniques can misleadingly inflate "success" because of the raised incidence of multiple births (which may, indeed, not be desirable). Although other reproductive outcomes such as biochemical or clinical pregnancy might prove useful indicators in the audit of a treatment unit's activities they are not relevant directly for public health [17].

In terms of the denominators used, reporting should include not only the average number of treatment cycles or time taken in order to achieve an outcome, but also the distribution. This may be displayed as the cumulative number of maternities by cycle or time period. To enable comparison within and between centres the characteristics of the patients selected must be clearly reported, as must cancelled cycles and people no longer being followed up for whatever reason. Patient characteristics of potential interest because they affect prognosis include age, parity, cause and duration of subfertility, severity of disease (e.g. when reporting outcomes of tubal surgery). There should be agreed criteria for diagnosing the causes of subfertility and agreed classification of severity of disease, where appropriate, to enable better comparison by using specific or adjusted rates.

Life-table or survival analysis is increasingly being used for the reporting of outcomes such as cumulative motherhood or conception rates. If the data are reported properly this can be a very powerful method for comparison [18]. However, such analysis is valid only if either those that are lost to follow-up are a random sample of those who continue to be observed (non-informative censoring) or if the variables which affect censoring and the probability of a successful outcome are fully known and recorded for those who are lost, so that adjustment can be made in the analysis. This is important because patients not infrequently drop out of treatment programmes and are not followed-up further. Many of those who drop out do so on medical advice because of poor prognosis (or perceived poor prognosis), while those who continue with treatment are those most likely to achieve a successful outcome. Without adjustment for the differences in characteristics between those who drop out of treatment programmes and those who continue, the result is an upwardly biased estimate of the effectiveness; in other words the cumulative outcome curve refers to an increasingly successful subset of those people along its length. Information on the characteristics of patients included in studies or treated in centres and on the nature of the interventions (e.g. number of embryos transferred, technique employed for tubal surgery, etc.) is also vital to be able to assess the generalisability or external validity of the results. If sufficient information of this sort is not available it is difficult to judge the degree to which results may be expected to be replicated in other settings.

There is a need for more information on the sorts of patients that are entering clinics and trials. It is likely that clinics in different parts of the country or the world are dealing with patients with different characteristics. Indeed part of the success of some centres is their decision to select for treatment only those who are most likely to conceive and carry a pregnancy to term. It would appear sensible to request all centres to follow up not only those undergoing active treatment but also those on the waiting list. In this way the spontaneous conception rate among those waiting can be estimated to allow comparison of the baseline fecundability of patients attending each centre and give additional information about the selection criteria in operation.

In terms of outcome measures, whilst the maternity rate is a vital reproductive outcome, broader outcome measures should also be included in line with the discussion above. One of the worrying features of the literature, and implicit in the outcome measures used, is the notion that only reproductive outcomes are valid. This ignores the fact that if treatment is provided in order to help people overcome the handicap of subfertility, this may be achieved either by facilitating conception and parenthood or by helping individuals to come to terms with their childlessness and overcome the feelings of low self esteem. Thus a couple who have decided not to go ahead with treatment or who have had a course of treatment but decided not to continue should not automatically be classed as "failures". As long as they are now better able to function socially, that is, their welfare has improved, the service may be regarded as effective.

In this area patient valuations of their quality of life or utilities before and after treatment – as measured, for example, by techniques such as standard gamble or time trade off [19] – would seem to be a potentially useful outcome measure. Alternatively, generic health profiles might be valuable.

Another outcome whose measurement should be encouraged is patient satisfaction with both the process and result of treatment [20]. This is particularly important given the considerable anxiety experienced by some patients, the nature of the treatment options and the fact that many patients will not conceive. More effort is needed to improve methods of obtaining and measuring patients' perceptions of their care.

All the preceding argument for better data, and improved recording, analysis and presentation of studies also applies to routine data collection for the purposes of clinical audit. If audit is to be a systematic and critical analysis of the quality of medical care, it must be carried out as a rigorous set of procedures [21].

It is also important that the longer-term outcomes of treatment on maternal and child health are monitored [22]. Thus the birthweights, growth and development of infants and children born after treatment should be recorded. Given the relatively short time that many of these treatments have been available it is important that good records are kept and retained for many years so that case-control studies may be carried out in the future if adverse outcomes for the child are thought to be associated with any particular treatment.

Lastly, not only is there a need for further and more rigorous studies on the effectiveness of interventions but there is an imperative to communicate the results accurately and honestly in a way that the purchasers and commissioners of health care services will understand and which is comprehensible to non-specialists and the public. In the UK at the moment there appears to be considerable confusion among those outside the subdiscipline.

A considerable proportion of purchasers assume that treatments are ineffective or cost-ineffective. This is to be expected in a new and rapidly developing area but more importantly it reflects the lack of clear standardised reporting mechanisms discussed above and the often inflated claims made by some. Good information is a necessary (though not sufficient) condition for rational policy making at a local level. One way of improving the situation is to produce comprehensive, rigorous and readable reviews or overviews. For example, Professor Hull from Bristol University has produced an accessible review of the area [23]. At the School of Public Health, University of Leeds, we are producing an issue of the *Effective Health Care* bulletin on the topic of the management of subfertility. About 28000 copies of this bulletin will be distributed free to health care policy makers in health authorities, hospitals and general practices around the country [7].

Nevertheless, we must be careful when using formal quantitative meta-analytic methods for overviewing the results of trials, even of RCTs [24], for as has been discussed already, there are huge variations in the outcome measures, the treatments, the subjects included and the reporting of those who drop out of treatment programmes. In addition, meta-analysis can actually hide the huge variations that exist between trials and centres [25].

The next issue which needs to be addressed is that of the most efficient way in which infertility services should be organised, delivered and financed.

Efficiency

One of the aims in organising the provision of services is to maximise allocative efficiency and technical efficiency.

Allocative Efficiency

By allocative efficiency we mean deciding on how resources will be allocated between competing areas or programmes of expenditure such that no greater benefit could be generated by alternative distributions of the same resources. This is important because the total resources available are limited and need to be divided between competing uses. This may be considered both within the health service (for example, in considering the allocation of resources for coronary bypass surgery or for renal dialysis as against those for subfertility treatments) and between the health service and other welfare-generating activities such as education and housing. For the former, cost utility analysis (CUA) is increasingly being carried out, and for the latter, one means of economic appraisal is cost–benefit analysis [26].

The total resources allocated to subfertility services compared to other health services must depend on an assessment of the relative benefit to be derived from the different services for a defined amount of money. Page and Brazier [27] have attempted to estimate the cost per Quality Adjusted Life Year (QALY) for in vitro fertilisation and to compare this with cost per QALY for other health programmes. However, the use of QALYs in this way has been severely criticised [28], on conceptual and statistical grounds. In the context of subfertility treatment it is even more unreasonable to use this approach because the QALY is specifically about health-related quality of life, but as discussed earlier the benefits from treatment are likely to affect general welfare rather than health (for a discussion of the limitations of QALYS in the context of allocative efficiency see reference [29].

Although the amount of resource devoted to this area must reflect information about the effectiveness of the treatments, estimated need for services and cost of meeting these needs, ultimately allocative decisions are political ones and should involve democratic debate and reflect the preferences of the public whilst taking into account the interests of minority groups. Initial attempts at involving the public in ranking medical services according to the priority they should have in public expenditure has indicated that IVF is very low down the list [30]. However, methods of eliciting public priorities have been rather crude and the results are very sensitive to the way questions are framed and the information available to the public.

Technical Efficiency

Having decided the amount of money to be allocated to this area of health care, the aim of planners should be to maximise the technical efficiency with which the resources are expended. In other words we want to use the techniques available to treat those patients who will benefit the most within the budget constraint imposed. The first difficulty, discussed above, is in defining those individuals who are most likely to benefit from treatment. Thereafter cost effectiveness analysis may be used to compare the costs per maternity achieved or other outcome by different techniques for different conditions; that is, to determine how, at least cost, to meet a particular objective, or, given a fixed budget, to maximise benefit.

There is a strong argument for collecting cost data along with clinical information and so carry out an economic analysis alongside clinical trials. Whilst cost per outcome comparisons have been made for particular conditions [31,32], they are unreliable for two reasons. First, there are difficulties in deciding which costs to include in the appraisal and second, as discussed above, there are problems in interpreting and comparing estimates of effectiveness. The "costs" associated with a treatment not only include the direct costs of the treatment but should also properly include opportunity costs such as the value of time used by the patient undergoing treatment as well as direct costs including psychological and social morbidity. It may also be appropriate to include knock-on costs such as those associated with increased use of antenatal care services or neonatal intensive care facilities and the costs of managing side-effects of treatments.

When comparing the cost effectiveness of interventions the role of prevention should not be ignored. It is possible that the rapid increase in sexually transmitted disease will, by causing tubal damage, increase the prevalence of subfertility. In particular there may be a case for exploring the rise of chlamydial infection rates to identify areas where prevention might be effective.

Towards a Rational Management Policy for Subfertility

Significant NHS resources are already devoted to the management of subfertility. The deployment of much of these is decided in an *ad hoc* fashion and they could be more efficiently deployed by a purchasing policy covering a broader range of services of proven effectiveness. For example, tubal surgery is carried out in many non-specialist units in district general hospitals, but there is strong evidence to indicate that this may not be as effective as surgery in specialist units or as effective as other treatments such as assisted conception [33].

As part of the discussion about ways to improve technical efficiency a case can be made for re-allocation of the resources currently used in the management of subfertility (though not always clearly identified as such) to interventions which are more cost effective and which are provided in an efficient organisational setting.

There are several aspects to this reorganisation. First, the potential role of primary care in the management of the preliminary investigations should be explored more thoroughly. Attention must also be given to the training requirements of general practitioners and the primary health care team to enable them to supervise the first phases of consultation with subfertile individuals, to make decisions about the timing and nature of preliminary investigations and to interpret their results. Protocols could be developed not only to aid the decision to refer but also to guide the appropriate investigation of patients and so reduce the need for unnecessary duplication of diagnostic tests. Consideration also needs to be given to the most appropriate setting in which to carry out laboratory investigations to maximise the accuracy of diagnosis.

If efficient choices about further diagnosis and treatment are to be possible the secondary/tertiary referral unit must have direct access to a comprehensive range of subfertility interventions so that patients can be directed towards the most effective treatment for their condition. It would be inefficient if tubal surgery were offered because it was the only service available at the local hospital, when another technique, for example, assisted conception, were more effective.

For optimal triaging, access to the full range of effective treatments is needed. It is unlikely that these would all be available within the district of residence so that there is a need for good information about the availability of services on a regional or national basis. The concentration of subfertility services in supradistrict specialty centres should be considered on the grounds of probable economies of scale and also a quality–volume relationship. It has been estimated that the most efficient treatment capacity for an IVF clinic is 750 started treatments (300–400 women) per year [15].

There is still the need for more cost effectiveness studies in this area. In order to maximise the welfare gain for each couple and to prevent unnecessarily prolonged treatment, Taylor [34] has suggested a more structured approach to the management of patients, which involves adequate provision of information on probabilities and the setting of personal goals. This could be extended by the use of a more formal decision analytical approach which has been used with some success in prenatal diagnosis [35,36]. By using the framework of a decision tree, staff can help patients to identify the treatment path which is likely to result in the greatest increase in their individual welfare (expected utility).

In summary, the preceding discussion examines technical efficiency and thus is relevant to the deployment of any reasonable budget for subfertility services. This is distinct from any argument about the total budget for these services for which competition with other health programmes must occur to secure allocative efficiency. In other words there is a strong case for arguing for a more effective use of the existing resources independent of any debate about the total amount allocated for the management of subfertility. There is a tendency for these arguments to get confused even though they are conceptually quite distinct. The former is based on the more technical arguments of effectiveness and cost effectiveness, the latter involves comparison of programmes with different outcomes, and the relative valuation of those outcomes.

Equity and Financing

The extent to which services for the diagnosis and treatment of subfertility are available to district health authority residents and are financed by the NHS varies between districts, the pattern of provision often being determined on an *ad hoc* basis, perhaps primarily as a result of individual local interest in the subspecialty. Many relatively effective elements of treatment, in particular assisted conception techniques, are not generally financed by the NHS and so are only accessible either to people who happen to be residents of a district which does purchase these services or to people with sufficient private financial resources. Thus the distribution of subfertility services in the UK is highly inequitable along both geographical and socioeconomic lines.

If a more integrated service were to be financed by the NHS this could improve the accessibility of services and possibly result in increased control over the costs and quality of services available nationally. There may be considerable advantages here in the NHS purchasing care from the private sector and exploiting the technical and organisational knowledge and skills which have been developed primarily in the private sector.

However, there is the risk of an explosion of costs if the NHS agrees to finance unlimited subfertility treatments. This risk comes, to a large extent, from the interplay between needs and supply which creates demand. Health care as a commodity is different from other sorts of products in that supply can create its own demand. This is the result of the "agency relationship" in which the doctor acts as the agent of the ill-informed patient [10]. This agency relationship is likely to be strong in the area of subfertility due to the complex nature of the area and the poor ability of the non-specialist to make informed judgements about prognosis and the appropriateness of treatment [37]. Thus it is the doctor who in some senses demands his own services on behalf of the patient.

It is no small wonder, then, that the amount of clinical activity tends to rise with an increase in service availability since doctors can, in good faith, convince patients of their need for services. The high prevalence estimated by some researchers, the variable definition of subfertility and the supply-created demand discussed above, when considered together, mean that the diagnosis and treatment of subfertility represents a potentially very large area of relatively expensive clinical activity. Currently in the UK the amount of activity is kept low because it is largely confined to the private sector. In several other countries where social insurance covers much of the cost of subfertility treatments, activity, and therefore total cost, are much higher. Stephenson and Svensson [15] estimate that already in some countries the resources devoted to IVF are most likely to be greater than is justified on the basis of need.

This presents a challenge: how can services be purchased to ensure maximum benefit is derived whilst preventing the unchecked proliferation of services and so containing costs? More research is needed to identify the purchasing structure and correct incentives to obtain the desired result. One option discussed by Haan and Rutten [38] in the context of practice in The Netherlands is to purchase outcomes rather than the actual services themselves. Thus rather than paying for operations or cycles of treatment where there are incentives to increase the amount of treatment their proposal is to pay only when treatment results in a birth. This should ensure that only couples who have a reasonable chance of "success" are treated. As long as a means of preventing the selection of patients who are likely to conceive spontaneously without intervention can be developed, this "no cure no pay" policy may establish the correct incentives. Units with good outcomes and low average costs will be rewarded at the expense of those who do not perform as well.

One major drawback to this approach is that it will reinforce the narrow clinical view that only a maternity can be regarded as a success rather than other outcomes which indicate that people have overcome or come to terms with the "handicap" arising from subfertility.

Conclusion

The main message of this chapter is that in parallel with the rapid pace of technological development there is a need to step back in order to look at subfertility from epidemiological and social perspectives to examine the size of the burden generated by subfertility and to consider the optimal organisation and delivery of services. Infertility care must be seen in the context of the competition for scarce resource by all other health services. Because there is a general undersupply of NHS services for the termination pregnancy, and the implications of some treatments for maternal and child health, the allocation of resources for the management of subfertility must be considered in the context of the overall health authority reproductive health strategy.

The broader social impact of reproductive technologies must be carefully considered as part of effectiveness. In particular the impact on maternal and child health, the danger of reinforcing the view that reproduction is necessary for fulfilment and the increased medical and technological control over reproduction. In addition to the ongoing search for technologies to ensure improved diagnosis and treatment there is a great need for systematic evaluation of the following areas:

1. The needs and demands for services for diagnosis and treatment of subfertility;
2. The impact in terms of relevant and valid measures of different patterns of delivery and variations in clinical practice;
3. The balance of and interface between primary, secondary and tertiary care in the management of subfertility;
4. The development and assessment of appropriate outcomes of treatment;
5. The assessment of the effectiveness and cost effectiveness of diagnosis and treatment of subfertility;
6. Examination of alternative ways to finance services;
7. Exploration of the social impact of the technologies especially on perceptions of childlessness, motherhood and family relationships.

The management of subfertility is not unique in generating this list of research questions; a similar list would be equally applicable to many other areas of clinical endeavour. There is a need for clinical researchers to work together with epidemiologists, health economists, medical sociologists and others in order to explore these questions. Only in this way can we begin to ensure that the finite resources available for health and social care are allocated equitably, efficiently and in an acceptable way in the pursuit of the improved health of populations and of individuals.

Acknowledgements We gratefully acknowledge Dr H. Page, Professor I. Cooke and Professor R. Lilford for helpful discussion of issues.

References

1. Greenhall E, Vessey M. The prevalence of subfertility: a review of the current confusion and a report of two new studies. Fertil Steril 1990; 54:978–83.
2. Office of Population Censuses and Surveys. General Household Survey 1989. London: HMSO, 1991.
3. Johnson G, Roberts D, Brown R, et al. Infertile or childless by choice? A multipractice survey of women aged 35 and 50. Br Med J 1987; 294:804–6.
4. Page H. Estimation of the prevalence and incidence of infertility in a population: a pilot study. Fertil Steril 1989; 51:571–7.
5. Page H. In-vitro fertilisation and the NHS: planning issues in in-vitro fertilisation. Unpublished MD thesis University of Leeds, UK, 1992.
6. Hull MG, Glazener CM, Kelly NJ et al. Population study of the causes, treatment and outcome of infertility. Br Med J 1985; 291:1693–7.
7. Leeds University, School of Public Health. The management of subfertility. Effective Health Care Bulletin no. 3 1992.
8. Edelmann RJ, Connelly KJ. Psychological aspects of infertility. Br J Med Psychol 1986; 59:209–19.
9. Stone J. Infertility treatment: a selective right to reproduce? In: Byrne P ed. Ethics and law in health care and research. Chichester: Wiley, 1991; 65–79.
10. Mooney G. Economics, medicine and health care, 2nd edn. London: Harvester Wheatsheaf, 1992.
11. Doyal L, Gough I.A theory of human need. Basingstoke: Macmillan, 1991.
12. Woods NF, Olshansky E, Draye MA. Infertility: women's experiences. Health Care Women Int 1991; 12:179–90.
13. World Health Organisation. International classification of impairments, disabilities and handicaps. Geneva: WHO, 1980.
14. Strickler J. The new reproductive technology: a problem or solution? Sociol Health Illness 1992; 14:111–32.
15. Stephenson PA, Svensson PG. In vitro fertilisation and infertility care: a primer for health planners. Int Health Sci 1991; 2:119–23.
16. Cochrane AL. Effectiveness and efficiency: random reflections on health services. Oxford: Nuffield Provincial Hospitals Trust, 1972.
17. Page H. Calculating the effectiveness of in-vitro fertilisation. A review. Br J Obstet Gynaecol 1989; 96:344–9.
18. Guzick DS, Bross DS, Rock JA. A parametric method for comparing pregnancy curves following infertility therapy. Fertil Steril 1982; 37:503–7.
19. Feeny DH, Torrance GW. Incorporating utility-based quality of life assessment measures in clinical trials. Medical Care 1989; 27:S190–204.
20. Fitzpatrick R. Measurement of patient satisfaction. In: Hopkins A, Costain D, eds. Measuring the outcomes of medical care. London: Royal College of Physicians, 1990; 19–26.
21. Russell IT, Wilson BJ. Audit: the third clinical science? Quality in Health Care 1992; 1:51–5.
22. MRC Working party on children conceived by in vitro fertilisation. Births in Great Britain resulting from assisted conception, 1978–87. Br Med J 1990; 300:1229–33.
23. Hull MG. Infertility treatment: needs and effectiveness. Department of Obstetrics and Gynaecology, University of Bristol, 1992.
24. Hughes EG. Meta-analysis and the critical appraisal of infertility literature. Fertil Steril 1992; 57:275–7.
25. Thompson S, Pocock S. Can meta-analysis be trusted? Lancet 1991; 338:1127–30.
26. Drummond MF, Stoddart GL, Torrance GW. Methods for the economic evaluation of health care programmes. Oxford: Oxford University Press, 1987.
27. Page H, Brazier J. Benefits of in vitro fertilisation. Lancet 1989; ii:1327–8.
28. Carr-Hill R. Allocating resources to health care: is the QALY (quality adjusted life year) a technical solution to a political problem? Int J Health Serv 1991; 21:351–63.
29. Mooney G, Olsen JA. QALYS: where next? In: McGuire A, Fenn P, Mayhew K, eds. Providing health care: the economics of alternative systems of finance and delivery. Oxford: Oxford University Press, 1991.
30. Dixon J, Welch HG. Priority setting: lessons from Oregon. Lancet 1991; i:891–4.
31. Page H. Economic appraisal of in vitro fertilisation: discussion paper. J R Soc Med 1989; 82:99–102.

32. Lilford RJ, Watson AJ. Commentary. Has in-vitro fertilization made salpingostomy obsolete? Br J Obstet Gynaecol 1990; 97:557–60.
33. Watson AJ, Gupta JK, O'Donovan P, Dalton ME, Lilford RJ. The results of tubal surgery in the treatment of infertility in two non-specialist hospitals. Br J Obstet Gynaecol 1990; 97:561–8.
34. Taylor PJ. Editor's corner: When is enough enough? Fertil Steril 1990; 54:772–4.
35. Pauker SP, Pauker SG. The aminocentesis decision: ten years of decision analytic experience. March of Dimes Birth Defects Foundation, birth defects: original article series 1987; 23:151–69.
36. Thornton JG, Lilford RJ, Johnson N. Decision analysis in medicine. Br Med J 1992; 304:1099–103.
37. Ryan M, Twaddle S. Economic issues in the evaluation of in-vitro fertilisation. Paper presented the UK Health Economists' Study Group meeting, Aberdeen 1991.
38. Haan G, Rutten F. No cure, no pay: an acceptable way of financing fertility treatment? Health Policy 1989; 13:239–49.

Discussion

Winston: I want to focus on the qustion of re-doing investigations.

There are two points about repeating investigations. For example, take laparoscopy. Many of these routine investigations are done by very good gynaecologists but without reproductive training, and very often there are subtle changes that need to be observed. The second point to consider is that disease is not static, it progresses, and investigations that might be negative initially may subsequently be positive.

We must always think in terms of making a diagnosis before we treat patients. It is fundamental. For example, one of the big errors we see in both the private and the public sectors is that we tend to give the patient clomiphene before we have made a diagnosis.

Drife: There is a conscious policy on the part of some gynaecologists to investigate slowly, because of the background pregnancy rate.

Winston: That is true of course.

Templeton: It is true that much that is not creditable is going on in the private sector, but it is wrong to focus too much on that. If we look at what we could do with present resources within the Health Service to optimise the management of patients, we should examine management at the secondary care level. All the building blocks are in place and without any additional resource an improvement in that part of the service would make a dramatic improvement to waiting times, to the general management of patients, and to proper diagnosis and treatment. We could make an enormous difference if we put in some work.

Hull: Mr Sheldon made a statement which I felt needed to be picked up. He said the problem with life-table analysis is that we are dealing with different populations. If we are dealing with different populations, then life-table analysis is invalid, but I want to re-emphasise what I said earlier, that it is valid as long as the characterisation of the group – the population under study – is well defined.

Sheldon: There are two issues and it may be that the way I presented it confused them.

If we are studying a group of people – following them up over time in the form of a life-table analysis or a survival analysis – if some of those people are dropping out of follow-up, those that remain are not representative of the group we began with. I did not mean falling out of treatment but falling out of follow-up.

What Professor Hull is discussing, which is also important, is comparisons of different groups in life-table analysis or different studies. They might be different populations and he is then absolutely right. As long as one can adjust for the key confounders, they can be compared. That is absolutely correct. But I was trying to make a different point.

Hull: But I believe the point is made wrongly because of an incorrect understanding of the fundamental theory of life-table or survival analysis. It depends on the assumption, which will have to be validated, that those who fail to continue as long in the study for whatever reason would have behaved in the same way as those who did. The only way that that assumption remains valid is if the characterisation of that group at the start was properly done. It is not true to say that because follow-up does not continue as long in some cases, they drop out for different reasons: so long as those reasons are not biased reasons in other words they are advised to drop out because for some reason they are less favourable, and they are only less favourable if they have been badly characterised.

Sheldon: That is exactly my point. It is called informative or non-informative censoring. If it is informative censoring, in other words if the reason they are dropping out is something that means that those who drop out are somehow different in their prognosis from those who stay in, then it is no longer valid.

Hull: Exactly so. And that is why it is so important for us to characterise accurately the groups that we are studying by life-table analysis accurately. They must be defined by various important parameters – diagnostic, age, perhaps duration of infertility, perhaps previous pregnancy – we do not have to go through the details. But it would be wrong to take away from this meeting the idea that life-table analysis is somehow inherently invalid. It is only the way that is applied that may invalidate it.

Sheldon: Absolutely.

Cooke: I wanted to follow-up the discussion about the full range of services, particularly at secondary level. The Fertility Sub-Committee of the RCOG attempted to do a national survey – I am afraid we did it very simplistically indeed and we had a second bite at it and did not do much better – but the important thing that came across was that although there is some sort of fertility provision quite generally around the country, the extent and quality of the service provision are enormously variable. The special interest that gynaecologists have may be manifest in one place by an ovulation induction service without any comment on quality, and in another place by a donor insemination service as the only reproductive medicine support. That means that if a patient or a couple want a spectrum of activity, they may have to cast their net over an enormously wide geographical area to get what is relevant for them. This is an unrecognised problem and Professor Templeton's plea to expand the secondary sectors would go a long way to improving that.

I am very nervous about John Dickson's league table; he did not use that phrase but that is what he is referring to. Fortunately the HFEA have eschewed this recommendation. Unless the volume, the experience, the case mix and the diagnostic categories are taken into account and presented, untold harm will be done in altering the perception of the quality. Only a large centre has the numbers of patients to break down into sufficiently large cells, and we have seen several examples of drawing conclusions from tiny cells in the past two days.

It is a mistake to call out for more money as the primary plea. What is most important is to emphasise quality. If we have quality, the money needs to follow that.

Drife: If one were to cost out the number of visits to an SHO and the number of duplicated investigations, this is a resource that could be reallocated from the current system of repeated, and often pointless, investigations. We might be staggered by the amount of resources already being spent that could be redirected. This would back up the idea that it is not more money, but better direction of resources that we need.

Dickson: I deliberately did not put in the answer to the problem: I was offering to try and work that out. My point is that if we leave the situation as it is now, that is untenable. I know there are problems, particularly with the smaller centres which have 200 patients or less per year. When they are split down to the various groups they are so small they are not statistically valid; I quite appreciate that. But I was referring to the situation where that is used as an argument to produce no data about success – where there is this vast difference between the top and the bottom. I cannot accept that, and my job is to put the patient's view, to say that that is unacceptable. I am told that by many many patients. They cannot understand why we cannot produce some form of information, and if it is not a league table then it must be some statement about those centres which are more successful.

Dame Mary Donaldson had the view of the three star, two star, one star system which was very much laughed at. But maybe that is the answer. Can we not have that debate? There should be a discussion.

Baird: This is an extremely important issue. We can take it as axiomatic that any purchaser, whether they be a private individual patient going to an IVF clinic and handing over the money, or a health authority, needs the information as to whether they are getting value for money. What I have considerable reservations about is that it should be the responsibility of the HFEA to provide that information. They are set up as a legislative body to carry out certain functions and it is not their function to provide a Consumer Association type guideline to where to get the best buy. It is obligatory, and indeed I suspect NHS purchasers would insist on some kind of information about the quality of the service they are buying, and similarly it should be obligatory that each individual private clinic that is charging individual purchasers should provide that sort of information. But the onus should be on the providers of that service not on an authority which was set up for a different purpose.

Braude: They are the only organisation that has a statutory duty to collect the data.

Baird: But not to draw up a league table which is available for purchasers.

Kerin: I empathise very much with Mr Dickson's point on this issue and the fact that he is being pressured by the consumer to give some sort of league table.

I have six years of personal experience in the United States, where a consumer protection policy was brought in for high-tech reproductive medicine. All programmes were asked to submit their data to a US Congress report and those results were published in a report issued by Congress and available to every consumer in the land. The biggest problem with this book was that the clinics were not standardised in the way they reported their figures, and because it is very much a dollar-driven society, people put in wrong results. They inflated their results when they submitted the data and it became a bigger problem because the consumer was misdirected to some programmes. So there was a bias there and it did not work.

One way this can be partly overcome would be to have some sort of accreditation body that sets minimum standards for all programmes, and independently reviews the programmes. If the consumer knew that a programme was accredited for a certain number of years by this independent group, that should be a safe programme to go to. But a league table will have several problems attached to it.

Abdalla: It is important for the patients to have the results of the different clinics available. If one is purchasing anything one wants to know how good it is. It may well be difficult. I would never believe what a clinic says! But there could be some kind of check – not necessarily that the HFEA should publish results, but they collect the data from all the clinics. If I were to claim tomorrow that my results were X, Y and Z, they have the right to question my analysis if I am giving out the wrong results. It is extremely important that they have this ability to prevent us massaging our data. In that context it would be extremely important to categorise the way any clinic in the land can publish its results. There has to be some standard set, so that those who publish their results have to publish them by age, by cause, by whatever. The HFEA has the data and if any centre is publishing wrong figures, they can counter-check on that and threaten them.

Hull: I have no difficulty with the idea of league tables, call them what you wish. If there are small units and small numbers are difficult to interpret, we can put confidence limits around them. The whole population is now very well aware – in this election week – of how we can put percentage points around proportions.

But I would emphasise the need to classify the data in a comparable way. In this College we have had problems with perinatal mortality. After generations of argument about how we can compare international and intranational figures, we all know what should be done, but it still has not been done and we can still argue about the meaning of perinatal mortality. We need at an early stage to define the way we present data so that they are truly comparable.

Nieschlag: I am not familiar with ISSUE and I should like to know more. It is a completely independent organisation or is it a part of a larger organisation? Is it supported entirely by its membership or does it get funding from other sources? How much is the membership fee, how many members are there, what is the growth of the membership, and what about fluctuations in membership? I can imagine that people may join when they do not have a child and that once their issue is resolved they leave ISSUE, or is that not the case?

Dickson: In terms of funding, our biggest sources of funding is from our membership. We get some sponsorship – Serono have helped us with the helpline – but also grant and trust money. The next largest source of money is the Department of Health grant.

We are independent, but also trying to work constructively in the area. What the change has been, or has tried to be over the last year, is that we are trying to work with the profession, with GPs, with anybody who has an interest in this area, whereas maybe 18 months ago we just used to kick out. If there was a complaint we would never try to understand the other person's point of view. The change from our point of view – which is why it is so exciting to be a part of a group like this, because it shows that that policy is working – means that we are better able to get the patient's view across.

Interestingly enough, membership has declined over the last 18 months. It has been a very difficult period for us. We are currently part of a larger Birmingham Charity called the Birmingham Settlement which we now wish to separate from, because we feel that we are big enough and we have no synergy with what they are doing. But financial difficulty has meant that people are less able to join, because our membership fee is quite high. We charge £30 to join and £15 per year after that, which is an attempt to deal with the fact that people join, strip out all the information and do not continue in membership, whereas we want people to stay in membership. Certainly when people succeed they do tend to drop out, because what we are doing is very much towards treatment – adoption, other forms of dealing with fertility – and once someone has solved their problem, what we are projecting is not particularly valid. That is something else I am trying to challenge by saying to people that they support charities and our charity should be the one they support even after they have solved their problem: their sympathy should still be with that group.

I am trying to change ISSUE from being a self-help group, very clearly focused on helping people, to be broader and more inclusive.

Drife: It is very important for us to have an alliance with the patients. It is difficult to prevent it becoming either too cosy or too confrontational. There is a right balance.

Hull: If I could return to Mr Sheldon's comments on audit. He quite rightly pointed to the need for rigour and completeness in studying groups undergoing treatment. But I would also suggest that we should pay attention not only to the groups that were treated, but to those who were not treated, because we need to see the whole population.

I want to give an example of problems that can arise. I can remember some years ago discussing with the authors of one of the early papers on the successful treatment by IVF of male infertility. The claim was that provided a certain minimum number of sperm could be recovered into culture medium, they did well. In my view they were not dealing with male infertility because they were demonstrating by test of function that those sperm were healthy, but when I asked about those they did not treat, I was told that they did not keep records of those and could tell me nothing about them. Really they were not treating male infertility.

We need to know as much about the couples we are not treating as those we do.

Drife: It occurred to me, when there was the suggestion that health authorities buy successful pregnancies rather than buy treatments, that the easiest way to cope with that funding problem is to treat people who do not need any treatment!

Whittall: One brief point on success rates. There are two possible sources for any publication of success rates – the centres themselves and the HFEA. In any event, the HFEA would have to have control over the nature of the figures that were published.

The Authority is about to go into discussion over that whole issue and I do not doubt it will be talking to people who hold all of the views that have been presented here. What the outcome will be I do not know.

Mr Dickson mentioned that he found no common understanding around the country of what is involved in counselling, and he said that the guidelines need to be set out. I have a difficulty here, and it has been said before, that there seems to be a view that the Authority's Code of Practice is not clear about what it expects in terms of counselling in centres. I find that rather surprising. We need to look at what is in the Code of Practice and look to develop counselling services in centres along those lines. Is what is in the Code of Practice not clear, or is it not meeting the requirements that people have?

Dickson: It would be better to say that there is a wide range of reaction to those standards. I would say that I am very interested to see what the Authority's first report will say about counselling.

Winston: We are facing colossal problems with provision. I wonder whether centres like the John Radcliffe and my own – and I have tried to initiate this – should not be offering transport IVF. We can form links with neighbouring District General Hospitals, particularly in places like Oxford, where they have trained a number of senior registrars who have then gone out to other hospitals in the district. They do the superovulation, a cigar lighter is used for the incubation, and we do the cultures and the transfers. There is something that probably should be explored here. The clinical results will probably not be as good as if one did the whole process oneself, but we will be able to spread IVF more widely on a relatively cheap basis.

Templeton: That is a useful suggestion. However, the HFEA regulations make that very difficult to do. We tried to do that in a simple way through setting up sperm banks in our region, which is very scattered geographically, we have got the Northern Islands to serve. We tried to set up sperm banks, as we have done in Orkney, and we are planning to do the same in Shetland. But those sperm banks, because of HFEA regulations, have to be visited and looked at. Not only that, but those satellite places have to fill in a complete set of application forms, pay the licence fee, and so on, and that is discouraging those smaller centres from collaborating with us. Bureaucracy is getting in the way of service.

Whittall: The Act itself makes clear that any centre that is storing frozen sperm does have to be licensed and has to comply with everything that goes along with that. The situation is different where transport IVF is involved, where the satellite centre is simply inducing ovulation, collecting the oocytes, and then transporting them to the major centre.

Winston: It is actually follicular fluid that is collected without necessarily knowing whether these is an egg in that fluid. That is one reason for the poor success rates, but it is better than nothing.

Whittall: That satellite centre is not carrying out any licensable activity.

Braude: If they are doing intravaginal culture they are. They put the sperm in the tube that goes into the vagina, and that has to be licensed.

Whittingham: That is mixing the gametes outside the body. If they are not doing that it does not come under the HFEA.

Braude: I wanted to pick up on service provision and whether infertility is a specialised case in that it is underfunded. It is a strange situation where we have the facilities and the abilities, but are not being given the funds to do it even when funds have been given for something similar. For example, a secondary centre or a tertiary centre can have funds to do tubal surgery, however badly it is being done. But the money is not avabilable to treat that same patient with superovulation for IVF. This is something I do not understand. Is there an analogous situation in any other part of medicine?

Drife: I do not think so. The solution that has been suggested is that there should be an infertility budget negotiated with the purchaser rather than a procedure-related budget.

Braude: But David Barlow would say that that is not the way to go because, as they have done for abortion, they can then set limits.

Barlow: It is a good idea if there is enough resource going towards it, but not if it is very limited.

Abdalla: I wanted to come back to the point about training and establishing standards. This is not an evangelistic point for the private sector; there is an advantage for the private sector to become involved in training and in development, and an advantage for the NHS when resources are scarce. But what we need is standards and these standards must be audited by a body, perhaps the College or the HFEA, or both.

Professor Winston raised a point about repeated investigations. Would it be wise in some patients in whom the investigations are being repeated to combine them with some kind of treatment? For instance, if I wanted to reassess a tube, would it not be wise to induce ovulation and collect one or two eggs?

Winston: Absolutely not.

Abdalla: And put them back? My point is that as one is repeating a procedure, and one wants to assess the situation, one could do a salpingogram, assess the state of the tube, and collect the eggs. One can do IVF on the same cycle and assess the situation, and if the patient needs tubal surgery in the future she can continue with that.

The third point I wanted to clarify is from the HFEA. Transport IVF is fine because they do not mix the gametes. But what about the communication that we do on the phone with the doctor who is doing the ovulation induction and who is looking after these patients?

Winston: There is no problem. It has worked extremely well in Montpelier, which is the prime model. The collaborating centres perforce have to meet with the professor, who directs it, and he ensures that there are standardisations of protocols and any centre that does not send a representative is not included.

It is effective, the French figures show that it can work and it is relatively inexpensive.

Abdalla: We are doing it.

Drife: For the record, there was some shaking of the heads when the possibility of intertwining investigation and treatment cycles was suggested.

Hull: To return to transport IVF, it would be quie wrong to go away with the assumption that it must be less expensive. The only logical conclusion is that is is likely to be more expensive. It must be less efficient. Costs may be lost or diffused but it cannot be less expensive.

Cooke: There is a halfway house. The North Carolina people reported having the ovulation induction in the peripheral centres and the patient travelling in to have the egg pick up in the main centre, so that the patient is transported pre-aspiration [1]. They looked at the efficiency, the effectiveness and the costs and there was no difference between that system and the person having all the work done peripherally. It is an interesting model that perhaps bears looking at.

Nieschlag: It is quite appropriate that I should be asked to make the final comment on this election day!

I am amazed to hear that patients under the National Health Service do not benefit from funding for IVF. In most European countries this is well regulated and patients have the right to get IVF under the health services. I am amazed, and almost shocked, to hear that in the country where in vitro fertilisation was invented and all the pioneering work was done, the patients do not benefit from this. You need to do something about it!

Reference

1. Talbert LM, Hammond M, Bailey L, Wing R. A satellite system for assisted reproductive technologies: an evaluation. Fertil Steril 1991; 55:555–8.

Recommendations

Epidemiology

1. National statistics are available for almost all aspects of reproduction except infertility and miscarriage. We recommend that national data are collected to fill this gap. This will provide information relevant to public health, and also to the planning and provision of medical services.

Investigation and Treatment of the Male

2. Andrology should be recognised as a relevant clinical and scientific discipline which forms a counterpart to gynaecology. Priority should be given to the development of mechanisms for the training and accreditation of specialists.
3. Emphasis should be given to the standardisation of the techniques and quality control criteria laid down in the "WHO Laboratory Manual for the Examination of Human Semen and Semen–Cervical Mucus Interaction" (Cambridge University Press, 1987).
4. Better methods of male infertility diagnosis and treatment need to be researched and evaluated in controlled prospective clinical trials.
5. There is a need to improve and standardise cryopreservation protocols in order to ensure uniform practice.
6. Although sperm micromanipulation techniques may be the only means of achieving pregnancies for some patients with deficient spermatogenesis,

the efficacy of these methods needs to be established. The techniques are still at the developmental stage and should be practised only in centres with appropriate facilities for evaluating the technical aspects on appropriate models.

Investigation and Treatment of the Female

7. The full value of reproductive surgery will be realised only if there is an improvement in surgical training and practice. There is a need to define grades of severity of tubal damage and hence the indications for surgery. Although it is likely that in vitro fertilisation will diminish the need for reproductive surgery, it will not replace this completely. It appears that tubal microsurgery remains the most effective treatment in properly selected cases.
8. It is unclear whether laparoscopic surgery is as effective as open microsurgery, although it is potentially less traumatic and requires less inpatient stay. Nor is there evidence that the use of lasers improves clinical results. Randomised prospective studies are needed for the evaluation of operative laparoscopic techniques.
9. In the management of induction of ovulation it is extremely important to prevent multiple pregnancy. This risk can be much reduced using modern regimens, for example low-dose gonadotrophin therapy in the management of clomiphene-resistant polycystic ovarian disease. In general, regimens which minimise or avoid the risk of multiple pregnancy, even at the expense of lower pregnancy rates, are recommended.
10. In clomiphene-resistant polycystic ovarian syndrome the results of ovarian cautery are promising but further studies are required to evaluate this treatment in comparison with gonadotrophins.
11. There is an established place for the treatment of the infertile woman by oocyte donation. However, there is considerable difficulty in obtaining eggs for donation. This problem could be helped if the recommended upper age limit of oocyte donors was raised to 38 years.
12. The transfer of embryos developed from cryopreserved oocytes after fertilisation in vitro should be permitted in clinics with appropriate expertise, provided that all reasonable steps are taken to exclude fetal abnormality early in pregnancy.

Reproductive Research

13. Properly designed randomised controlled trials of sufficient size are necessary for the formal evaluation and comparison of many forms of infertility treatment. Audit and research reports should clearly state the inclusion criteria, particularly with respect to age, diagnosis and duration of infertility. Results should be presented according to a standard format, including where appropriate, time-specific cumulative conception and birth rates and outcomes per cycle of treatment.

14. Authors of scientific papers in the field of infertility should be encouraged to adopt more rigorous study design and interpretation, and journal editors should reflect this trend in the editing and acceptance of papers.
15. There have been significant developments in preimplantation diagnosis resulting in normal live births. Further research in this field should continue with the use of appropriate animal models and donated human embryos. It is essential that any centre embarking on these difficult techniques should have access to an appropriate animal model to establish the basic skills and have an established high quality clinical IVF programme.

Service Provision

16. There is a need to develop national and regional strategies for provision and audit of care to infertile couples. The RCOG has produced a booklet "Infertility – Guidelines for Practice" and the relevant recommendations in this booklet indicate the way forward in this respect.
17. Infertility investigation and treatment is a continuum in which assisted reproduction is an important and effective component. The NHS should use its limited resources more appropriately, and provide a comprehensive fertility service which includes assisted reproduction in tertiary centres.
18. The RCOG should develop, where appropriate in collaboration with patient support groups, expert fact sheets on infertility treatment. These should be specifically targeted at different groups, including the gynaecologist, the general practitioner and the patient.
19. The RCOG and patient support groups should communicate with the Family Planning Association (FPA) about making infertility part of family planning education and services.
20. There should be a substantial increase in the number of consultant posts for subspecialists in reproductive medicine.

Training in Infertility

21. Reproductive Medicine subspecialty training programmes should be made accessible to more trainees by provision of more funded fellowships.
22. There is a need for far greater exposure to reproductive medicine during training for the MRCOG.
23. The opportunities for training in the private sector should be utilised. Training posts should be created combining private sector and NHS experience.
24. We support the Royal College of Nursing recommendation that appropriate training that fulfils postregistration educational requirements is implemented for nurses in the subspecialty of infertility.
25. All nurses who wish to undertake counselling must have appropriate training. Where there is a full counselling role, conditions of practice should

include the need for counselling to take place in an appropriate setting and within a clearly defined code of ethics and practice.

26. There should be training, accreditation and professional affiliation for embryologists, and other laboratory workers handling gametes and zygotes in a clinical setting. An association of clinical embryologists should be formed under the auspices of the RCOG, to set appropriate standards for clinical embryology and to offer registration and protection to this branch of the profession.

Legal Aspects

27. The confidentiality clause (33,5) in the Human Fertilisation and Embryology Act (1990) is unnecessarily restrictive and severely impedes patient care. We recommend that urgent attention is given to altering this part of the Act.

Subject Index